Drawn by Fire 4

DRAWN BY FIRE

FIRE SERVICE EDITORIAL CARTOONS BY

PAUL COMBS

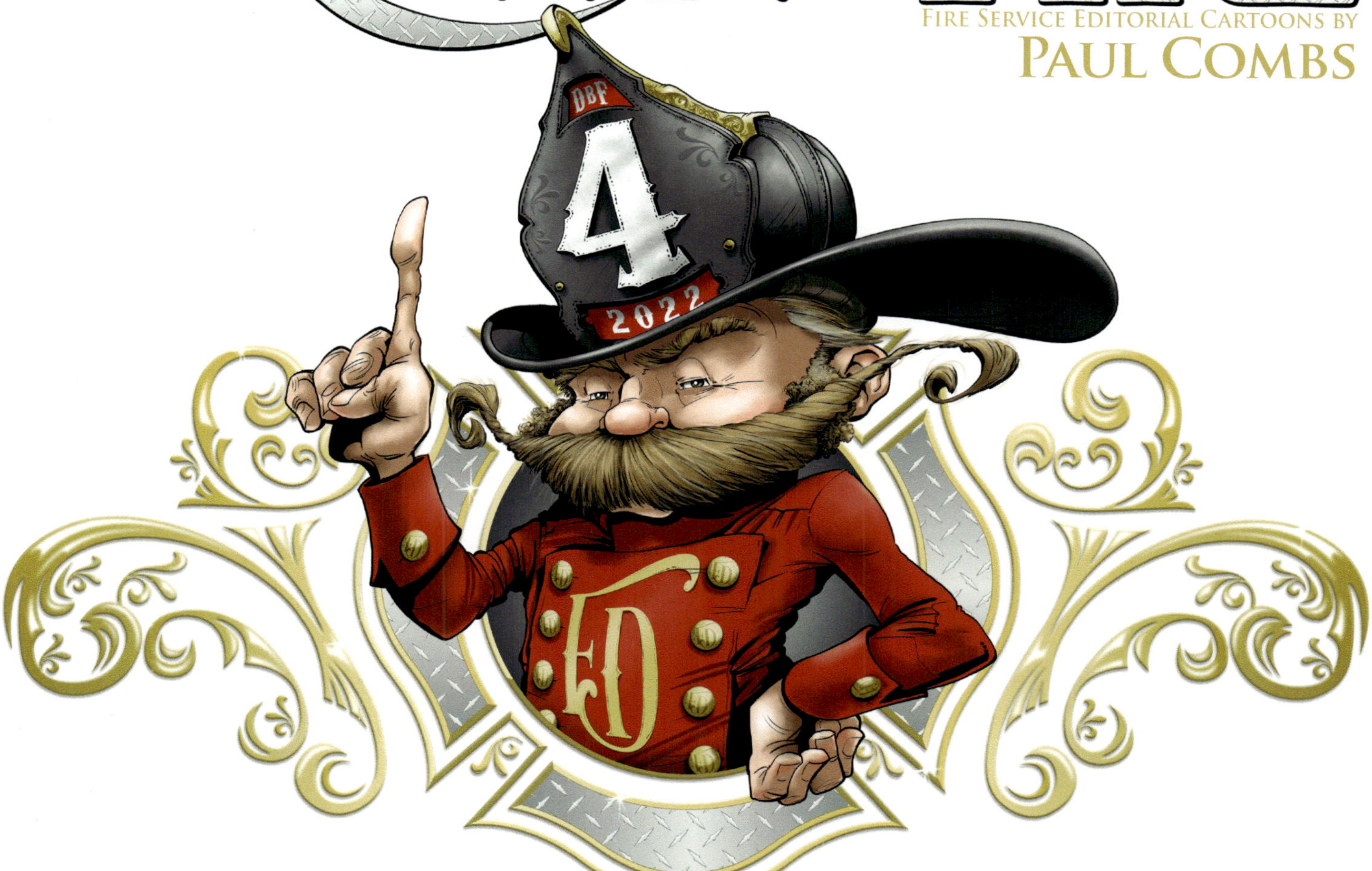

Fire Engineering
BOOKS & VIDEOS

Fire Engineering Books & Videos
110 S. Hartford Ave., Suite 200
Tulsa, Oklahoma 74120 USA

800.752.9764
+1.918.831.9421
sales@pennwell.com
info@fireengineeringbooks.com
www.Fire EngineeringBooks.com

Senior Vice President: Eric Schlett
Vice President: Amanda Champion
Operations Manager: Holly Fournier
Sales Manager: Joshua Neal
Managing Editor: Mark Haugh
Production Manager: Tony Quinn
Developmental Editor: Chris Barton
Cover Designer: Paul Combs

Library of Congress Cataloging-in-Publication Data available upon request

Disclaimer

The recommendations, advice, descriptions, and the methods in this book are presented solely for educational purposes. The author and publisher assume no liability whatsoever for any loss or damage that results from the use of any of the material in this book. Use of the material in this book is solely at the risk of the user.

Printed in the United States of America

1 2 3 4 5 26 25 24 23 22

For Sheryl and Brittney—you are the breeze in my sail.

Sometimes supple, sometimes a gale, but always pushing me forward.

Contents

FOREWORD

When we look back through the story of our life, there are always a few key teachers who helped shape and mold us, who helped us find our own true north. Everybody has these people in their life, but those who are smart and successful listened to them. What common traits do the great teachers share?

The first rule of teaching is that you must engage your audience. To do so, you must capture and entertain them. The best educators are always the most creative and most informed, and their lessons are full of relevant topics. Lastly, the greatest writers and teachers are inspirational. Their lessons make us believe in ourselves and inspire us to do the right thing. When you bundle up these traits and combine them with humor and amazing artwork, you are describing the work of Paul Combs.

Paul has the unique ability to make us laugh at ourselves and learn a powerful lesson at the same time. Through his unbelievable artwork and creative genius, we are able to see through the smoke and into the heart and mind of Paul Combs. Many of the lessons are designed to modify the behavior of the reader and encourage them to do the right thing: It's not a sign of weakness to ask for help. It's okay to call a mayday if you're lost, trapped, or injured. You are not a tough guy because you fail to wear your PPE properly.

Paul Combs's books are saving lives, modifying behavior, and helping change the culture of the fire service.

I've been a proud member of the fire service for more than 40 years. I retired from the Chicago Fire Department as a deputy district chief after a 36-year career. I've been a fire service instructor for more than 35 years, and I am blown away by Paul's approach to the mission of educating the noble breed of men and women that fill the ranks of first responders. Education through humor not only opens the eyes and mind of the reader, but it also engages them and makes it fun to learn.

Why is humor so important to firefighters and medics? Humor has always has been a big part of the fire service and always will be. The stressful nature of our job and the dark memories that come from the pain, suffering, and death firefighters and medics witness far too often makes humor a necessity as a blow-off valve. Long before critical incident stress debriefings existed, it was humor that kept us sane. It's better to send them home laughing than crying.

This book should be on the coffee table in every firefighter's home and on the kitchen table in every firehouse. Paul Combs's books don't require you to sit down and read them from cover to cover. His books are like magnets. They can be picked up and opened to any page at any time, and the reader will be entertained. More importantly, they makes you think and encourage you to do the right thing.

You already did the right thing by picking up this book. Enjoy. I did.

Steve Chikerotis, Deputy District Chief
Chicago Fire Department (ret.)

Producer, NBC Chicago Fire

Introduction

For years I read the cartoons by Paul Combs in the trade magazines and online. I had never met the man, but always enjoyed the insight he brought to his work. His artwork was easy to appreciate, especially for someone like me who has no artistic ability. Fast-forward a few years, and a mutual friend, the late EJ Mascaro, introduced me to Paul at FDIC. EJ was a tough judge of character and had a keen eye for good people and a better eye for good firefighters. He described Paul as a top-shelf firefighter and instructor who really "got it." High praise coming from our late friend!

As we travel through our fire department careers as well as life in general, we all experience encounters with people who leave us feeling many ways. We feel joy, sorrow, empathy, sympathy, anger, fear, and love, to name a few. There is no doubt that I have felt many of these things in a single shift, yet I lacked the ability to easily express such feelings in a meaningful and productive way. Paul has no such dilemma. From the eye and the hand of a seasoned fireman comes insight we all need.

His art and editorial content provide us with constant reminders of what is good in the fire service as well as what we really need to improve. At times he reminds us of the simple things like buckling our seatbelts, wearing our gear properly, and stepping up for the firehouse chores. He doesn't sugarcoat it! Paul isn't afraid to peel back the layers of the fire service and expose things many of us are unwilling or uncomfortable to discuss. Leadership issues, poor communication, physical fitness, suicide, cancer prevention, bullying, and lack of training come to mind. Paul's ability to entice his readers to take a hard look in the mirror is uncanny.

What makes this book, and all of Paul's work, uniquely special is that it comes from the heart. His pride in the profession is second to none. Having watched him train and communicate the nuances of the job at training venues across the globe, it is clear that this amazing artist has a tremendous passion for the fire service, and that is something so very special to witness. As our pal EJ said, Paul "gets it."

To Paul's readers, buckle up, hold on tight, and keep your arms and legs inside the car at all times, because you're about to go on a fantastic fire service ride.

Paul, I am blessed to call you my friend and honored to call you my Brother! Thank you for giving our beloved fire service such an important means of communication.

Aaron Heller, Deputy Chief
Hamilton, NJ, Fire Department (ret.)

Owner and Lead Instructor for On Scene Training Associates, LLC

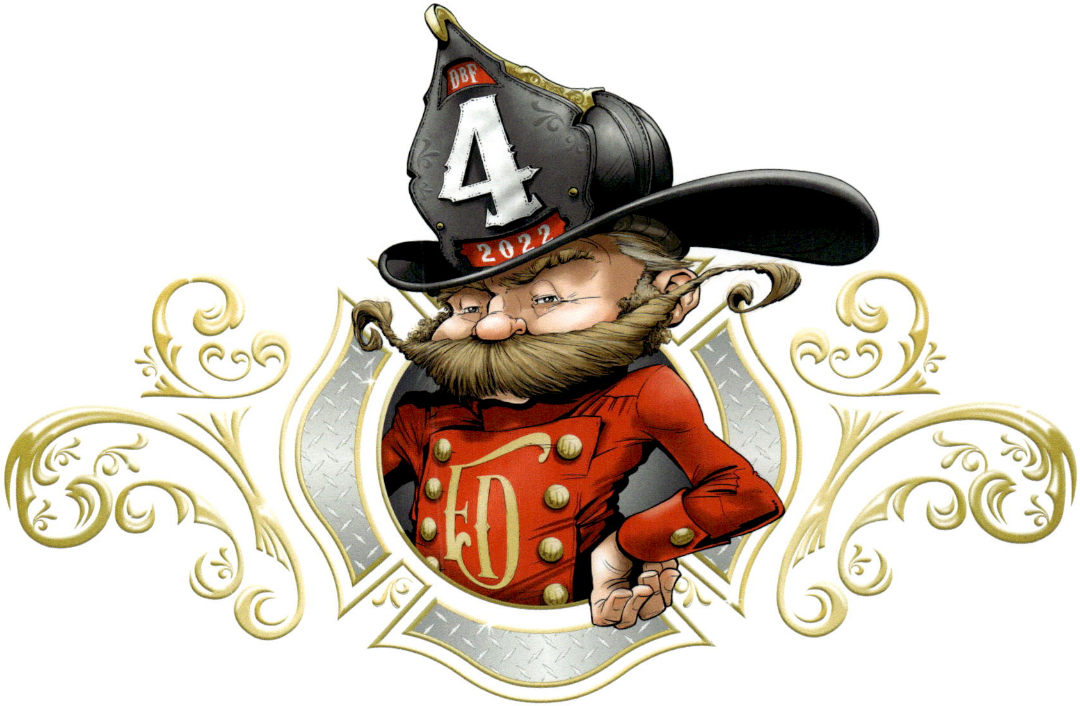

Death, Taxes, and Bad Leaders

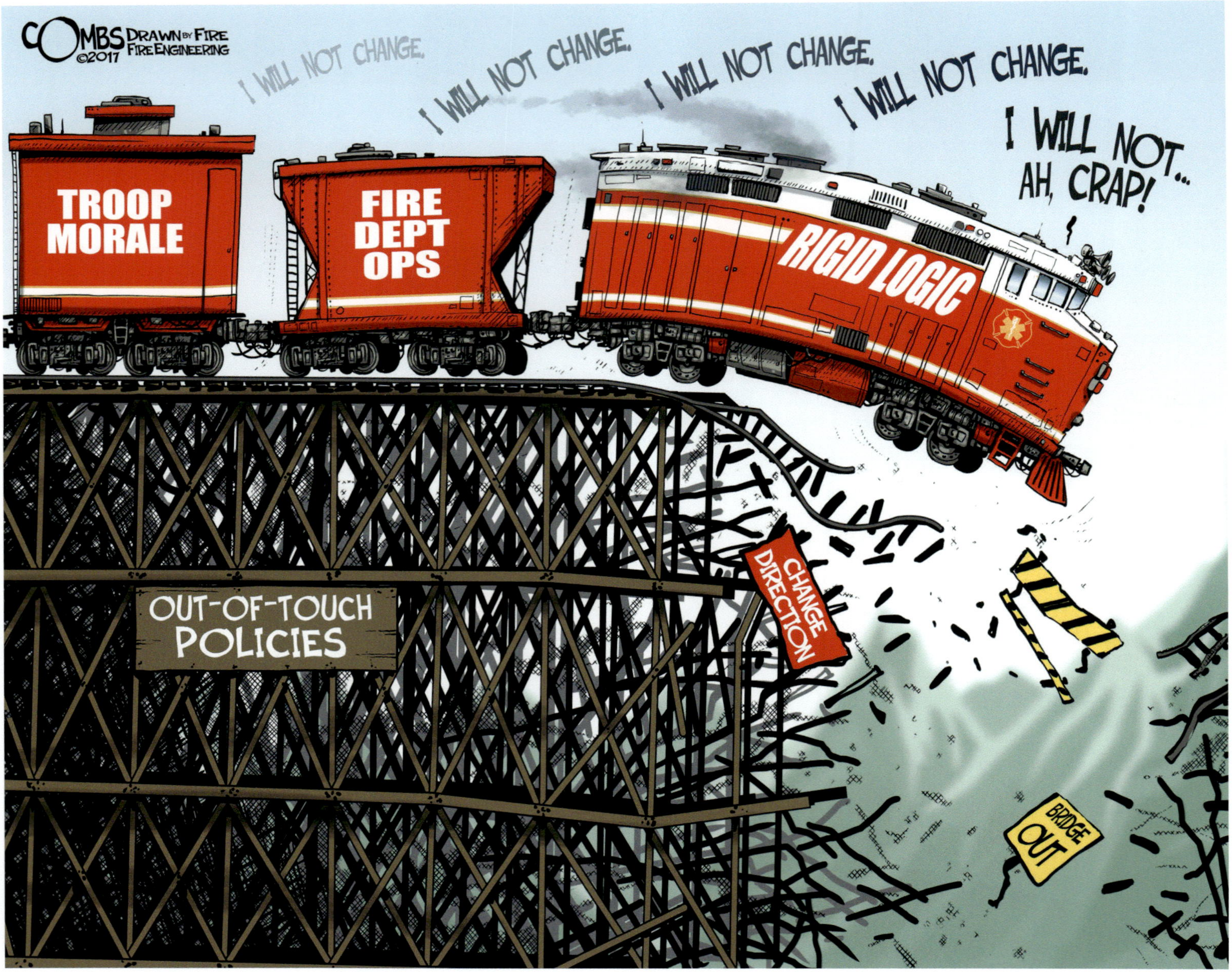
COMBS DRAWN BY FIRE
FIRE ENGINEERING
©2017
I WILL NOT CHANGE.
I WILL NOT CHANGE.
I WILL NOT CHANGE.
I WILL NOT CHANGE.
I WILL NOT... AH, CRAP!
TROOP MORALE
FIRE DEPT OPS
RIGID LOGIC
OUT-OF-TOUCH POLICIES
CHANGE DIRECTION
BRIDGE OUT

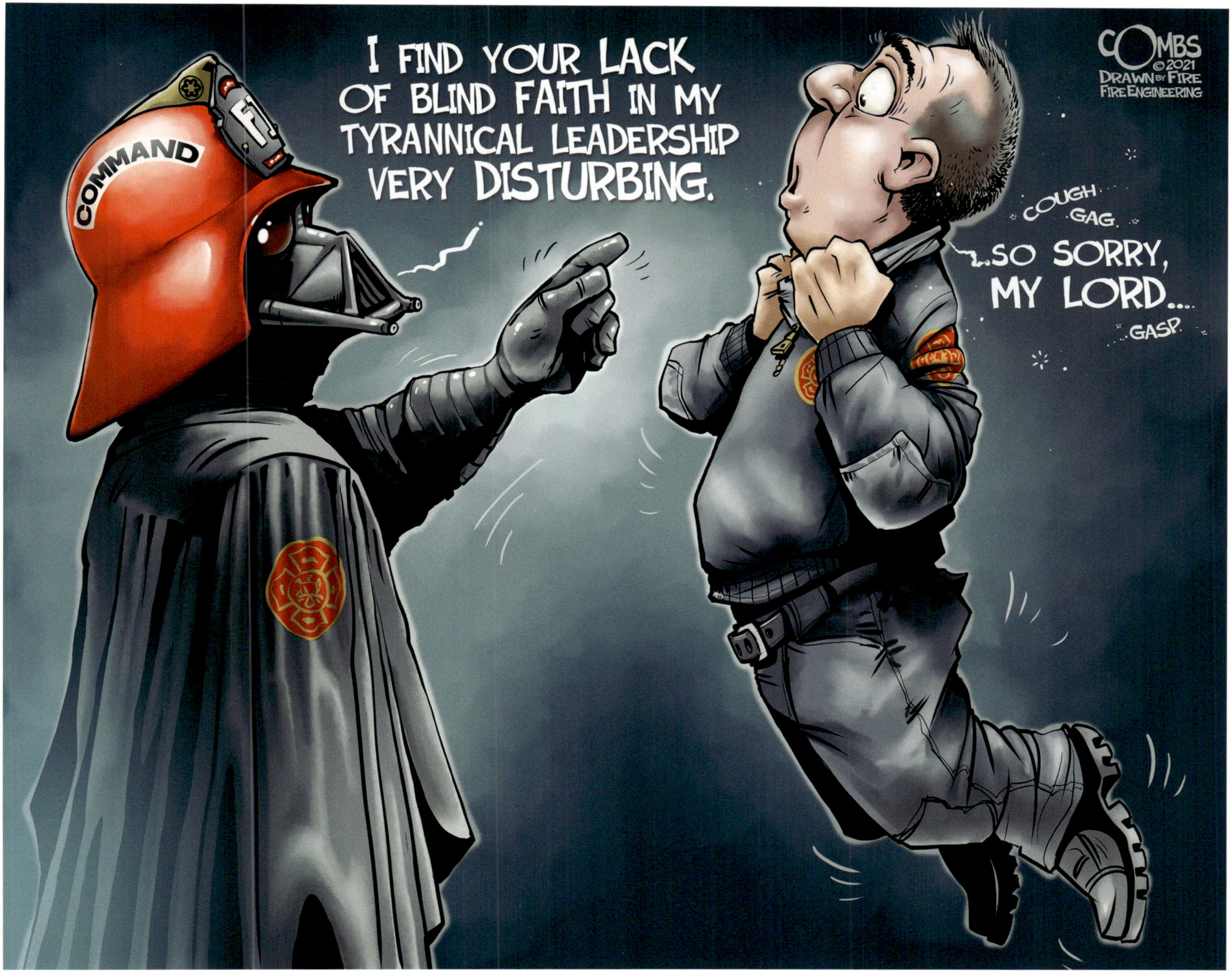
I FIND YOUR LACK OF BLIND FAITH IN MY TYRANNICAL LEADERSHIP VERY DISTURBING.
COUGH GAG
...SO SORRY, MY LORD...
GASP
COMMAND
COMBS ©2021 DRAWN BY FIRE FIRE ENGINEERING

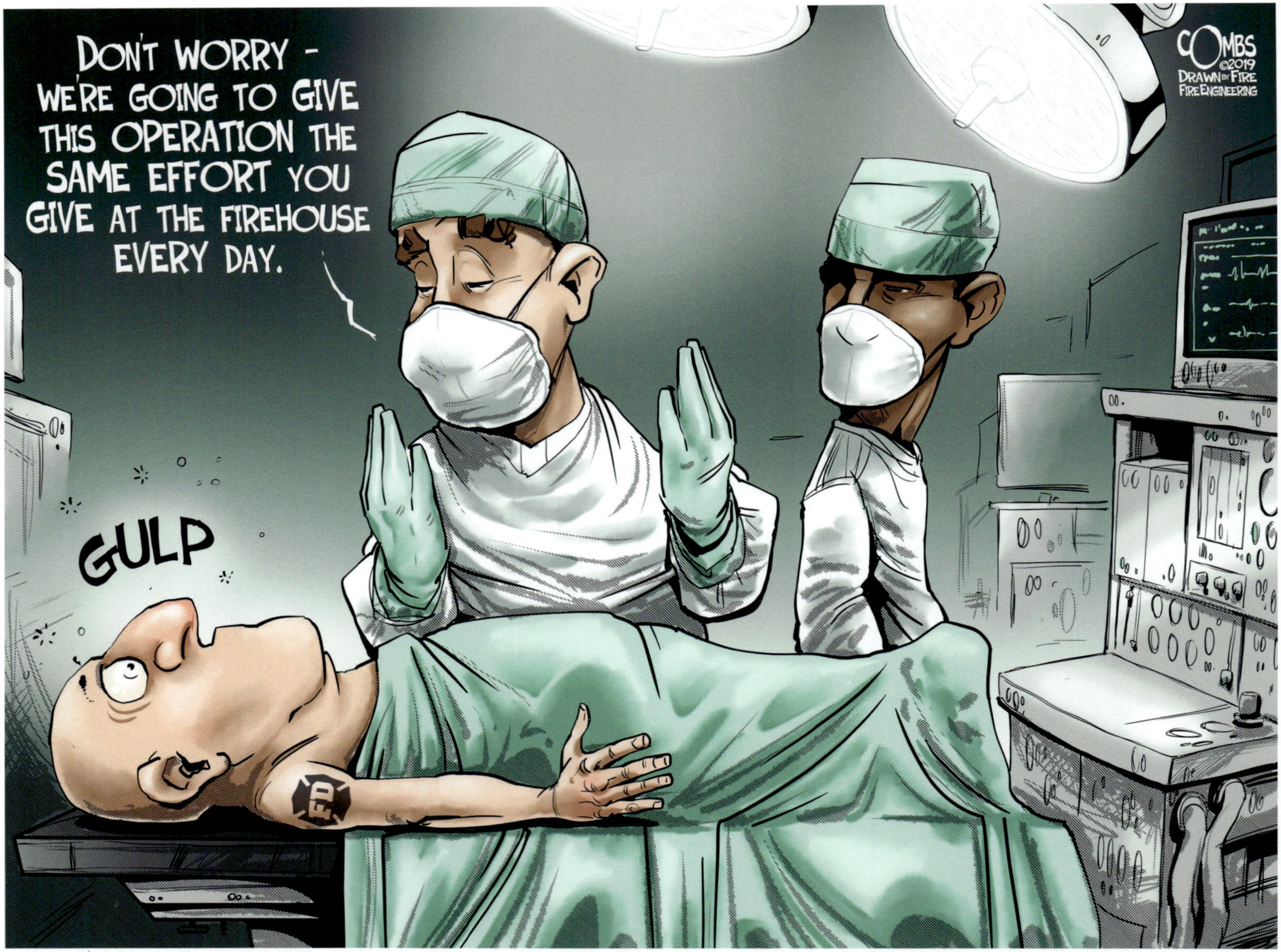

ARE YOU GIVING AN ALL OUT EFFORT?

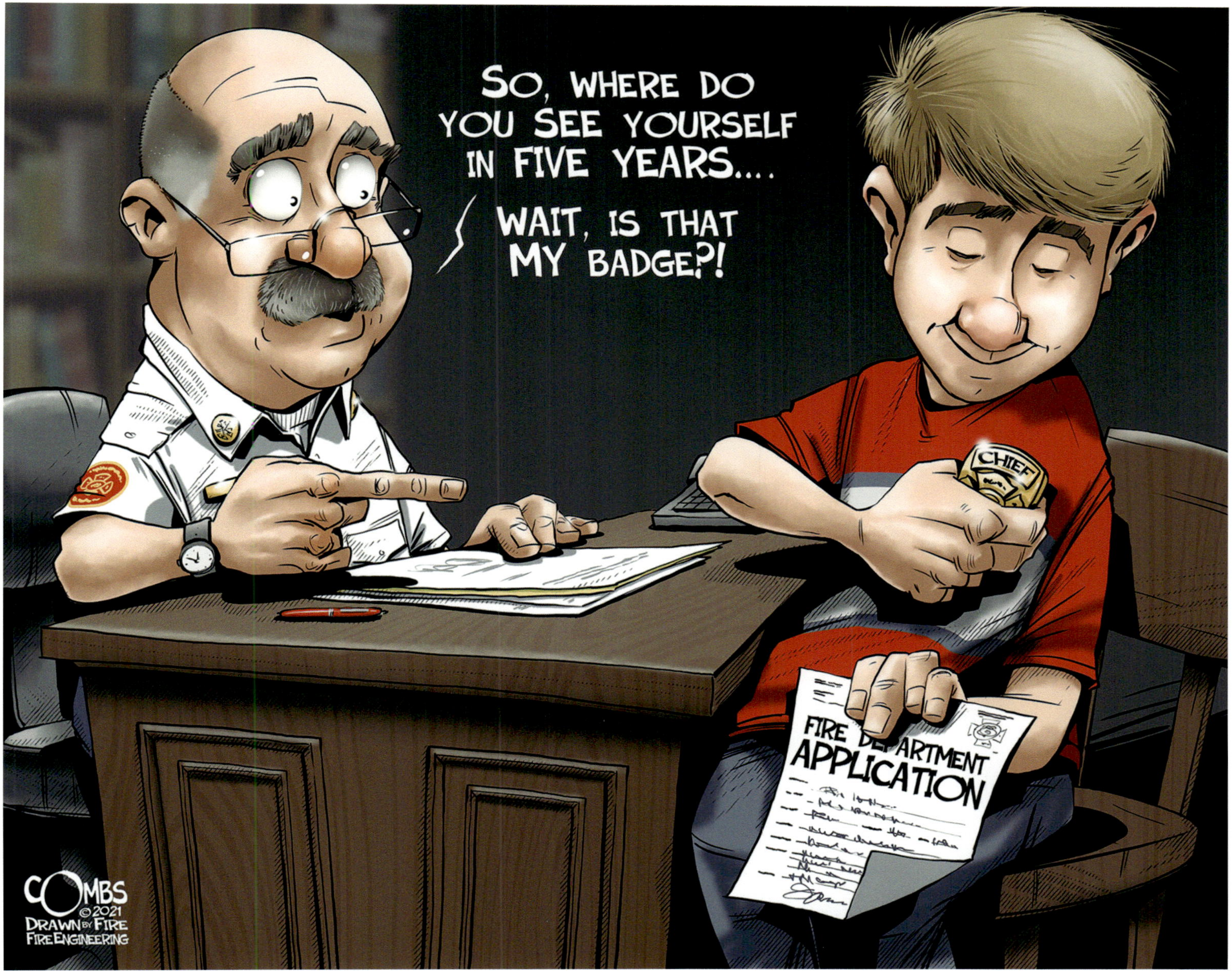
SO, WHERE DO YOU SEE YOURSELF IN FIVE YEARS....
WAIT, IS THAT MY BADGE?!
CHIEF
FIRE DEPARTMENT APPLICATION
COMBS ©2021 DRAWN BY FIRE FIRE ENGINEERING

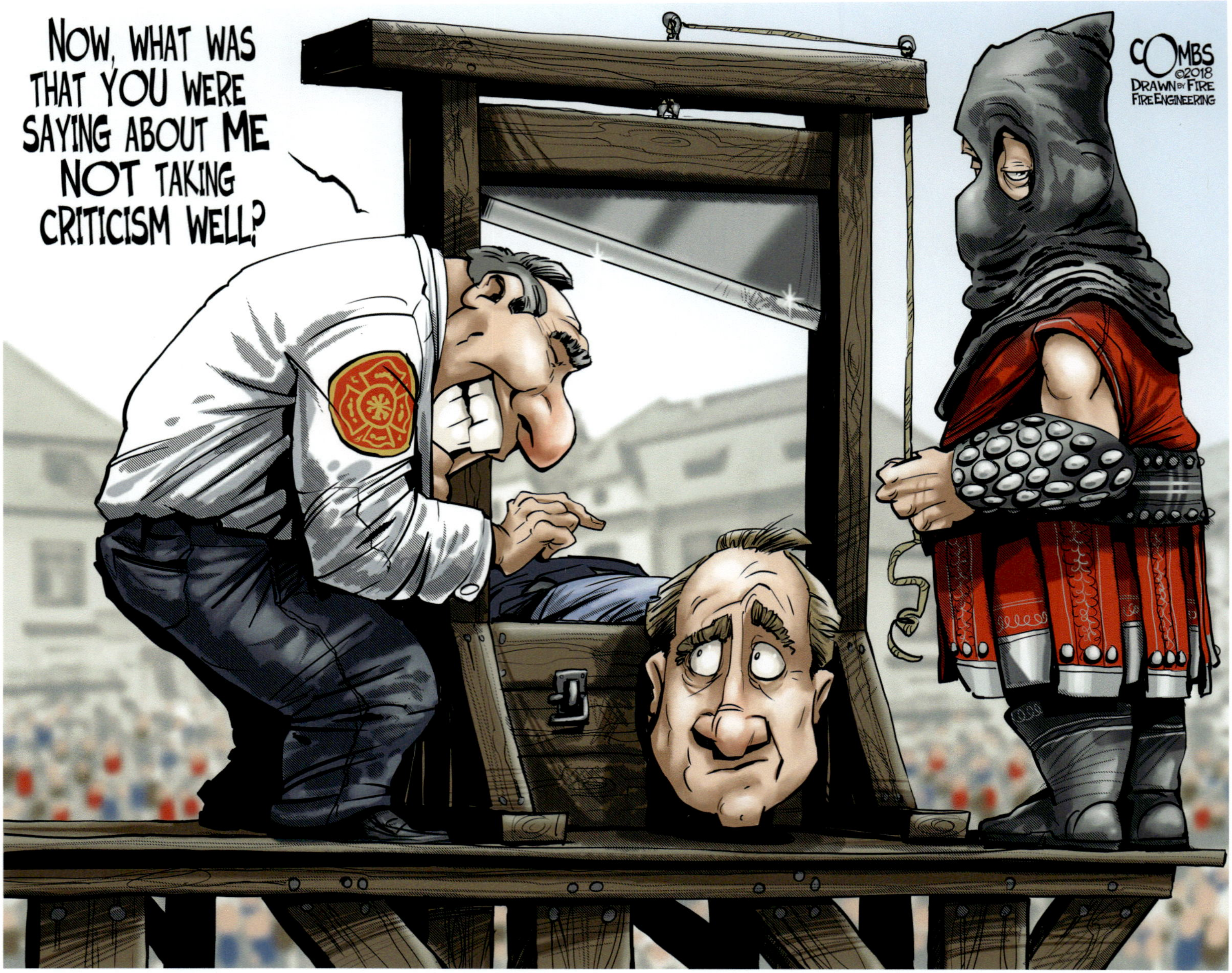
NOW, WHAT WAS THAT YOU WERE SAYING ABOUT ME NOT TAKING CRITICISM WELL?
COMBS ©2018 DRAWN BY FIRE FIRE ENGINEERING

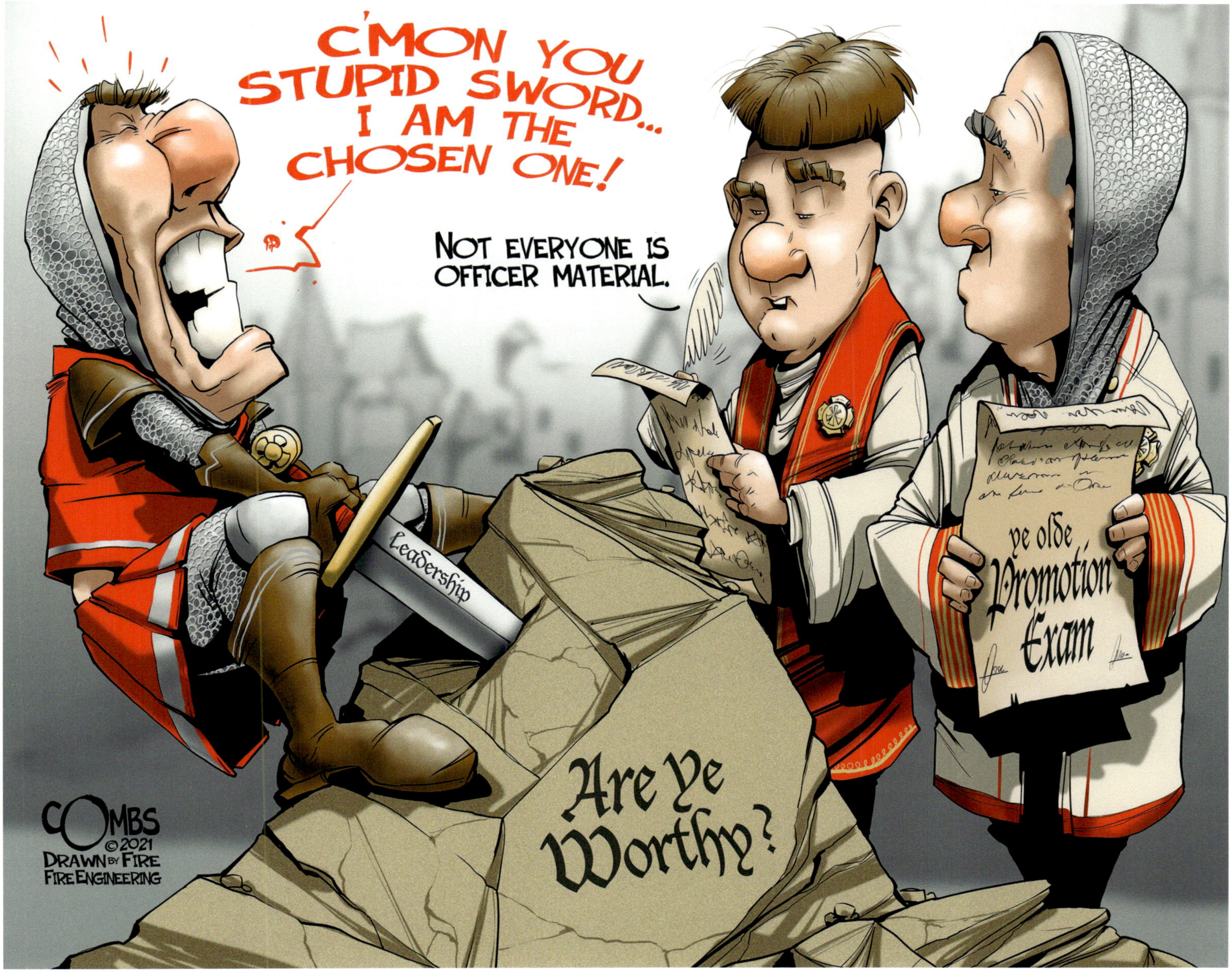
C'MON YOU STUPID SWORD... I AM THE CHOSEN ONE!
NOT EVERYONE IS OFFICER MATERIAL.
Leadership
ye olde Promotion Exam
Are ye Worthy?
COMBS ©2021 DRAWN BY FIRE FIRE ENGINEERING

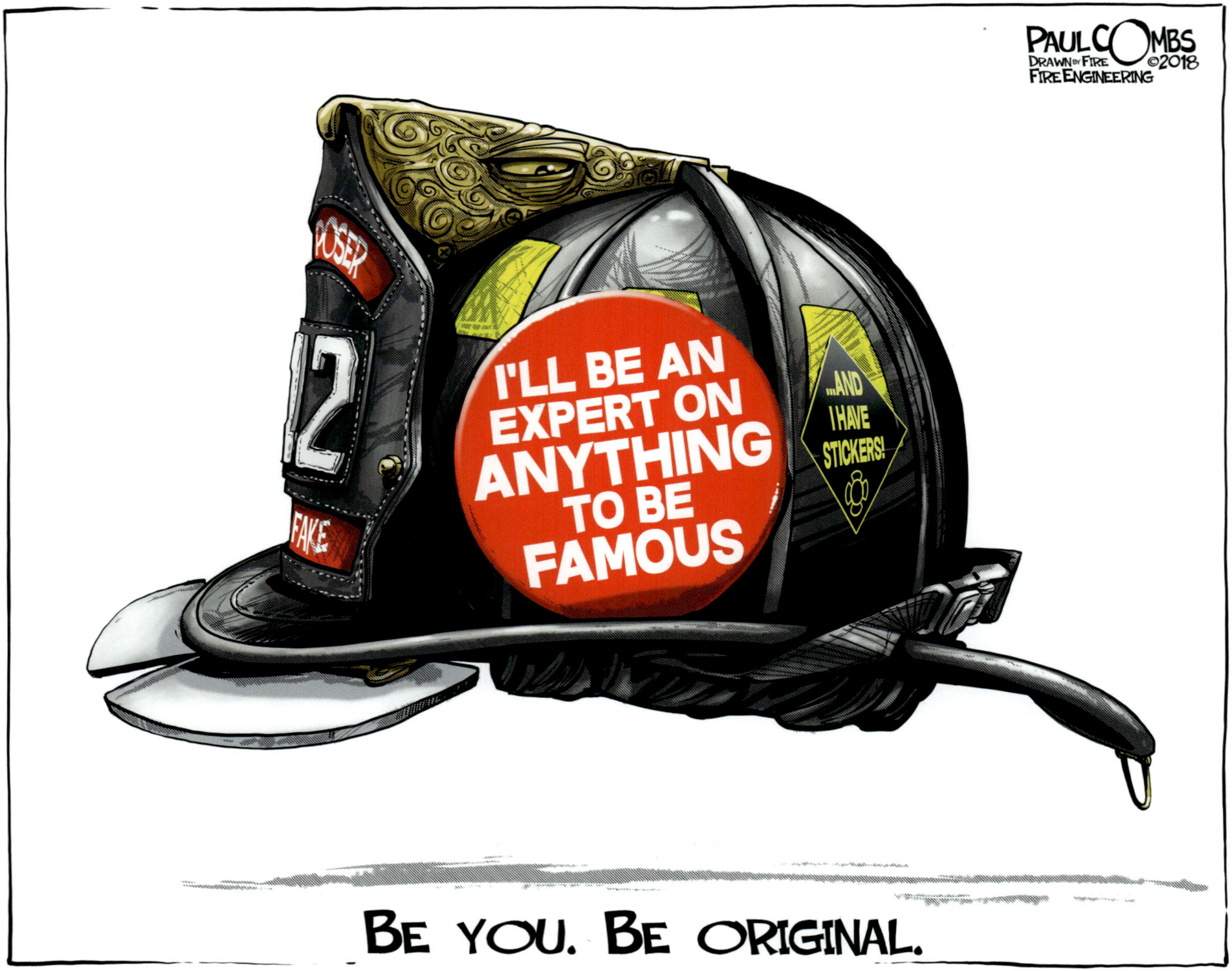
PAUL COMBS
DRAWN BY FIRE ©2018
FIRE ENGINEERING
POSER
12
FAKE
I'LL BE AN EXPERT ON ANYTHING TO BE FAMOUS
...AND I HAVE STICKERS!
BE YOU. BE ORIGINAL.

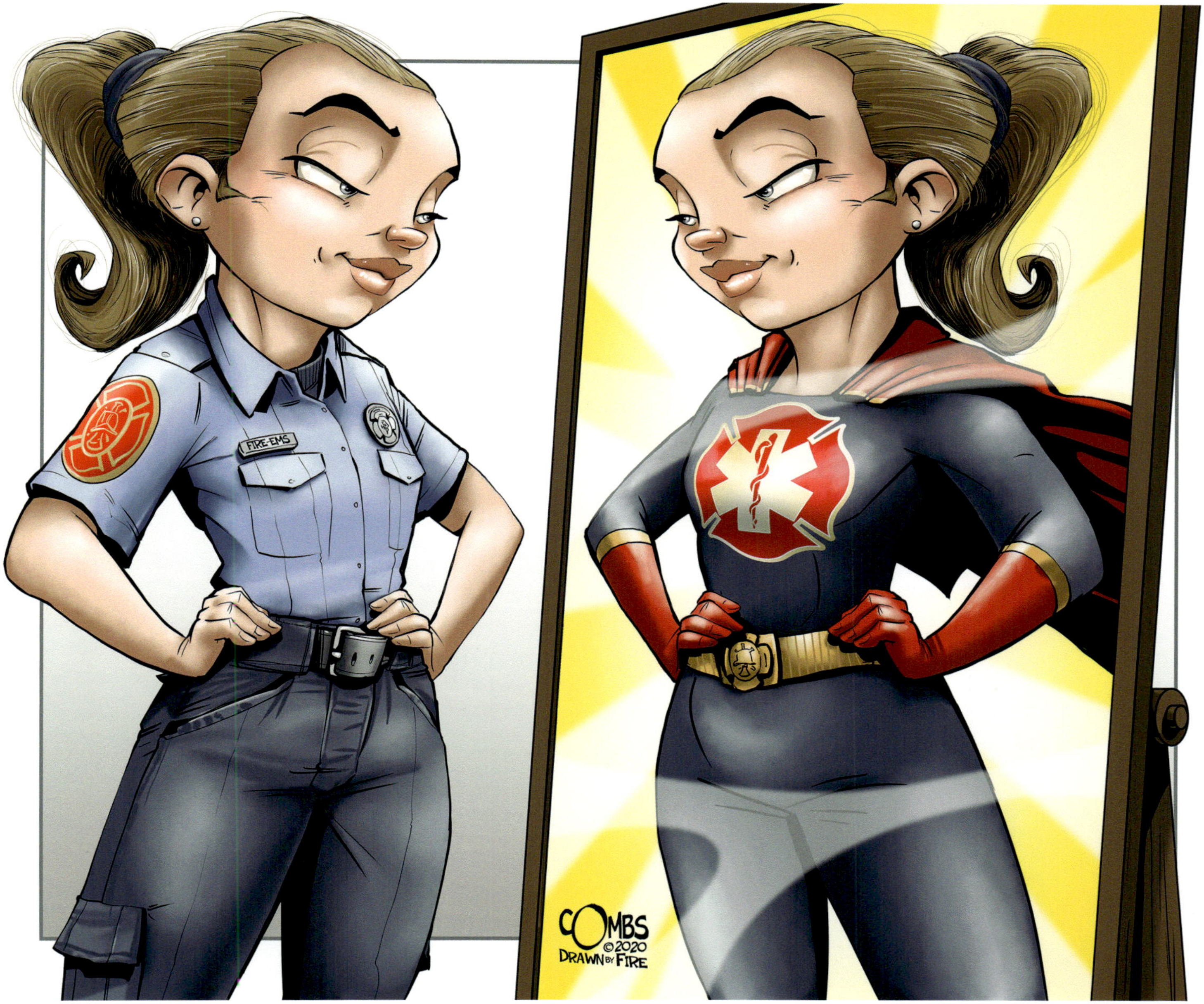
FIRE-EMS
COMBS
©2020
DRAWN BY FIRE

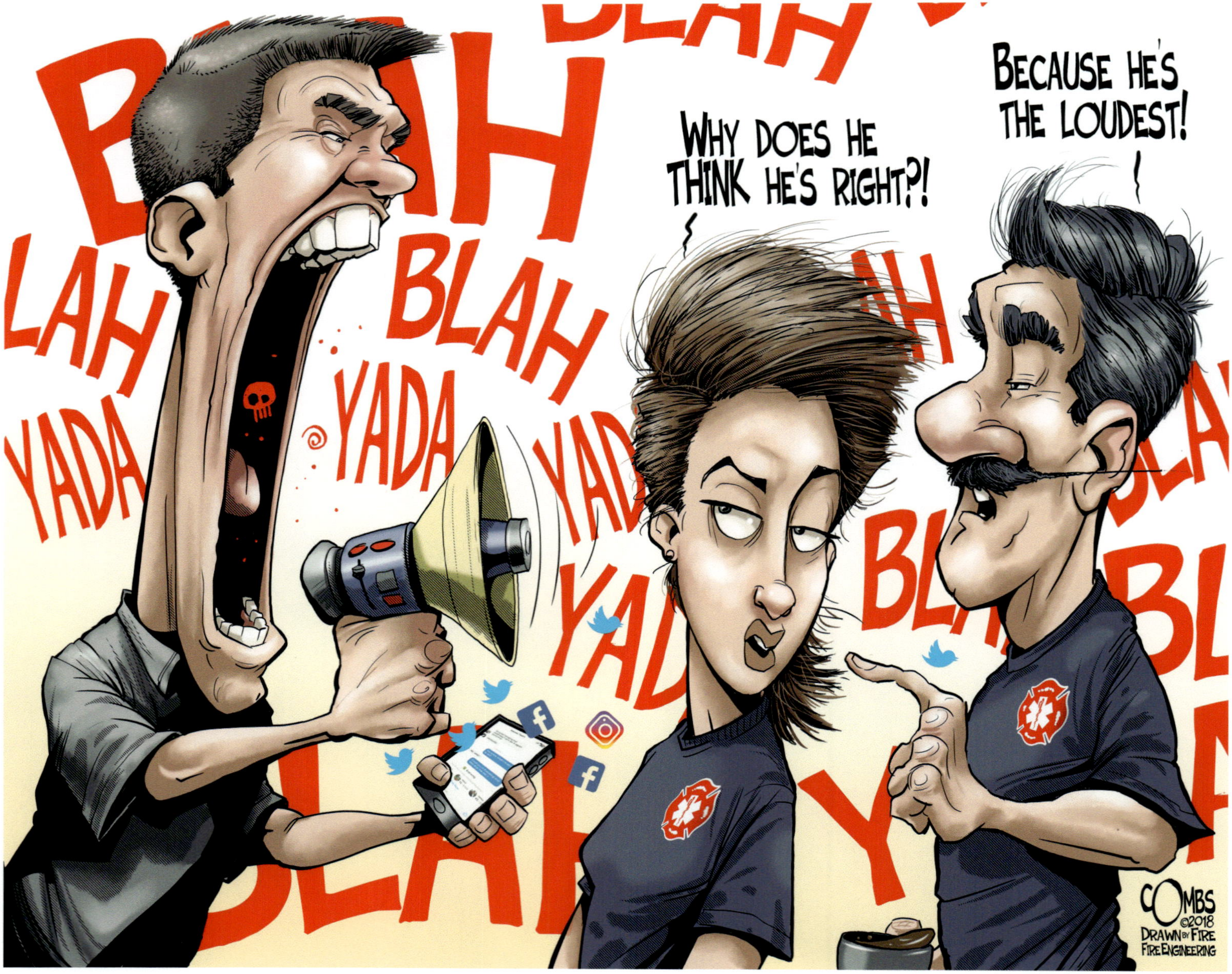
BLAH
BLAH
YADA
WHY DOES HE THINK HE'S RIGHT?!
BECAUSE HE'S THE LOUDEST!
COMBS ©2018 DRAWN BY FIRE FIRE ENGINEERING

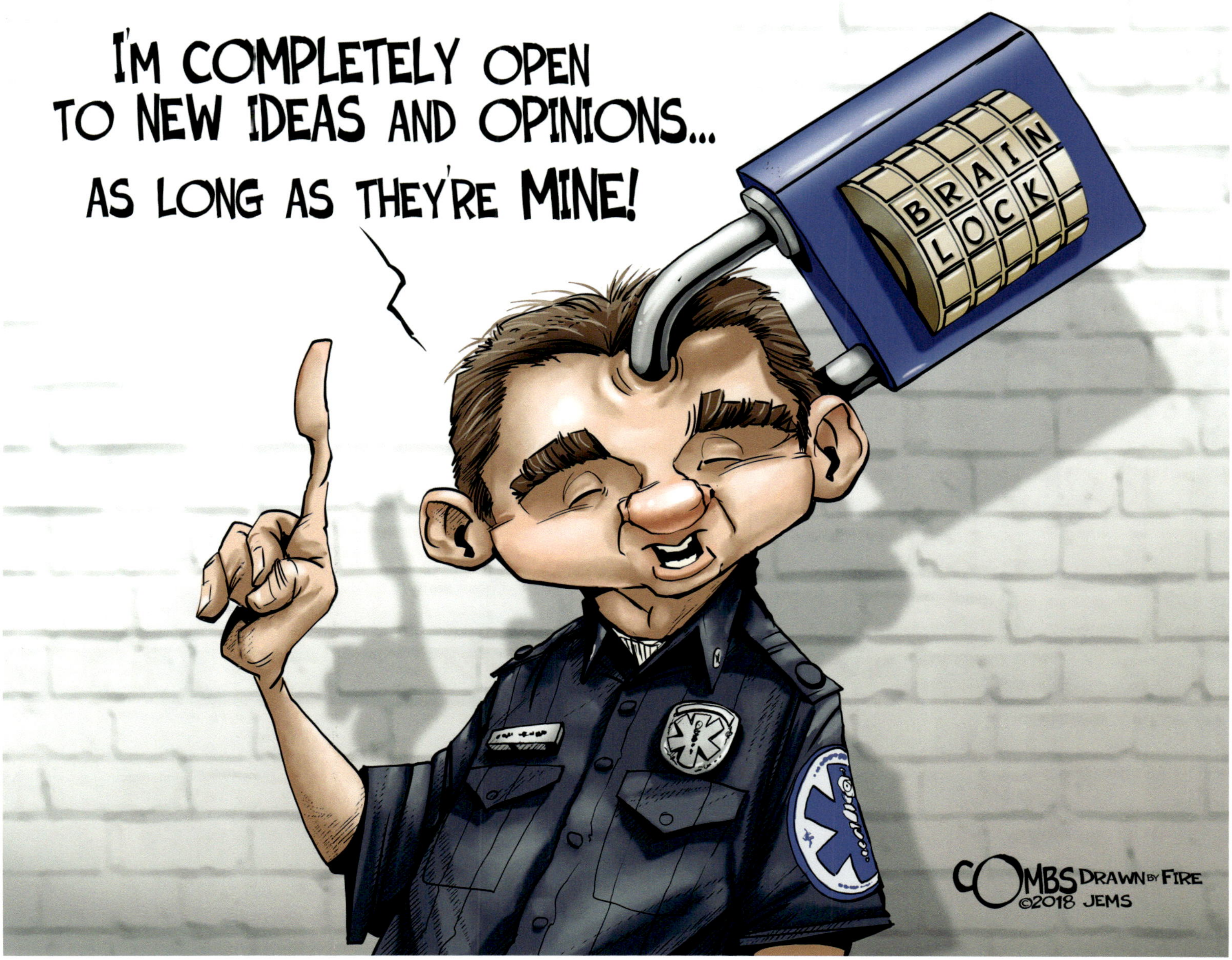
I'M COMPLETELY OPEN TO NEW IDEAS AND OPINIONS...
AS LONG AS THEY'RE MINE!
BRAIN
LOCK
COMBS DRAWN BY FIRE ©2018 JEMS

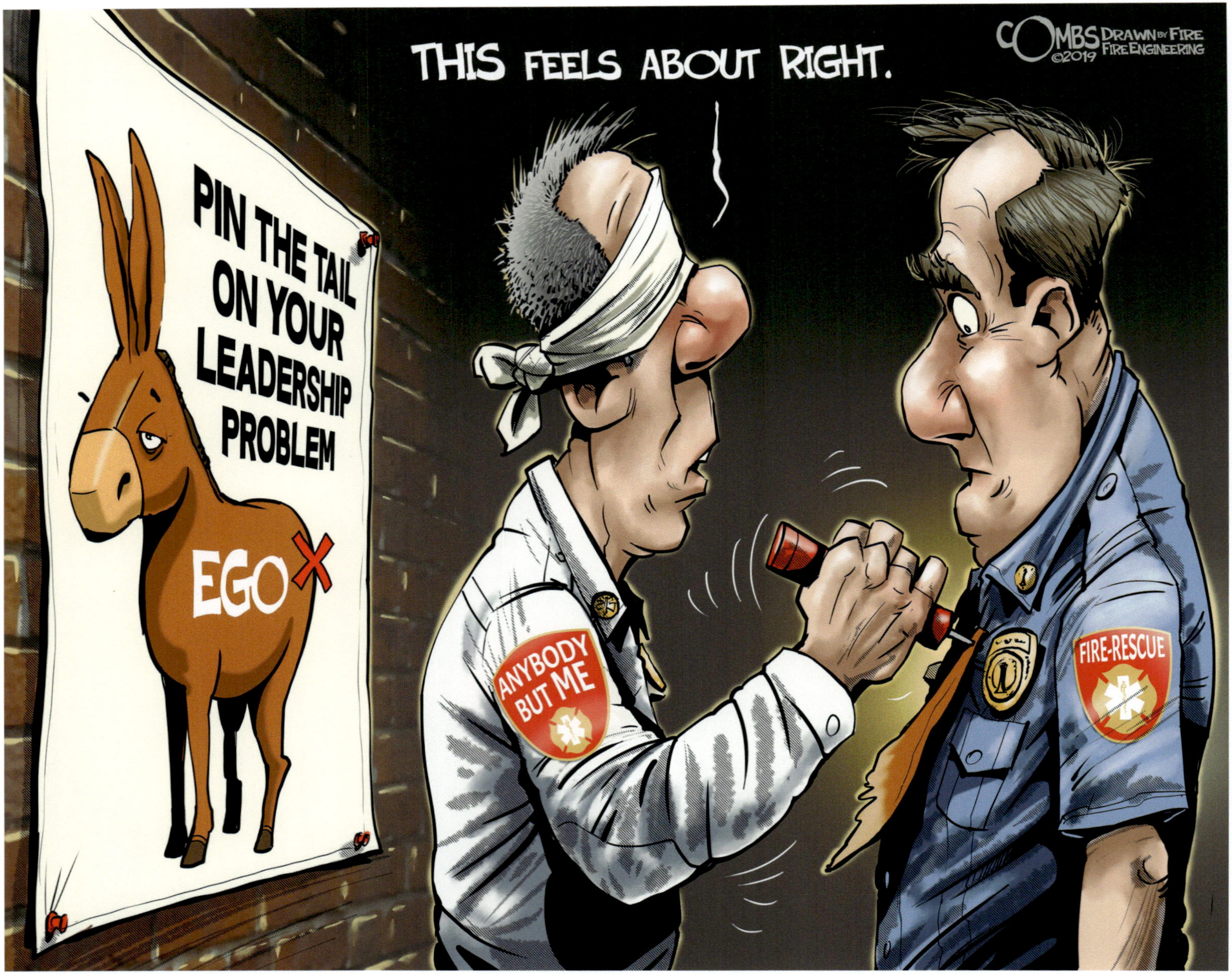
THIS FEELS ABOUT RIGHT.
COMBS DRAWN BY FIRE ©2019 FIRE ENGINEERING
PIN THE TAIL ON YOUR LEADERSHIP PROBLEM
EGO
ANYBODY BUT ME
FIRE-RESCUE

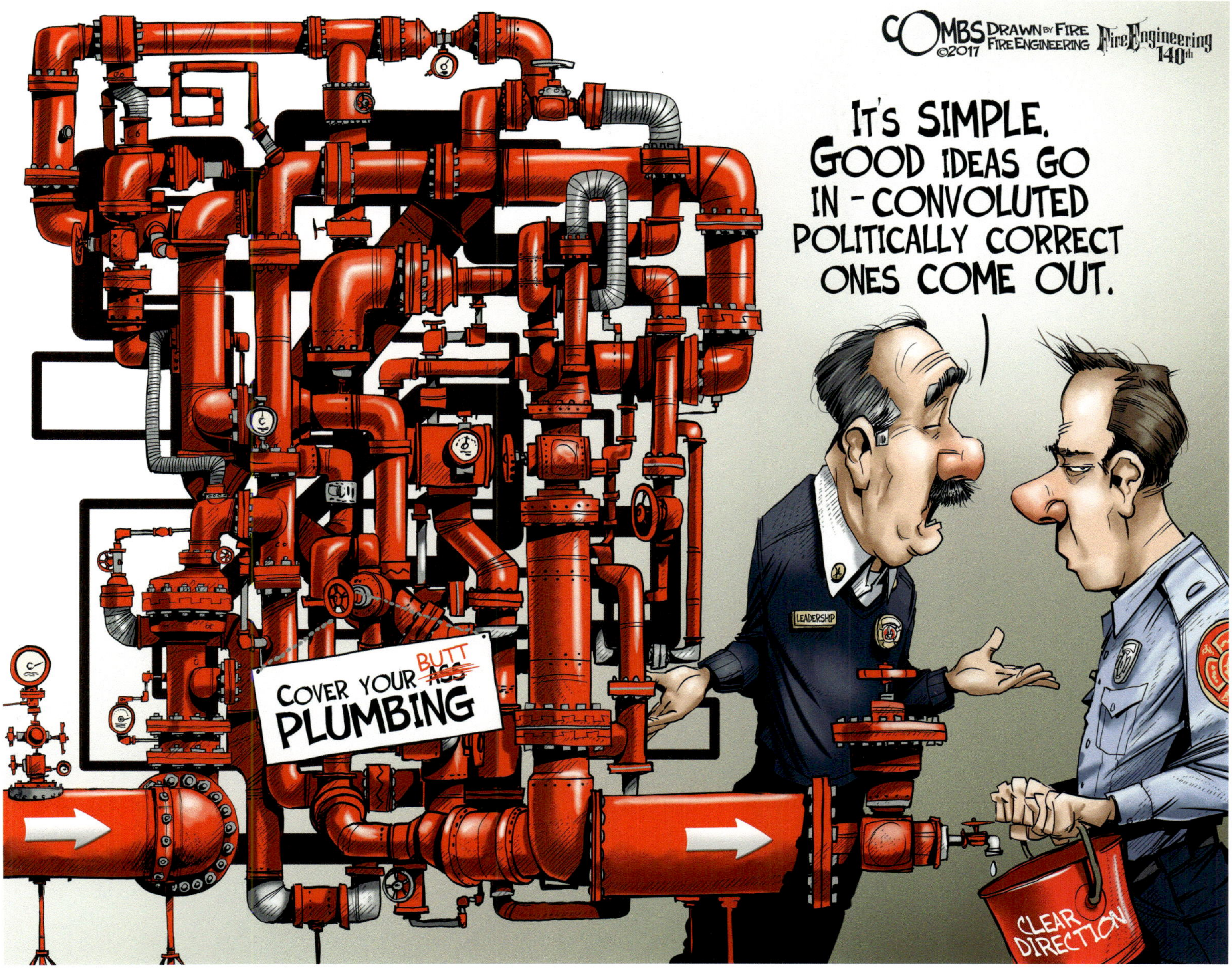
COMBS ©2017
DRAWN BY FIRE
FIRE ENGINEERING
Fire Engineering 140th
IT'S SIMPLE. GOOD IDEAS GO IN - CONVOLUTED POLITICALLY CORRECT ONES COME OUT.
COVER YOUR ~~ASS~~ BUTT PLUMBING
LEADERSHIP
CLEAR DIRECTION

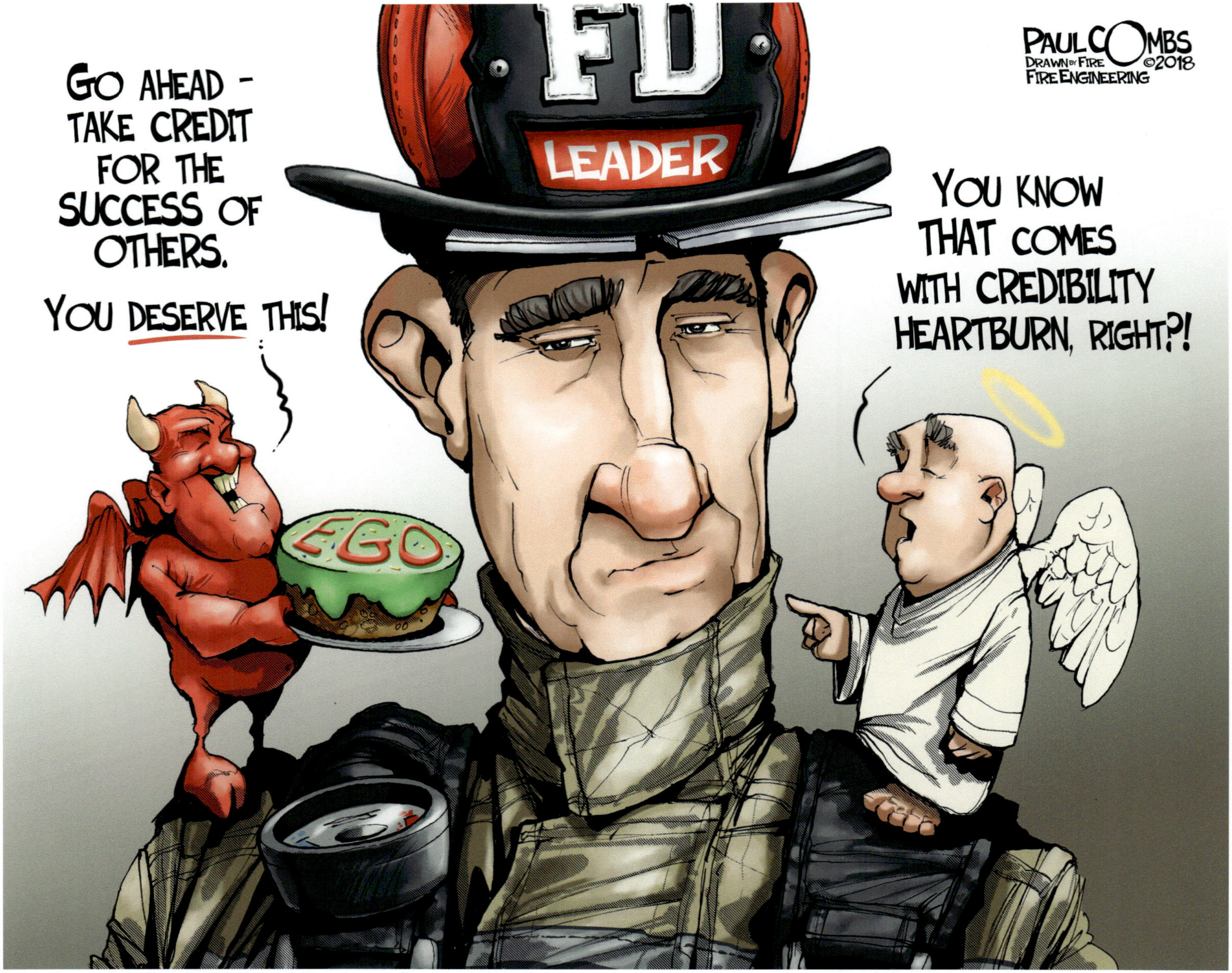
FD
LEADER
GO AHEAD - TAKE CREDIT FOR THE SUCCESS OF OTHERS.
YOU DESERVE THIS!
YOU KNOW THAT COMES WITH CREDIBILITY HEARTBURN, RIGHT?!
EGO
PAUL COMBS
DRAWN BY FIRE ©2018
FIRE ENGINEERING

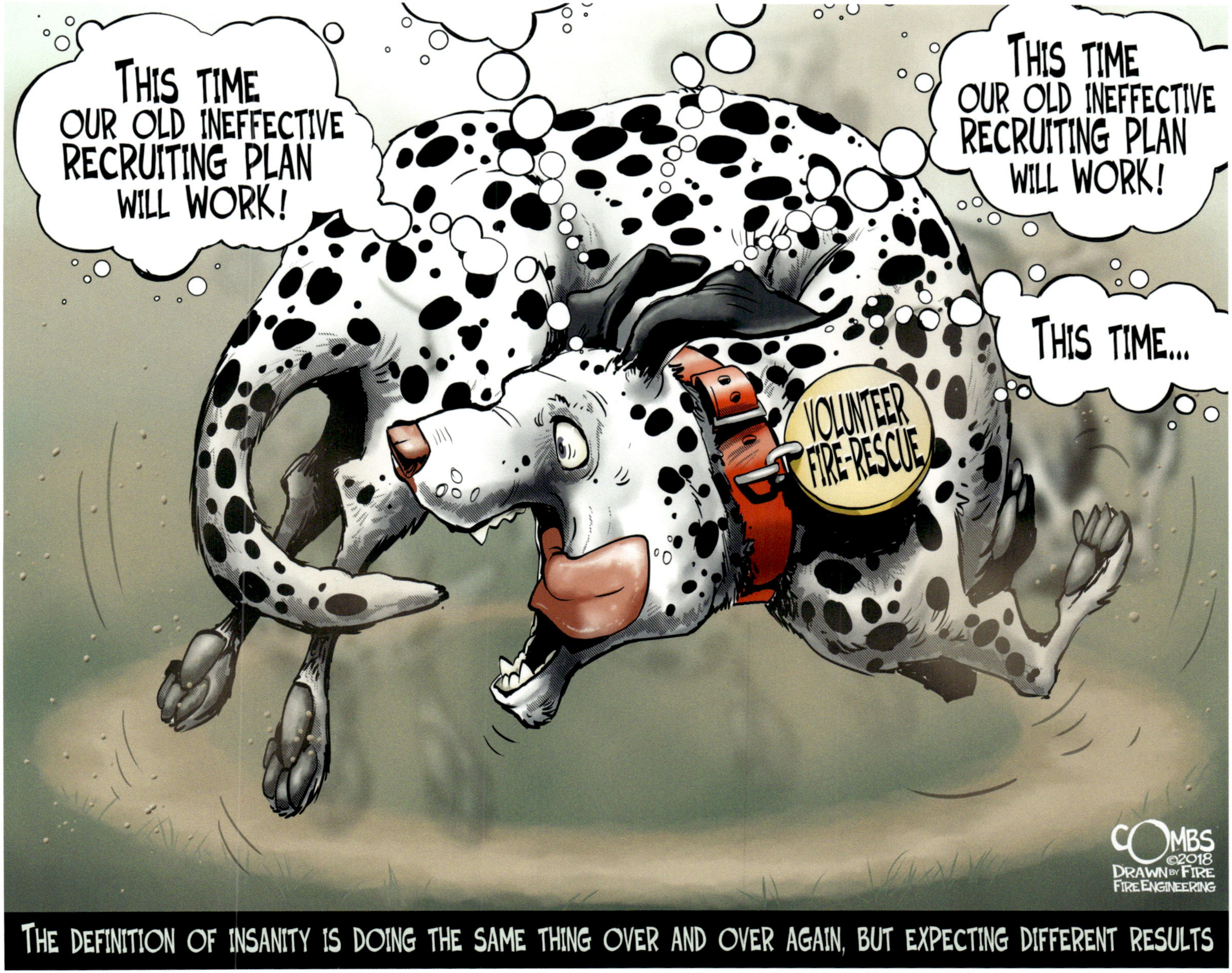
THIS TIME OUR OLD INEFFECTIVE RECRUITING PLAN WILL WORK!
THIS TIME OUR OLD INEFFECTIVE RECRUITING PLAN WILL WORK!
THIS TIME...
VOLUNTEER FIRE-RESCUE
COMBS ©2018 DRAWN BY FIRE FIRE ENGINEERING
THE DEFINITION OF INSANITY IS DOING THE SAME THING OVER AND OVER AGAIN, BUT EXPECTING DIFFERENT RESULTS

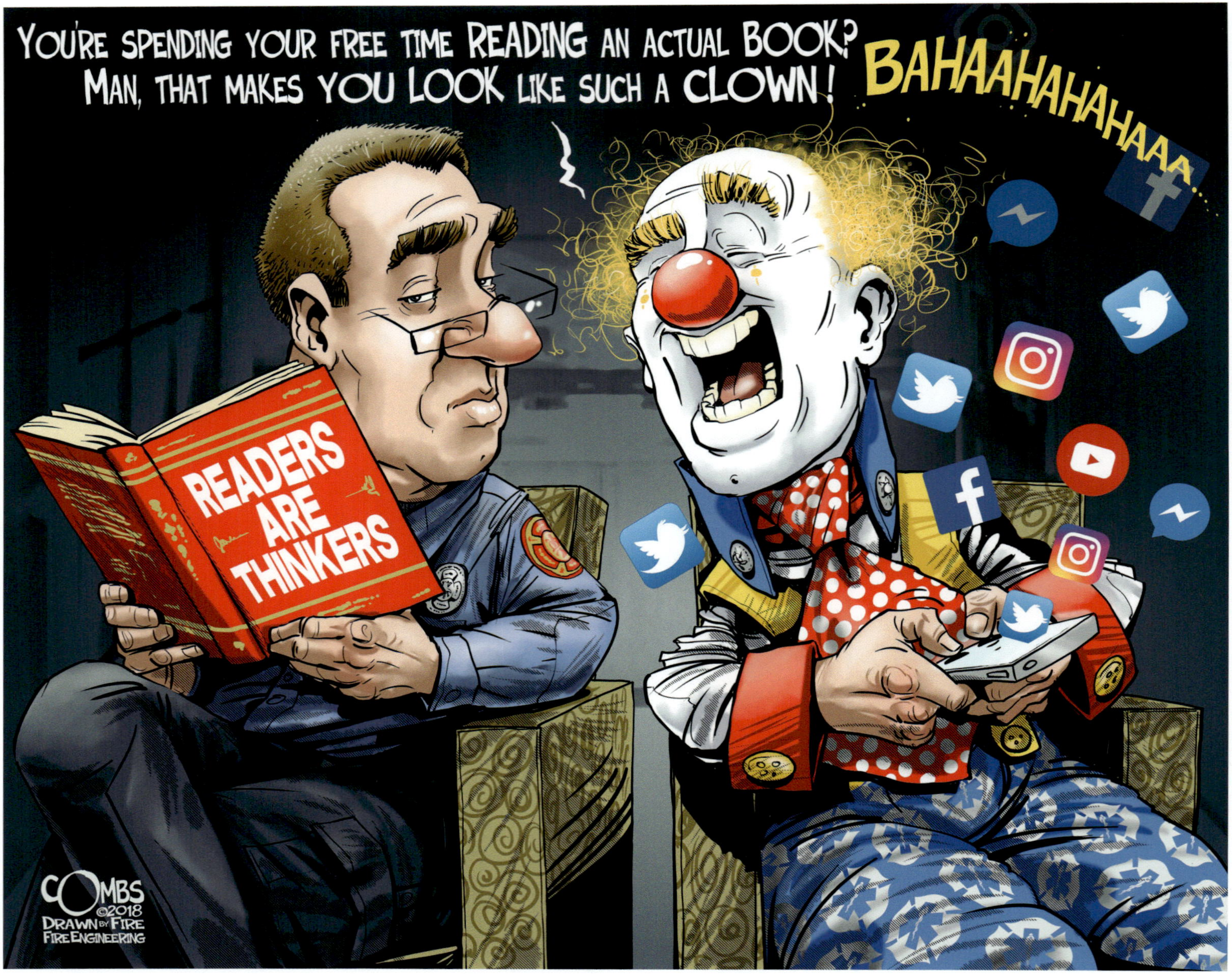

YOU'RE SPENDING YOUR FREE TIME READING AN ACTUAL BOOK? MAN, THAT MAKES YOU LOOK LIKE SUCH A CLOWN!
BAHAAHAHAHAAA
READERS ARE THINKERS
COMBS ©2018 DRAWN BY FIRE FIRE ENGINEERING

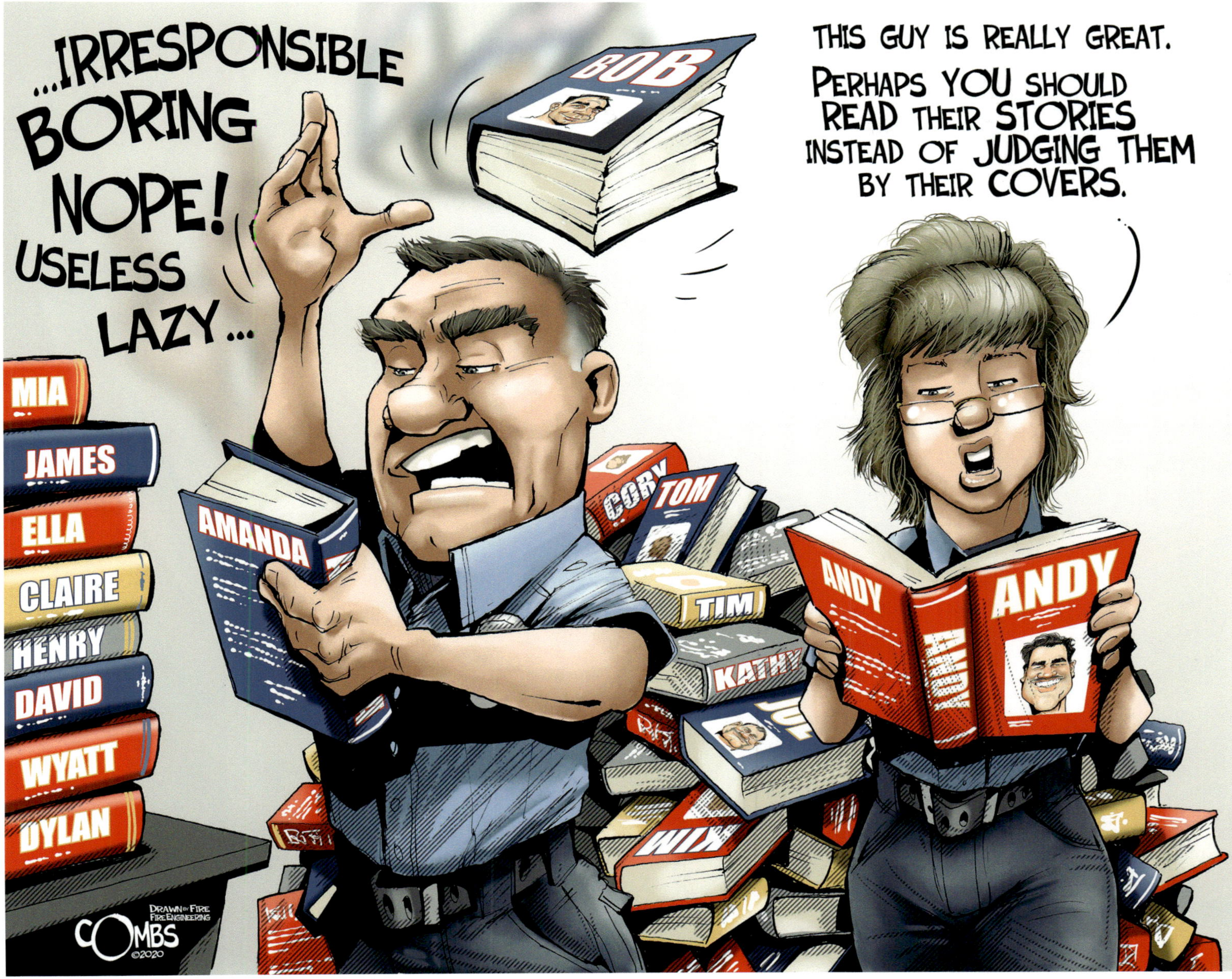
...IRRESPONSIBLE
BORING
NOPE!
USELESS
LAZY...
BOB
THIS GUY IS REALLY GREAT.
PERHAPS YOU SHOULD READ THEIR STORIES INSTEAD OF JUDGING THEM BY THEIR COVERS.
MIA
JAMES
ELLA
CLAIRE
HENRY
DAVID
WYATT
DYLAN
AMANDA
CORY
TOM
TIM
KATHY
MIX
ANDY
ANDY
ANDY
DRAWN BY FIRE
FIRE ENGINEERING
COMBS
©2020

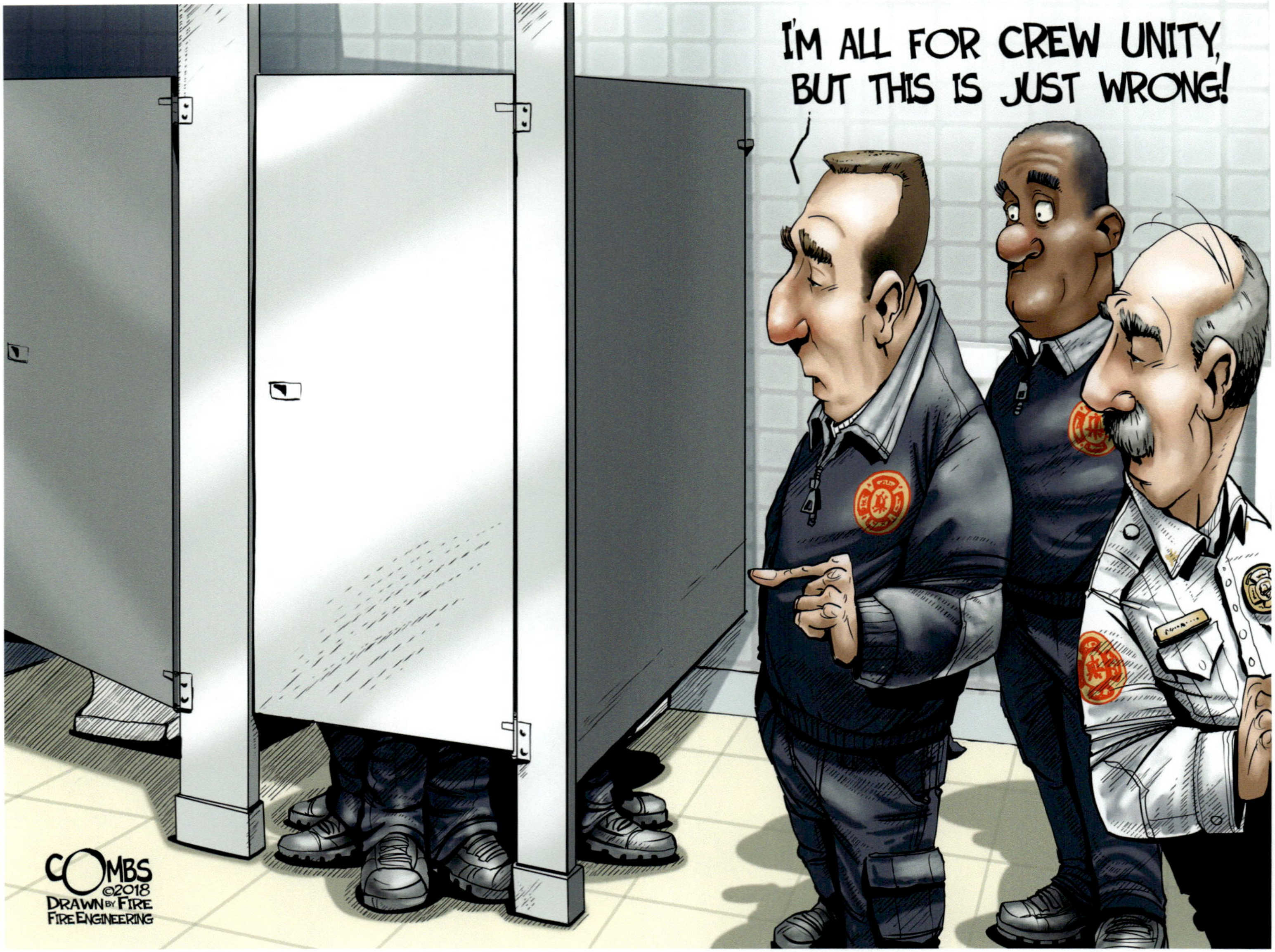

HOW CLOSE IS YOUR CREW?

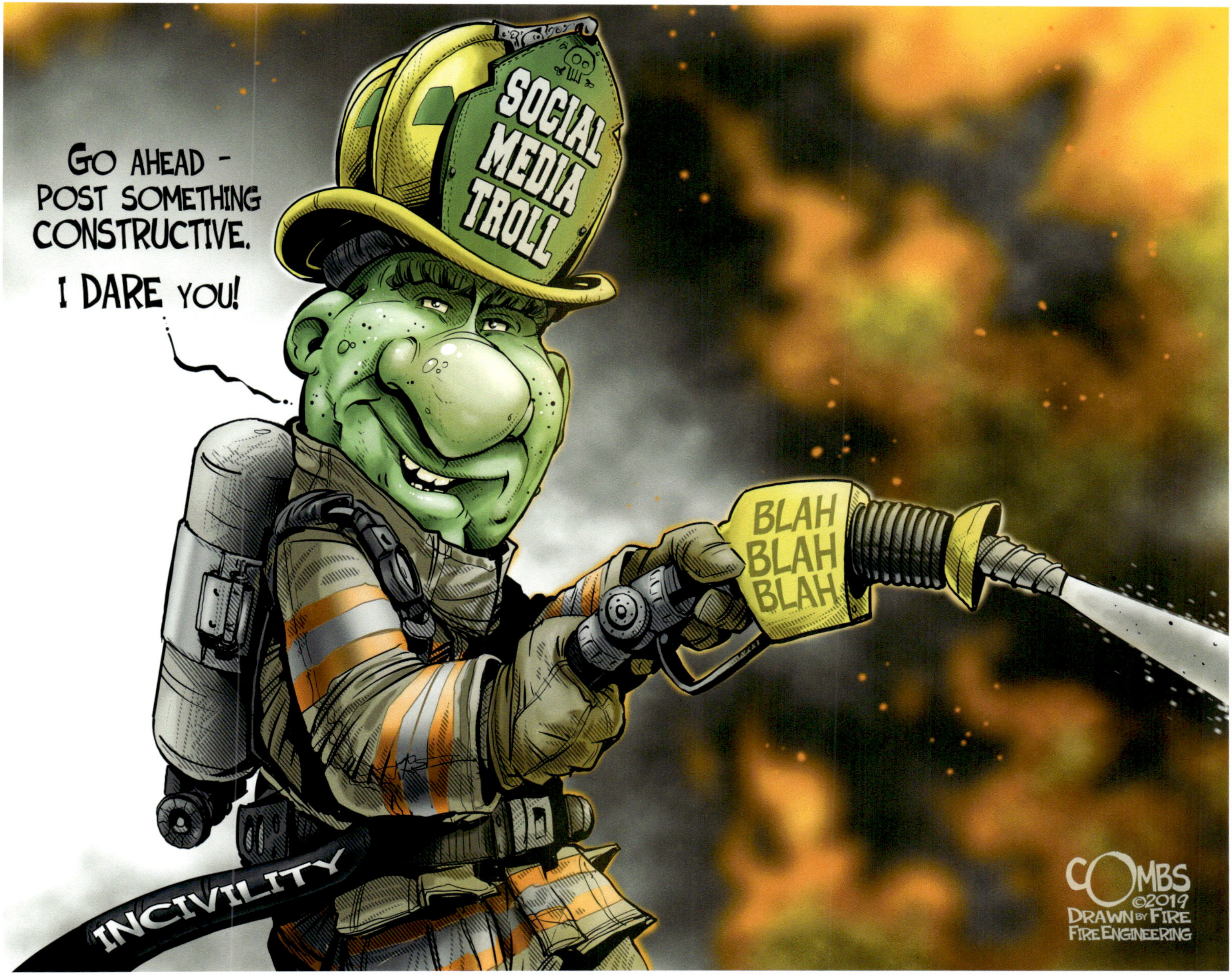
GO AHEAD -
POST SOMETHING
CONSTRUCTIVE.
I DARE YOU!
SOCIAL
MEDIA
TROLL
BLAH
BLAH
BLAH
INCIVILITY
COMBS
©2019
DRAWN BY FIRE
FIRE ENGINEERING

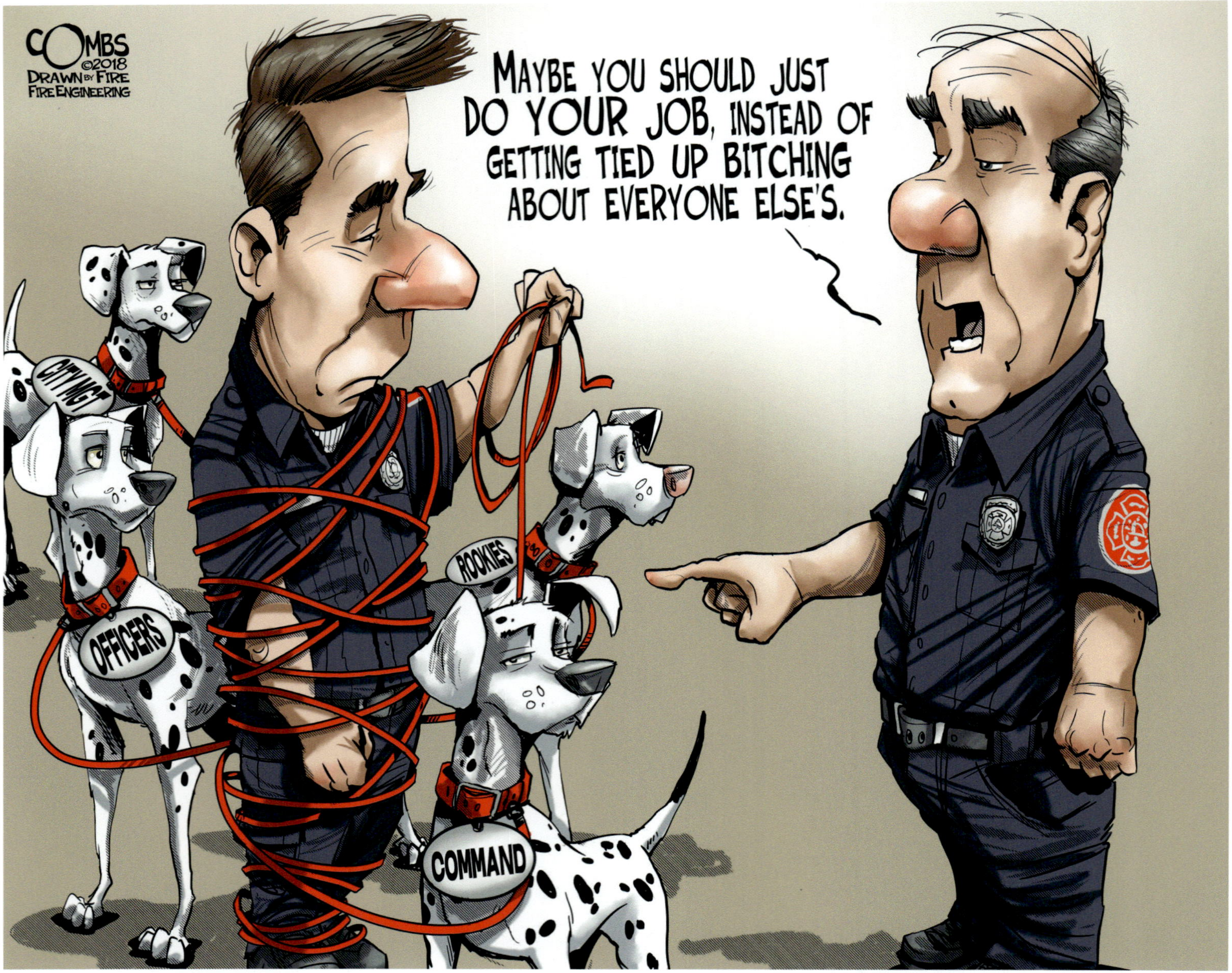
COMBS ©2018 DRAWN BY FIRE FIRE ENGINEERING
MAYBE YOU SHOULD JUST DO YOUR JOB, INSTEAD OF GETTING TIED UP BITCHING ABOUT EVERYONE ELSE'S.
CITY MGT
OFFICERS
ROOKIES
COMMAND

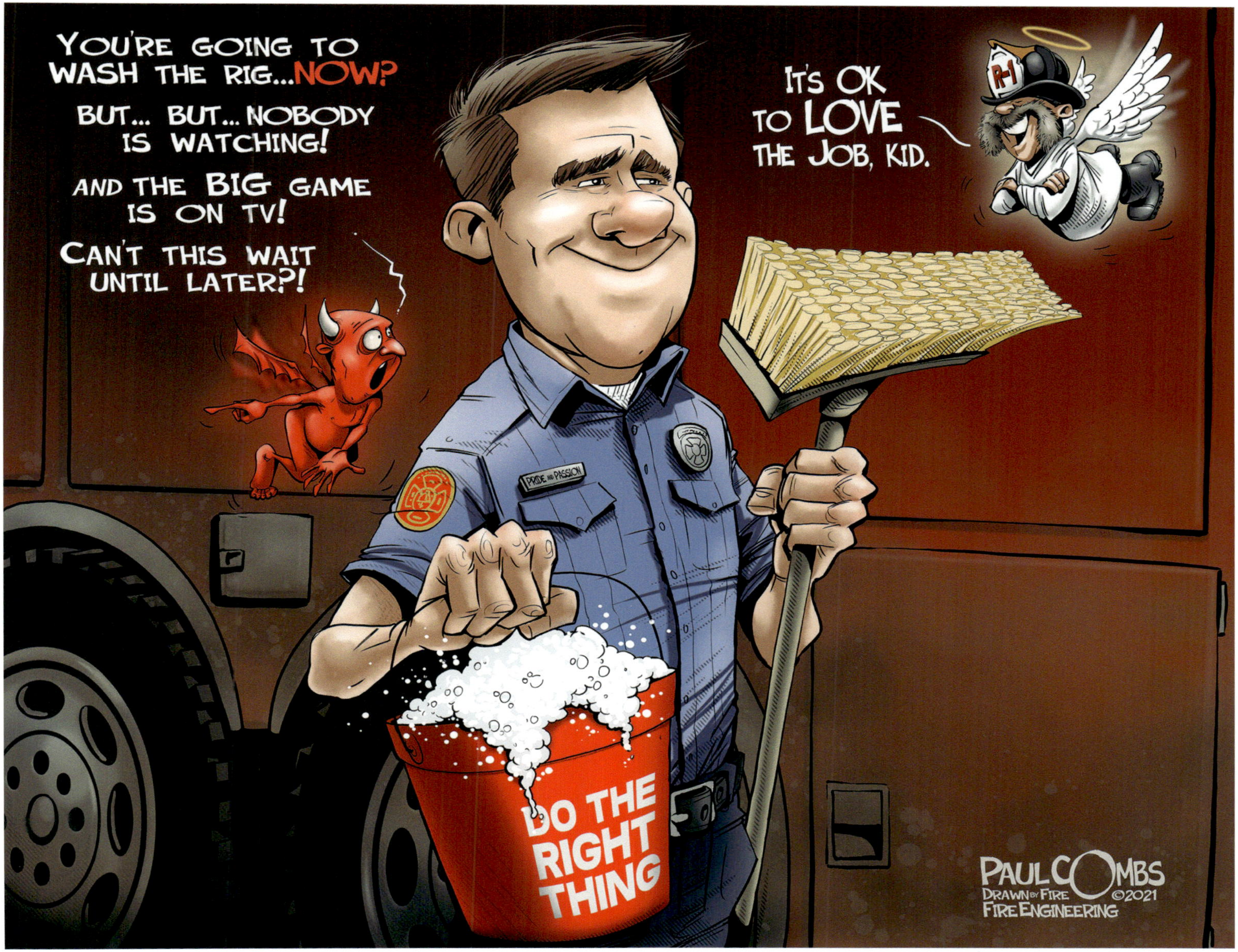
YOU'RE GOING TO WASH THE RIG...NOW?
BUT... BUT... NOBODY IS WATCHING!
AND THE BIG GAME IS ON TV!
CAN'T THIS WAIT UNTIL LATER?!
IT'S OK TO LOVE THE JOB, KID.
R-1
PRIDE AND PASSION
DO THE RIGHT THING
PAUL COMBS
DRAWN BY FIRE ©2021
FIRE ENGINEERING

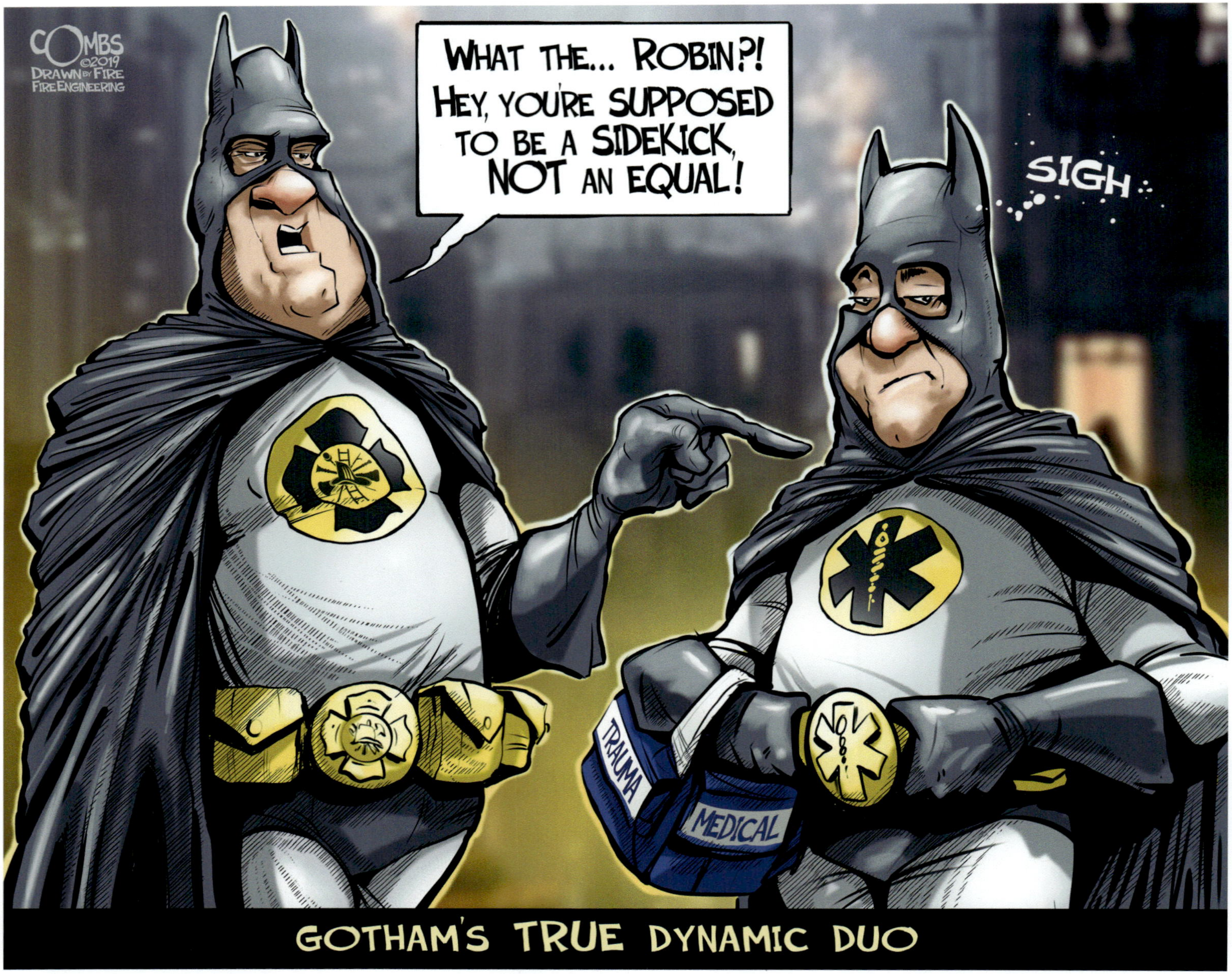

GOTHAM'S TRUE DYNAMIC DUO

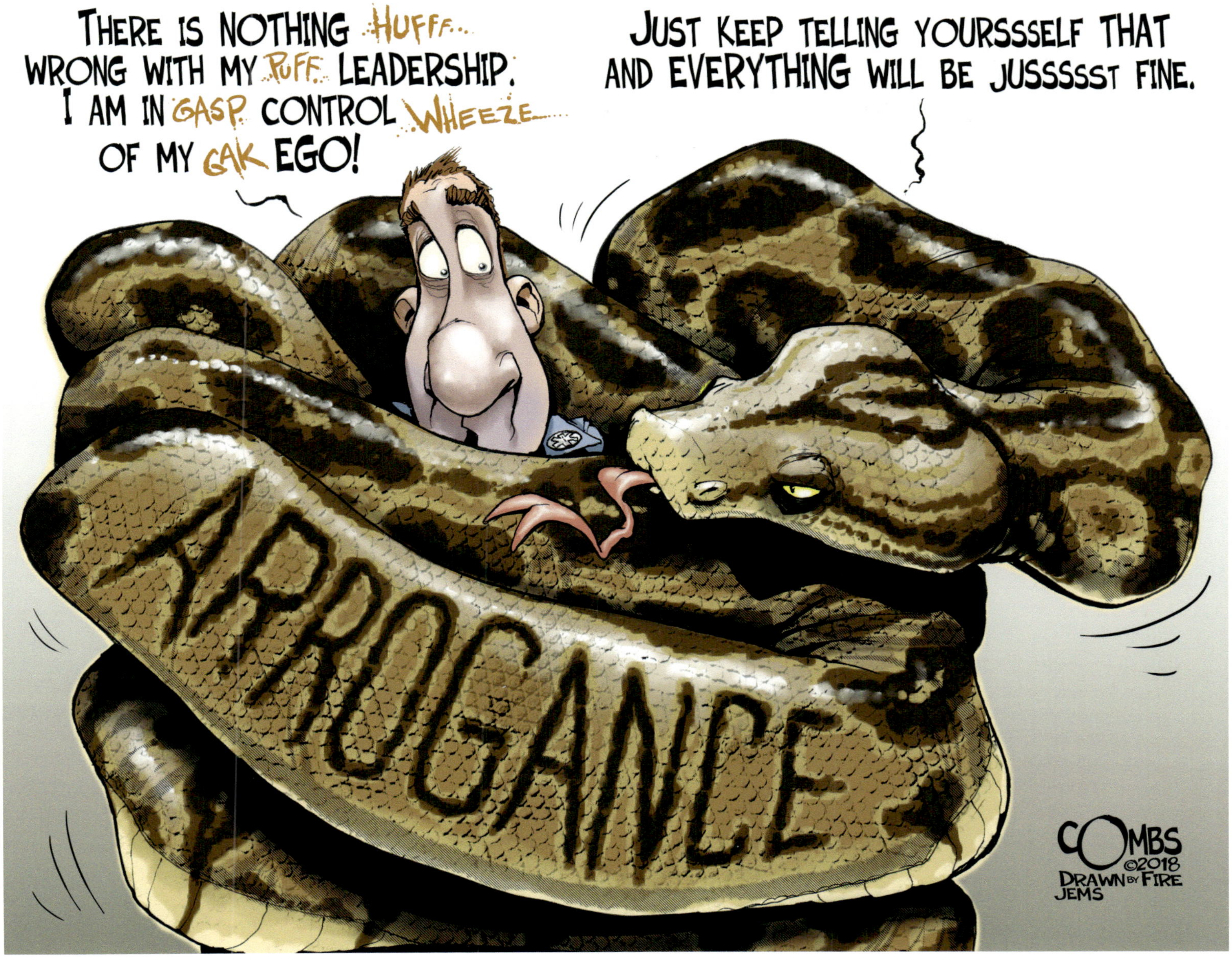
THERE IS NOTHING HUFFF... WRONG WITH MY PUFF LEADERSHIP. I AM IN GASP CONTROL WHEEZE... OF MY GAK EGO!
JUST KEEP TELLING YOURSSSELF THAT AND EVERYTHING WILL BE JUSSSSST FINE.
ARROGANCE
COMBS ©2018 DRAWN BY FIRE JEMS

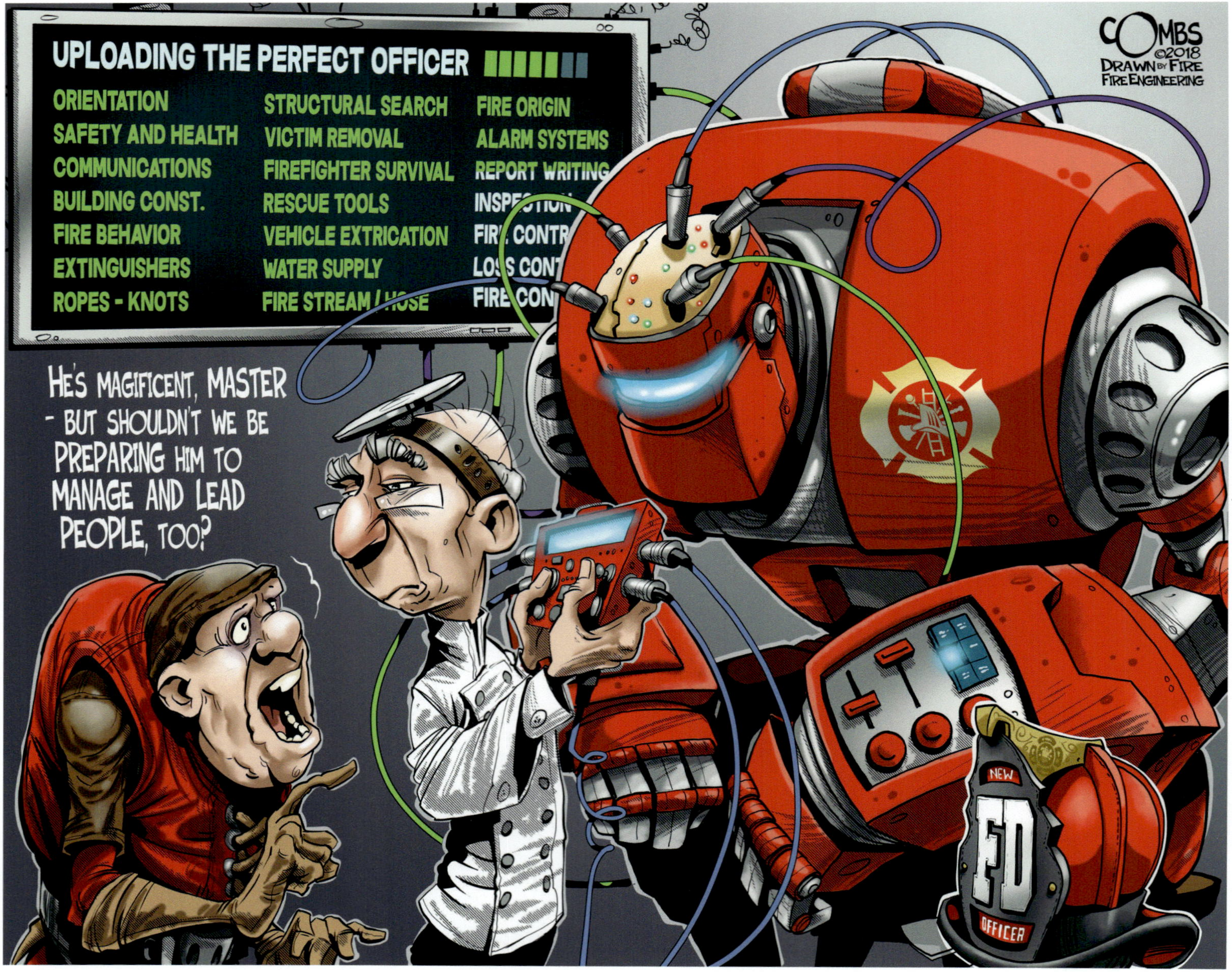

UPLOADING THE PERFECT OFFICER
ORIENTATION
SAFETY AND HEALTH
COMMUNICATIONS
BUILDING CONST.
FIRE BEHAVIOR
EXTINGUISHERS
ROPES - KNOTS
STRUCTURAL SEARCH
VICTIM REMOVAL
FIREFIGHTER SURVIVAL
RESCUE TOOLS
VEHICLE EXTRICATION
WATER SUPPLY
FIRE ORIGIN
ALARM SYSTEMS
REPORT WRITING
COMBS ©2018 DRAWN BY FIRE FIRE ENGINEERING
HE'S MAGIFICENT, MASTER - BUT SHOULDN'T WE BE PREPARING HIM TO MANAGE AND LEAD PEOPLE, TOO?
NEW
FD
OFFICER

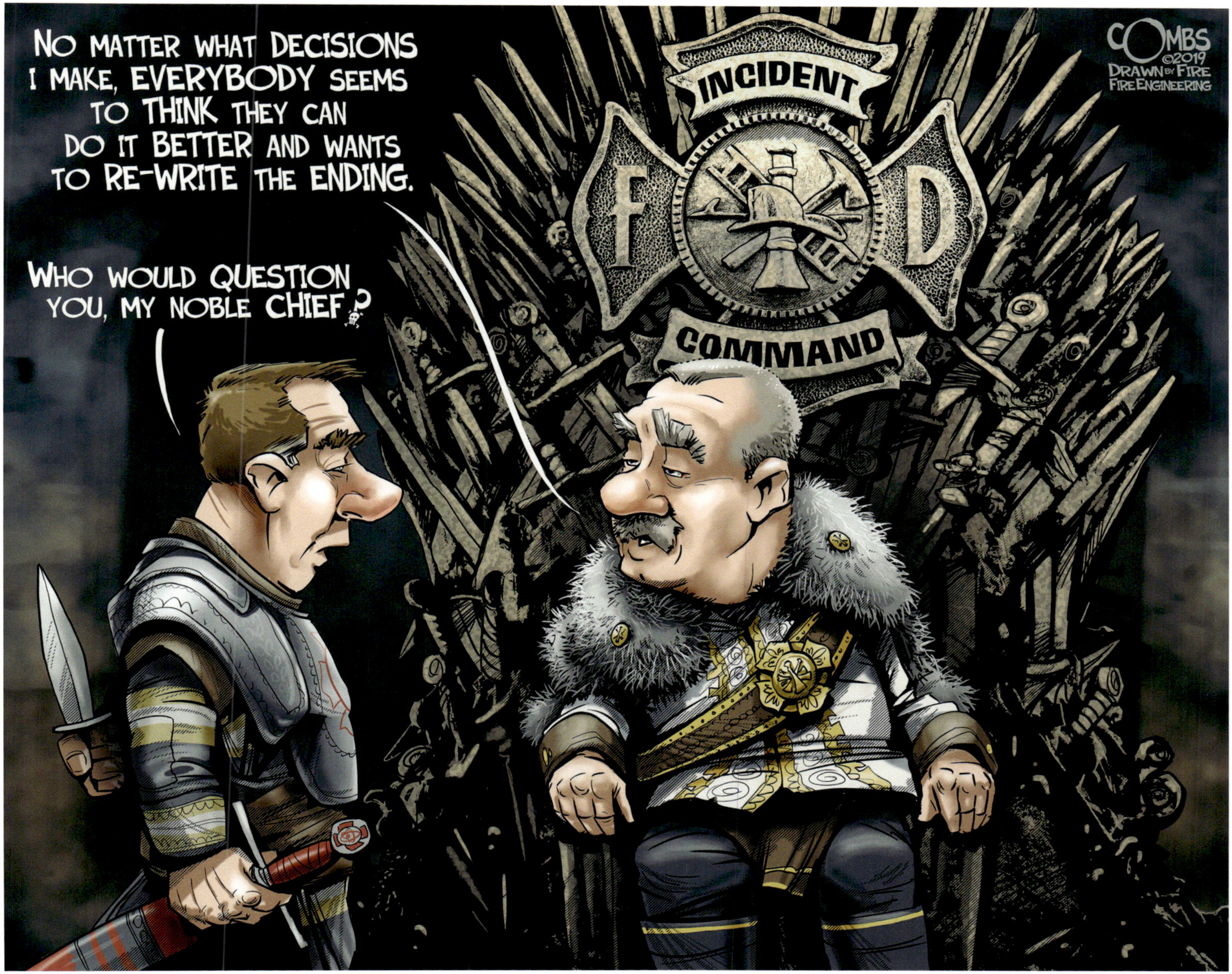
NO MATTER WHAT DECISIONS I MAKE, EVERYBODY SEEMS TO THINK THEY CAN DO IT BETTER AND WANTS TO RE-WRITE THE ENDING.
WHO WOULD QUESTION YOU, MY NOBLE CHIEF?
INCIDENT
F D
COMMAND
COMBS ©2019 DRAWN BY FIRE FIRE ENGINEERING

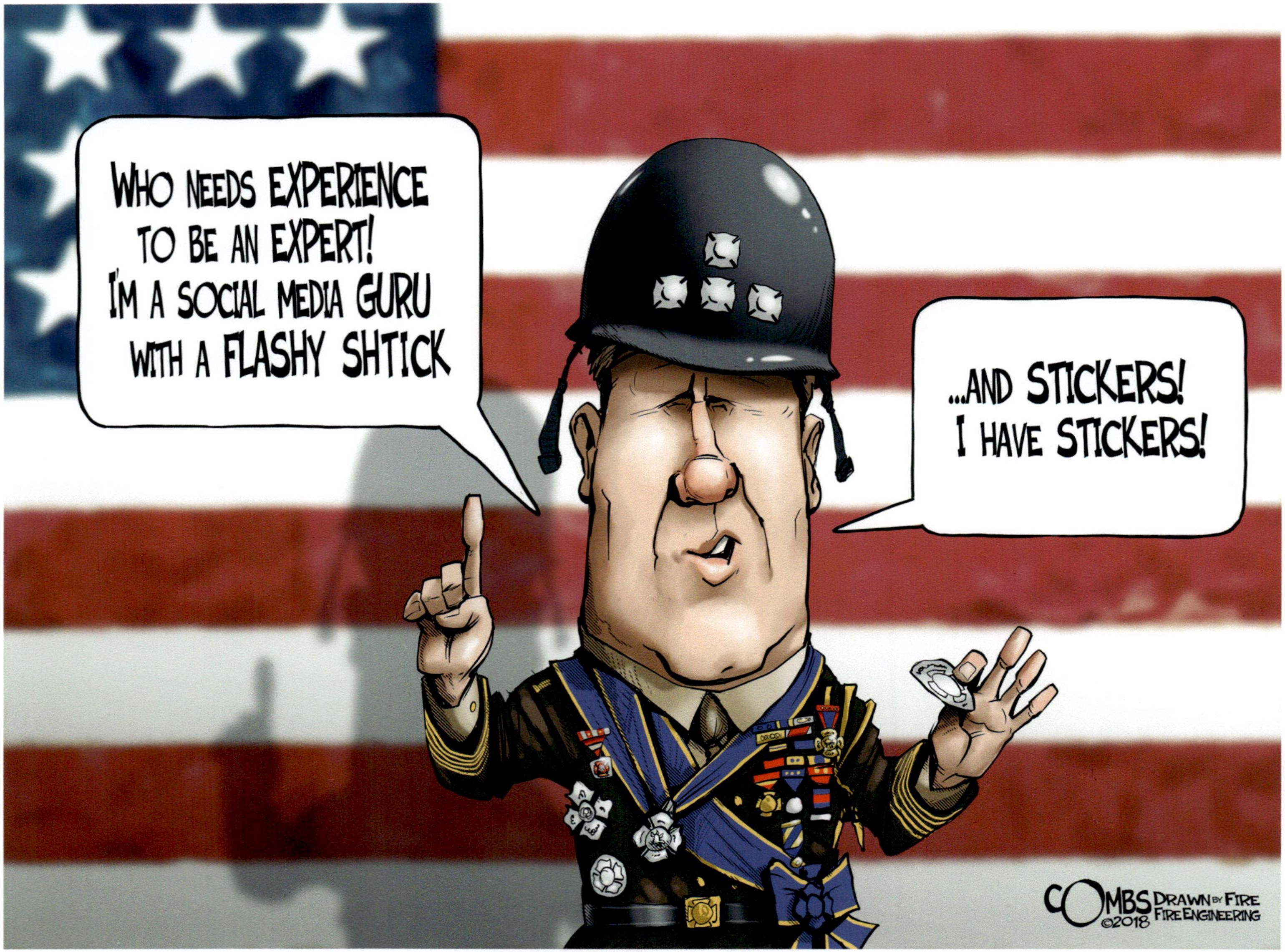

GENERAL POSER

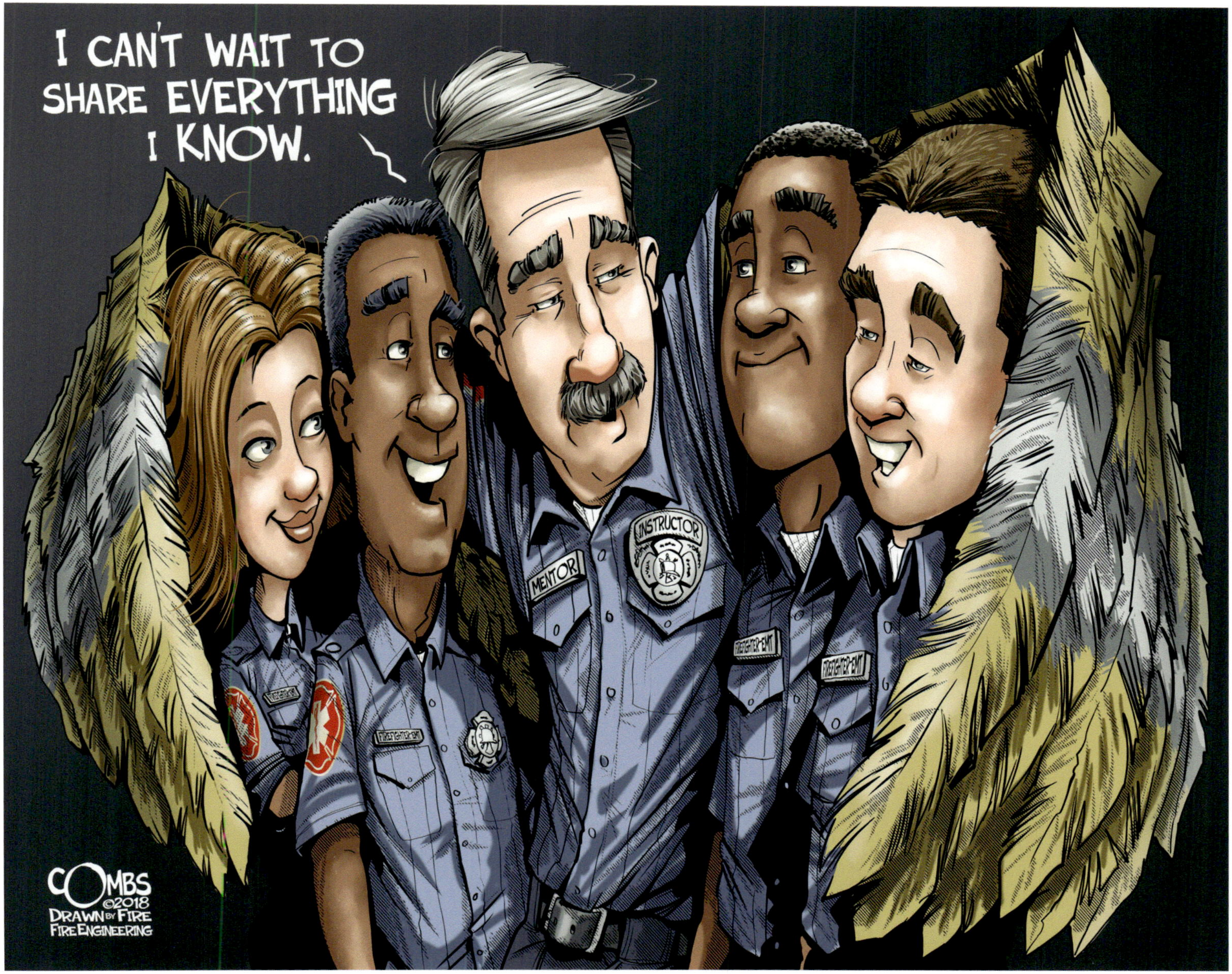
I CAN'T WAIT TO SHARE EVERYTHING I KNOW.
INSTRUCTOR
MENTOR
COMBS ©2018 DRAWN BY FIRE FIRE ENGINEERING

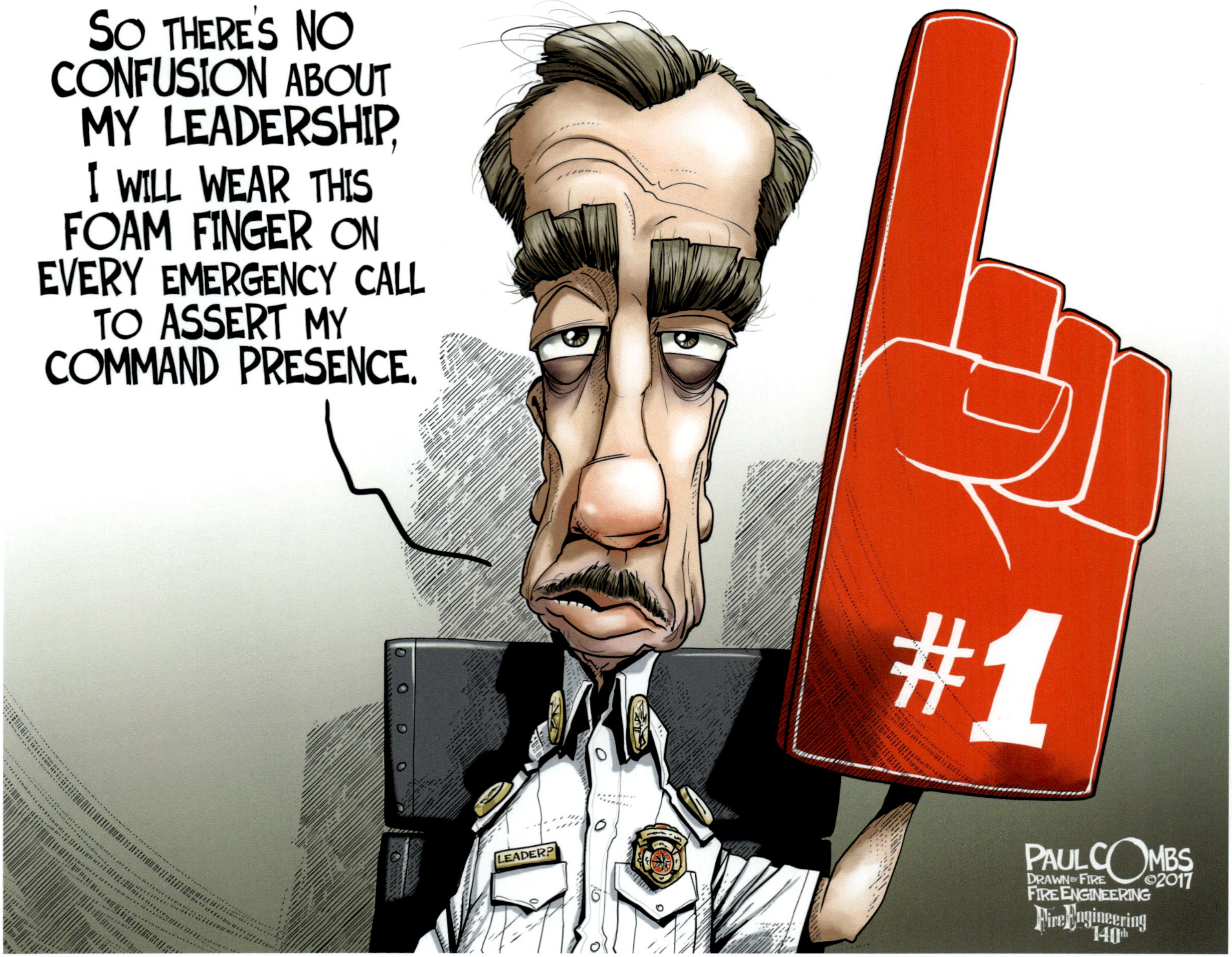
SO THERE'S NO CONFUSION ABOUT MY LEADERSHIP, I WILL WEAR THIS FOAM FINGER ON EVERY EMERGENCY CALL TO ASSERT MY COMMAND PRESENCE.
#1
LEADER?
PAUL COMBS
DRAWN BY FIRE ©2017
FIRE ENGINEERING
Fire Engineering 140th

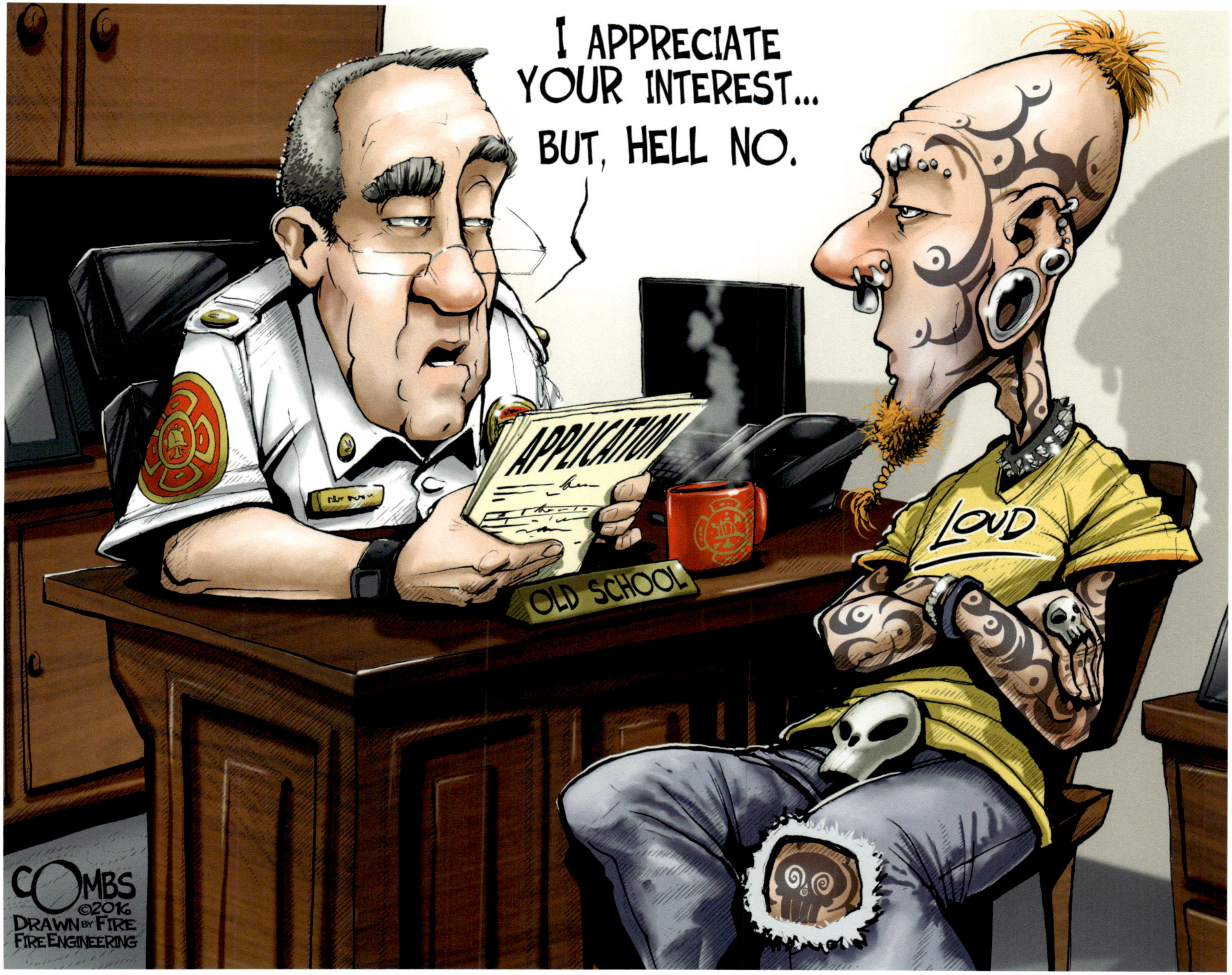
I APPRECIATE YOUR INTEREST... BUT, HELL NO.
APPLICATION
OLD SCHOOL
LOUD
COMBS ©2016 DRAWN BY FIRE FIRE ENGINEERING

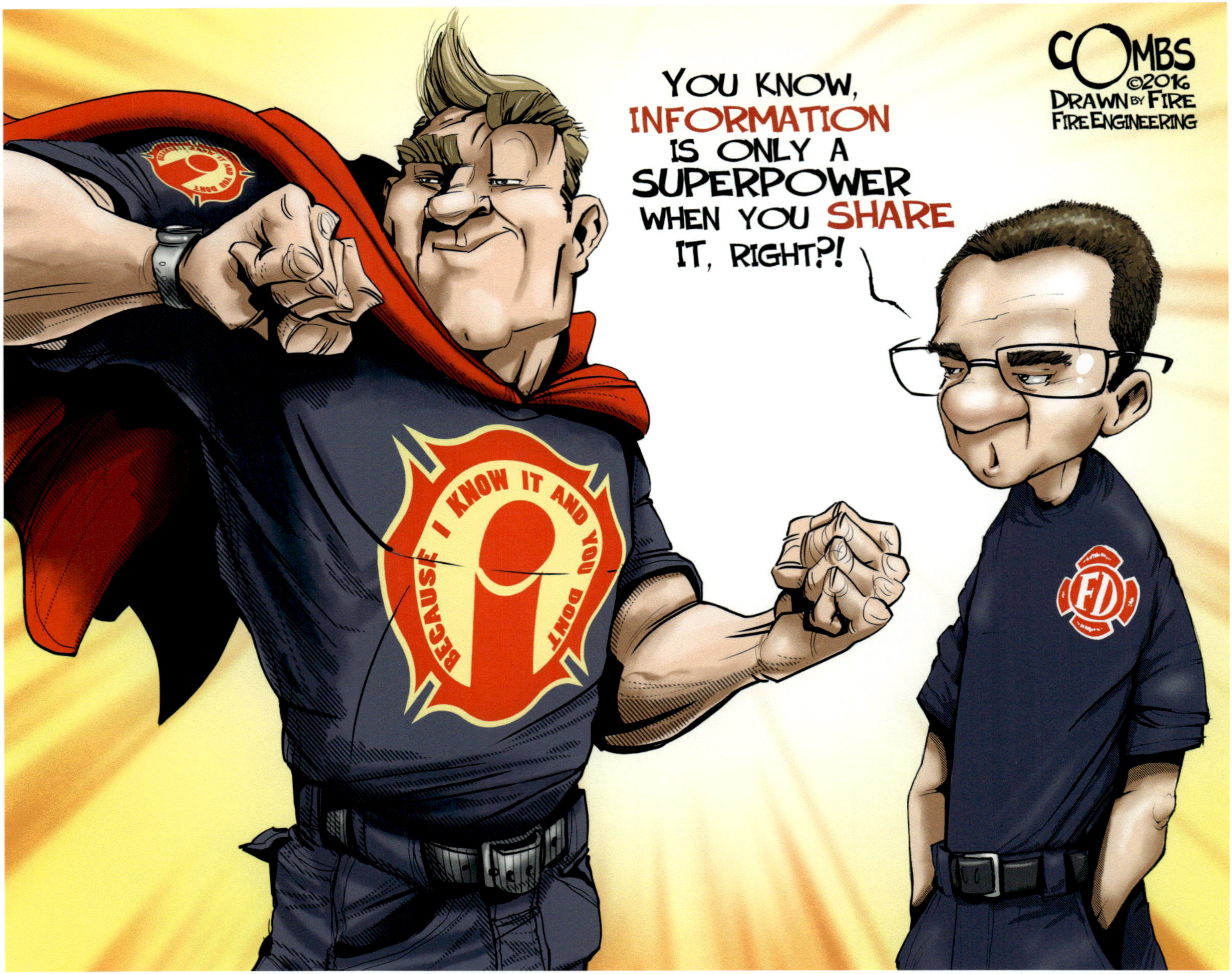
COMBS
©2016
DRAWN BY FIRE
FIRE ENGINEERING
YOU KNOW, INFORMATION IS ONLY A SUPERPOWER WHEN YOU SHARE IT, RIGHT?!
BECAUSE I KNOW IT AND YOU DON'T
FD

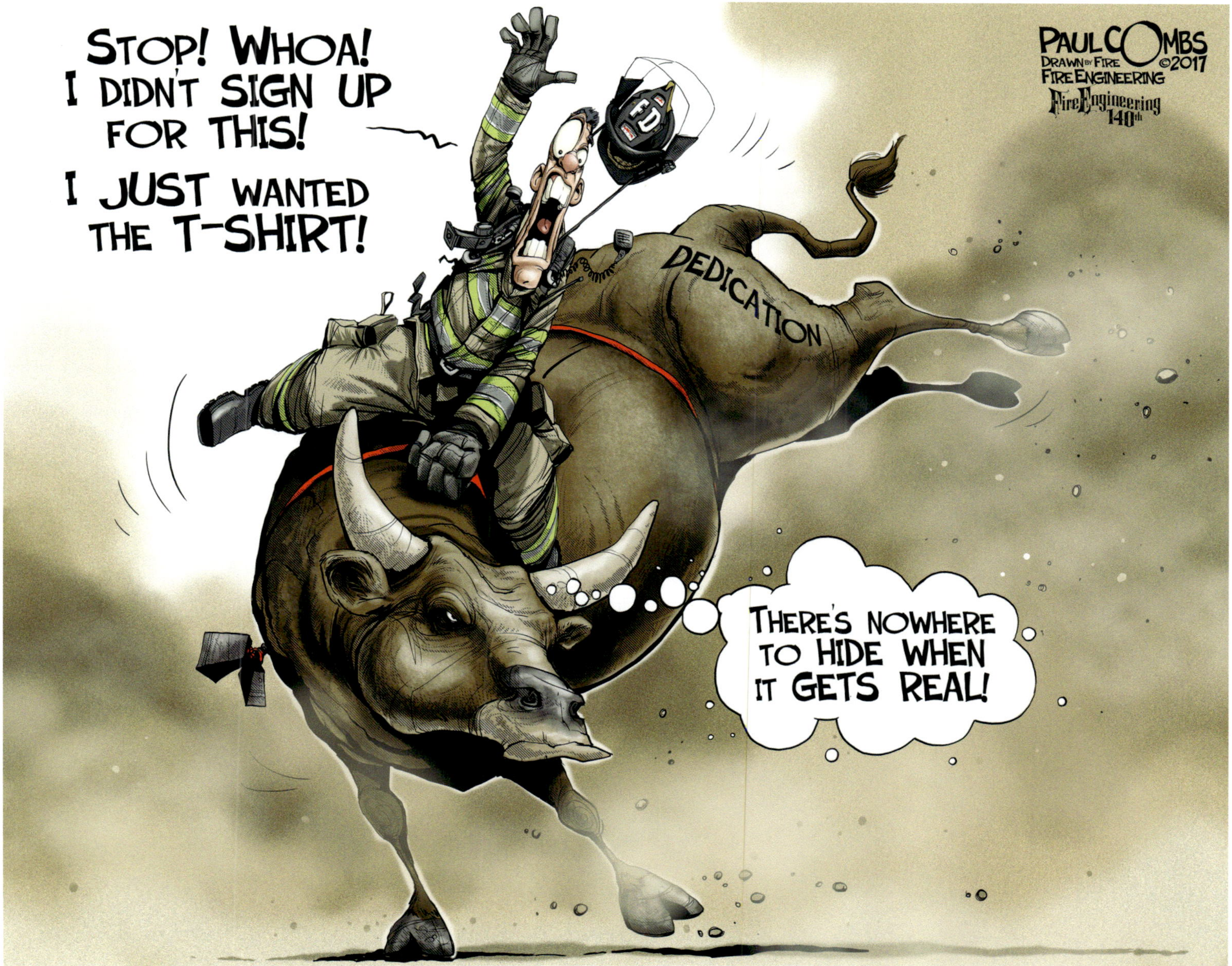
STOP! WHOA!
I DIDN'T SIGN UP FOR THIS!
I JUST WANTED THE T-SHIRT!
FD
DEDICATION
THERE'S NOWHERE TO HIDE WHEN IT GETS REAL!
PAUL COMBS
DRAWN BY FIRE ©2017
FIRE ENGINEERING
FireEngineering 140th

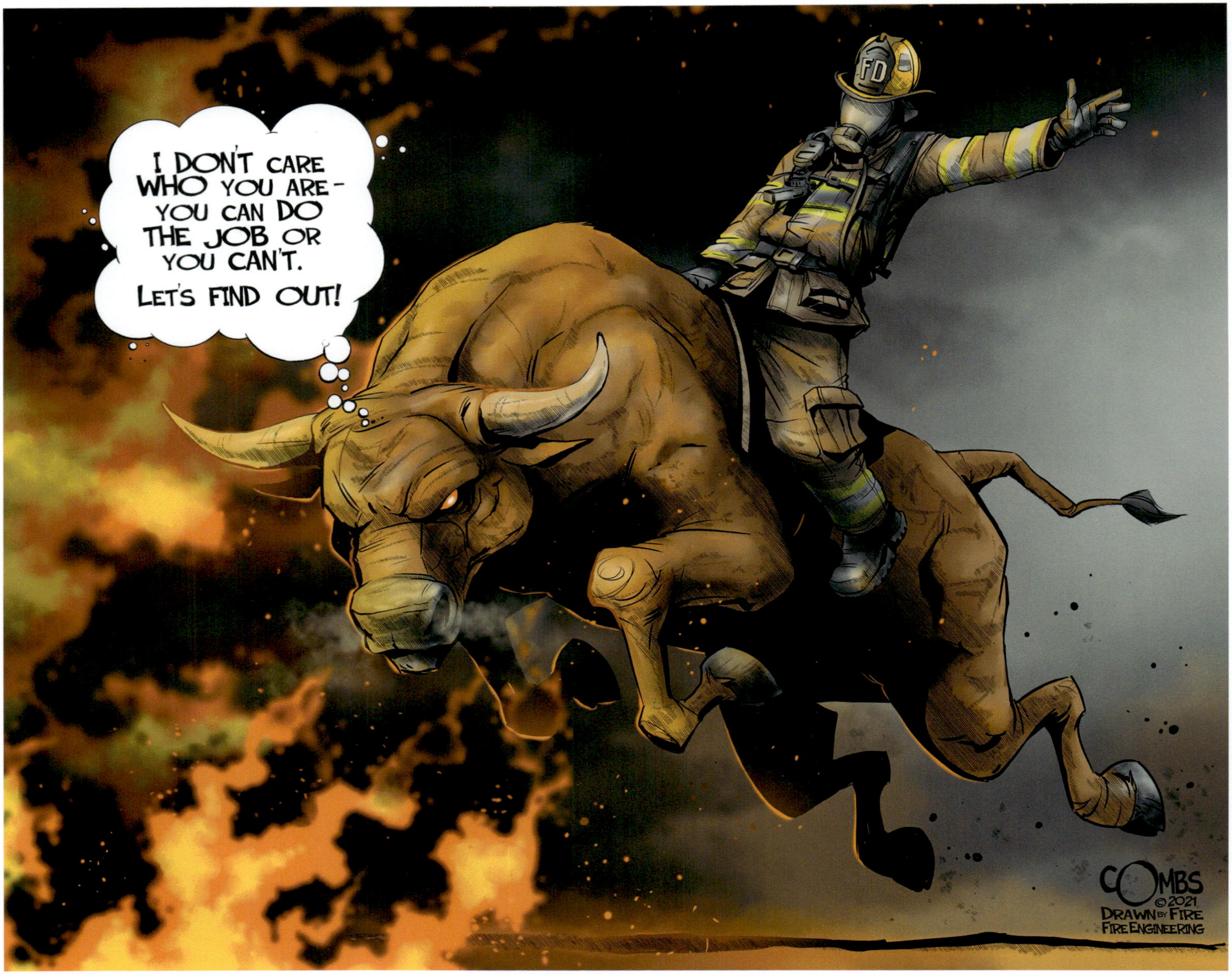
I DON'T CARE WHO YOU ARE - YOU CAN DO THE JOB OR YOU CAN'T.
LET'S FIND OUT!
COMBS ©2021 DRAWN BY FIRE FIRE ENGINEERING

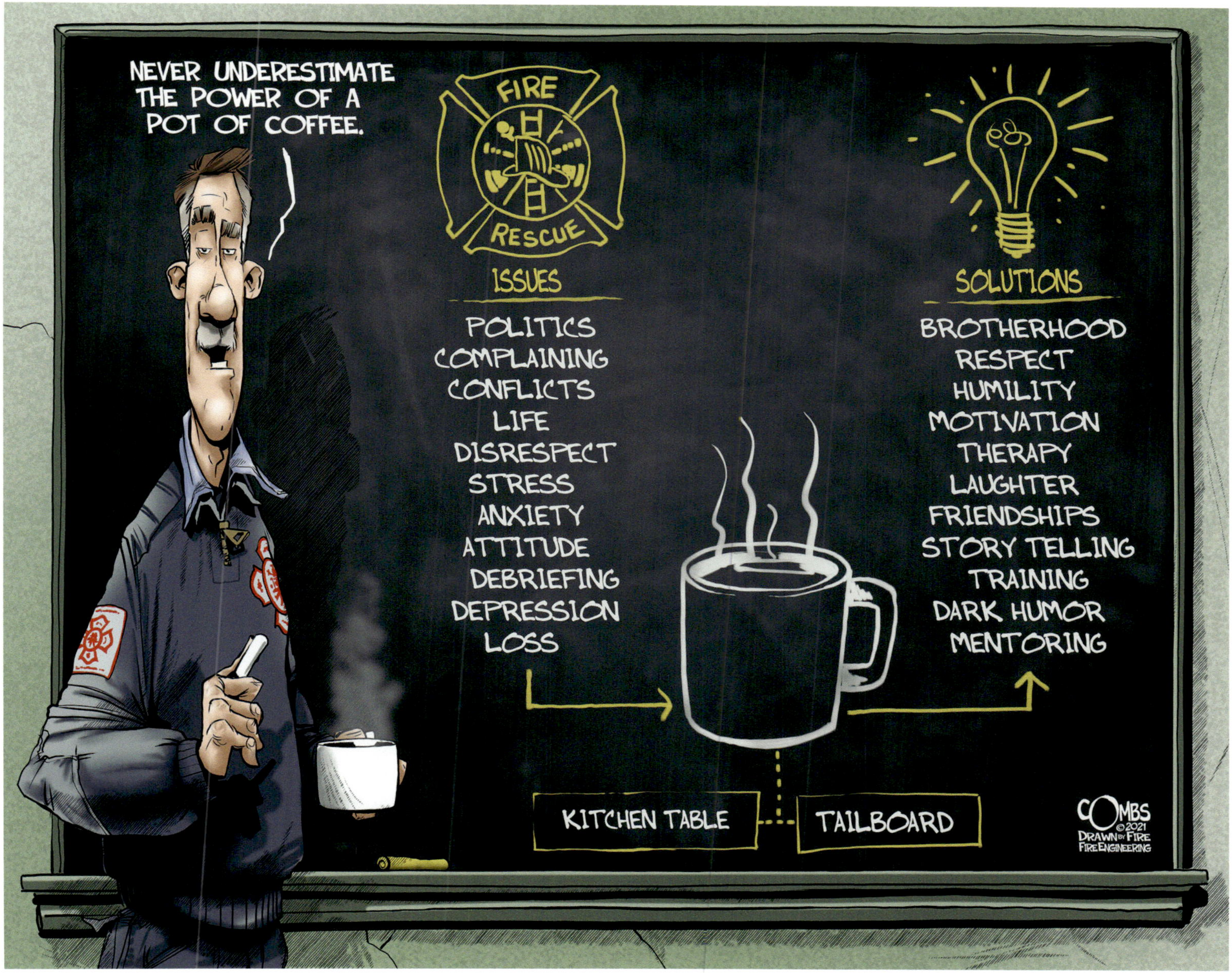
NEVER UNDERESTIMATE THE POWER OF A POT OF COFFEE.
FIRE
RESCUE
ISSUES
POLITICS
COMPLAINING
CONFLICTS
LIFE
DISRESPECT
STRESS
ANXIETY
ATTITUDE
DEBRIEFING
DEPRESSION
LOSS
SOLUTIONS
BROTHERHOOD
RESPECT
HUMILITY
MOTIVATION
THERAPY
LAUGHTER
FRIENDSHIPS
STORY TELLING
TRAINING
DARK HUMOR
MENTORING
KITCHEN TABLE
TAILBOARD
COMBS ©2021 DRAWN BY FIRE FIRE ENGINEERING

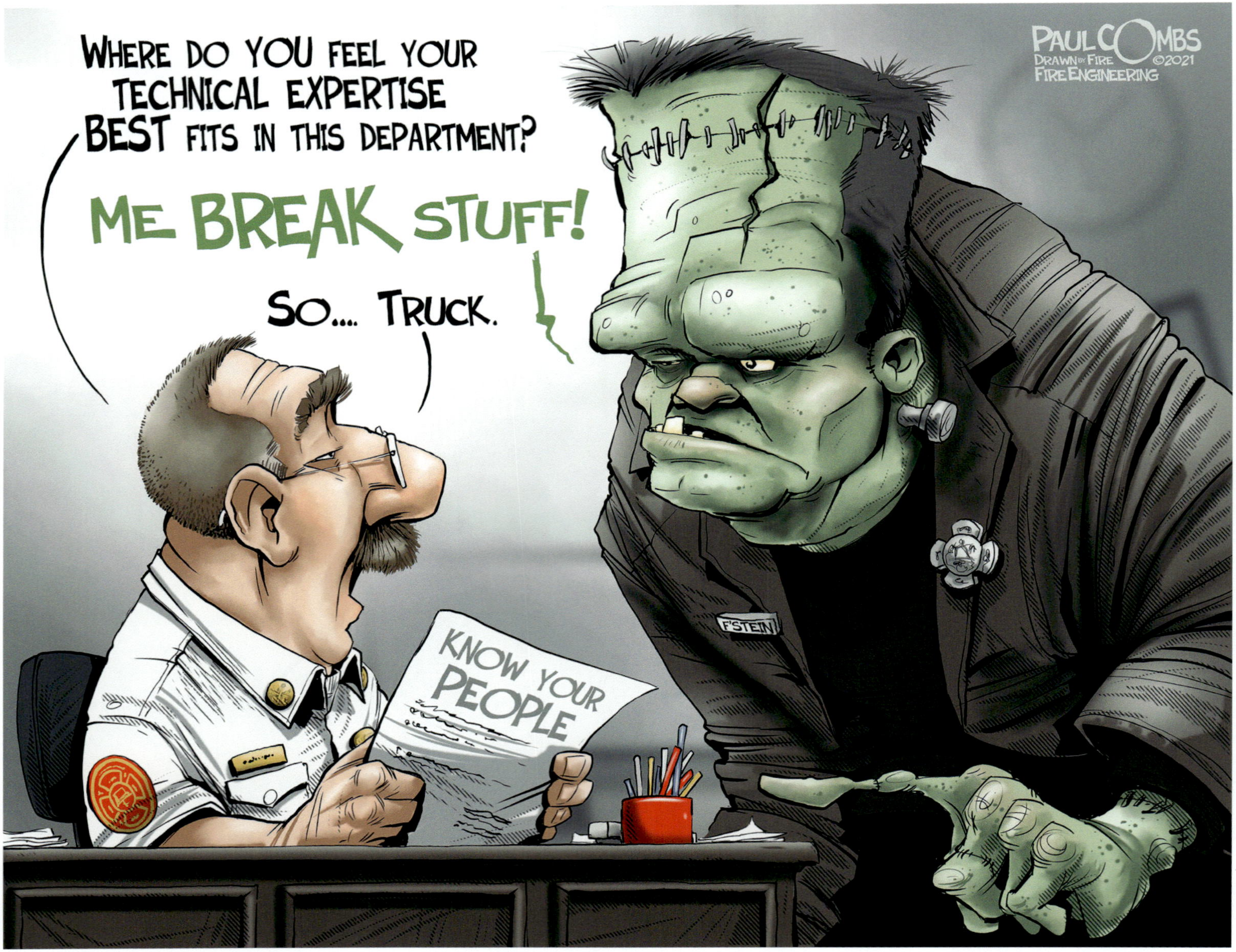
PAUL COMBS
DRAWN BY FIRE ©2021
FIRE ENGINEERING
WHERE DO YOU FEEL YOUR TECHNICAL EXPERTISE BEST FITS IN THIS DEPARTMENT?
ME BREAK STUFF!
SO.... TRUCK.
KNOW YOUR PEOPLE
F'STEIN

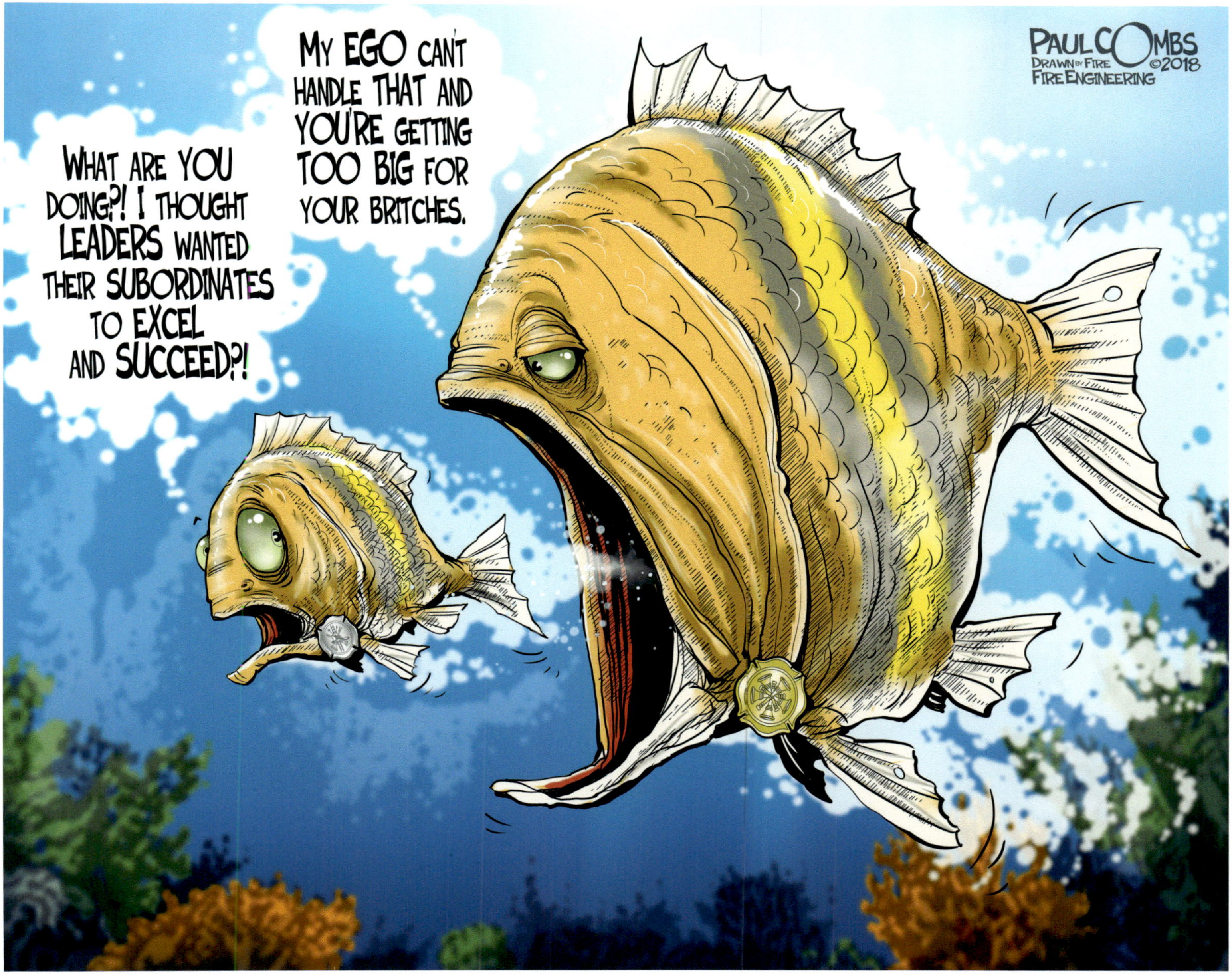
MY EGO CAN'T HANDLE THAT AND YOU'RE GETTING TOO BIG FOR YOUR BRITCHES.
WHAT ARE YOU DOING?! I THOUGHT LEADERS WANTED THEIR SUBORDINATES TO EXCEL AND SUCCEED?!
PAUL COMBS
DRAWN BY FIRE ©2018
FIRE ENGINEERING

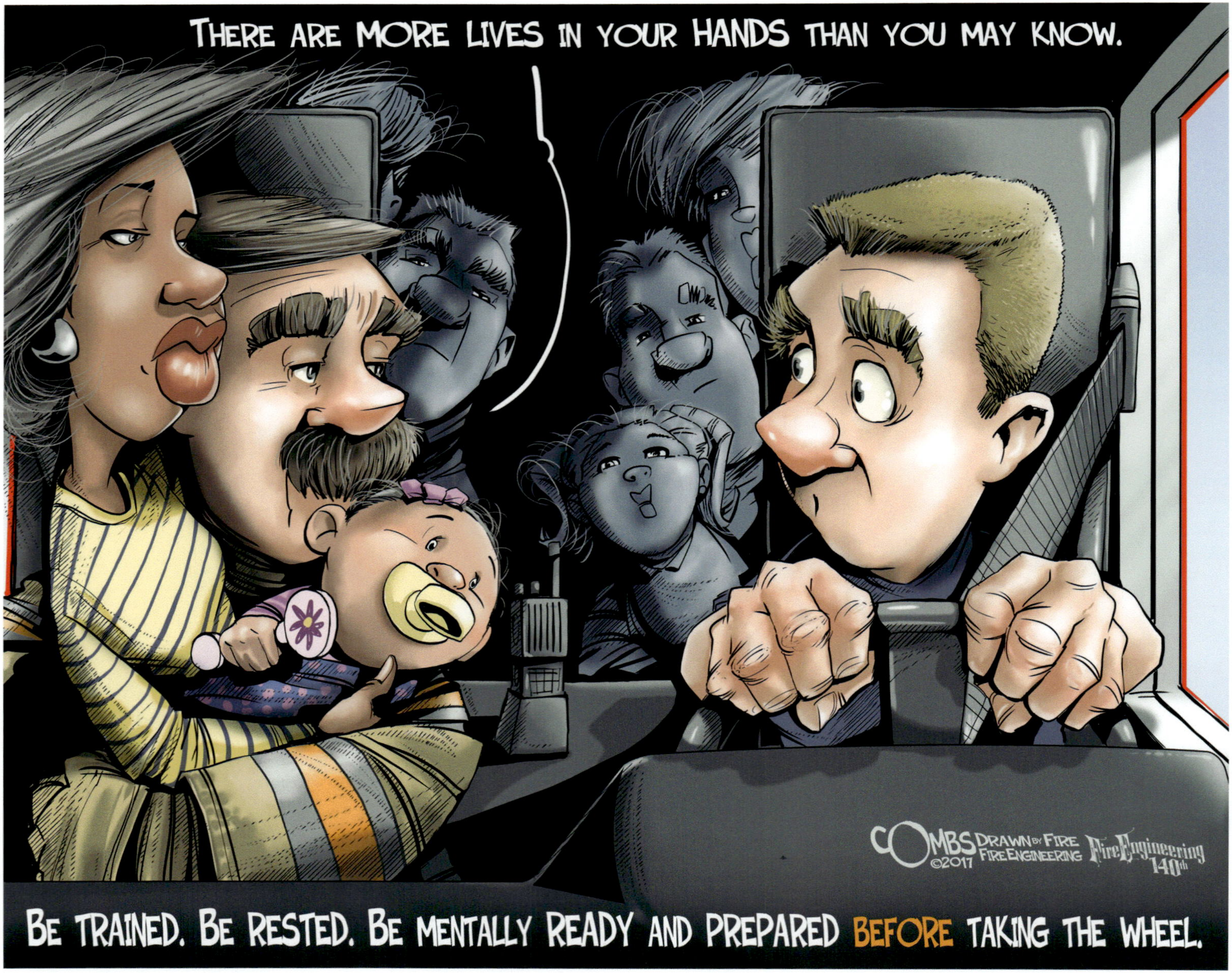
THERE ARE MORE LIVES IN YOUR HANDS THAN YOU MAY KNOW.
COMBS DRAWN BY FIRE FIRE ENGINEERING ©2017
Fire Engineering 140th
BE TRAINED. BE RESTED. BE MENTALLY READY AND PREPARED BEFORE TAKING THE WHEEL.

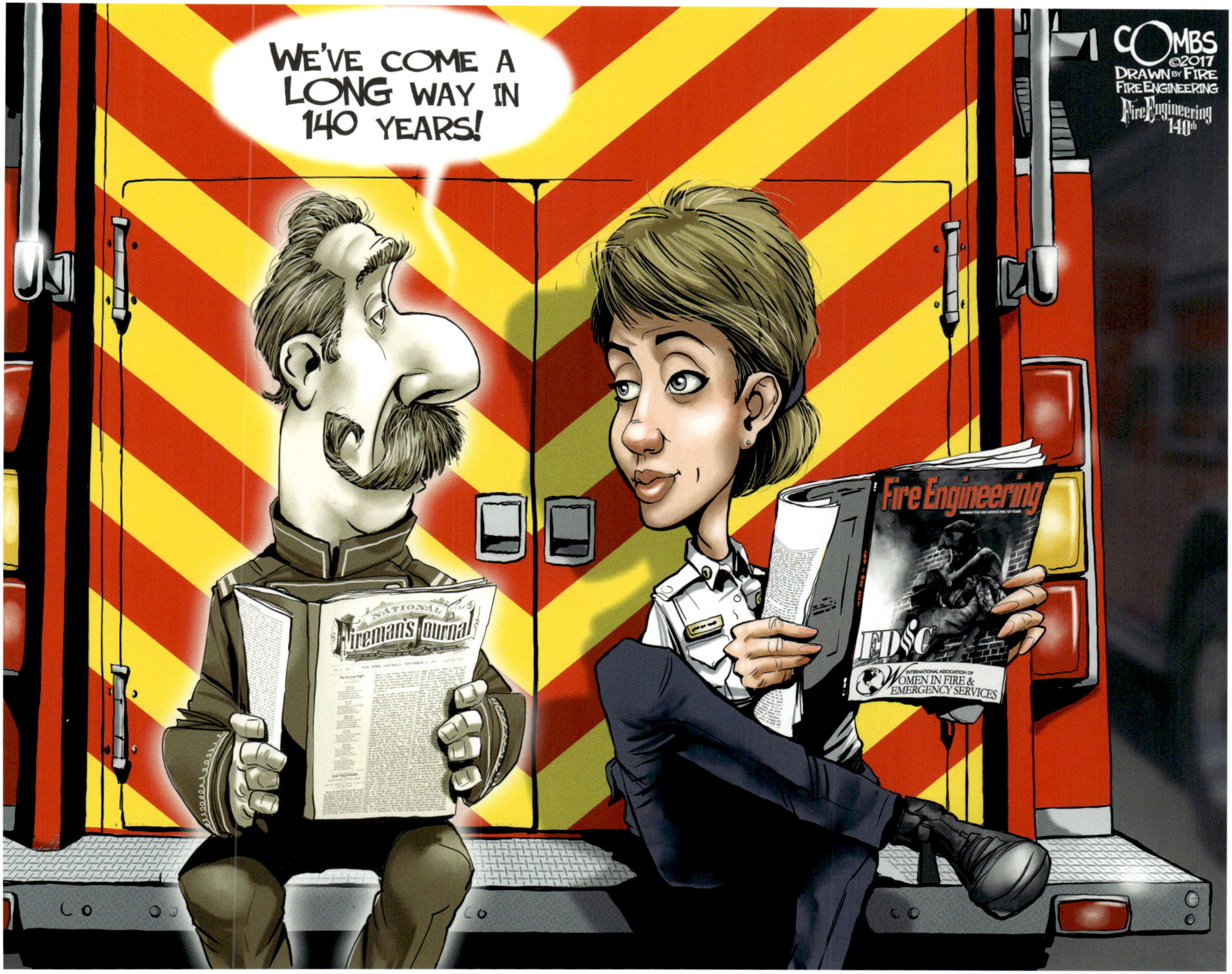
WE'VE COME A LONG WAY IN 140 YEARS!
COMBS ©2017 DRAWN BY FIRE FIRE ENGINEERING
Fire Engineering 140th
National Fireman's Journal
Fire Engineering
FDIC
INTERNATIONAL ASSOCIATION OF WOMEN IN FIRE & EMERGENCY SERVICES

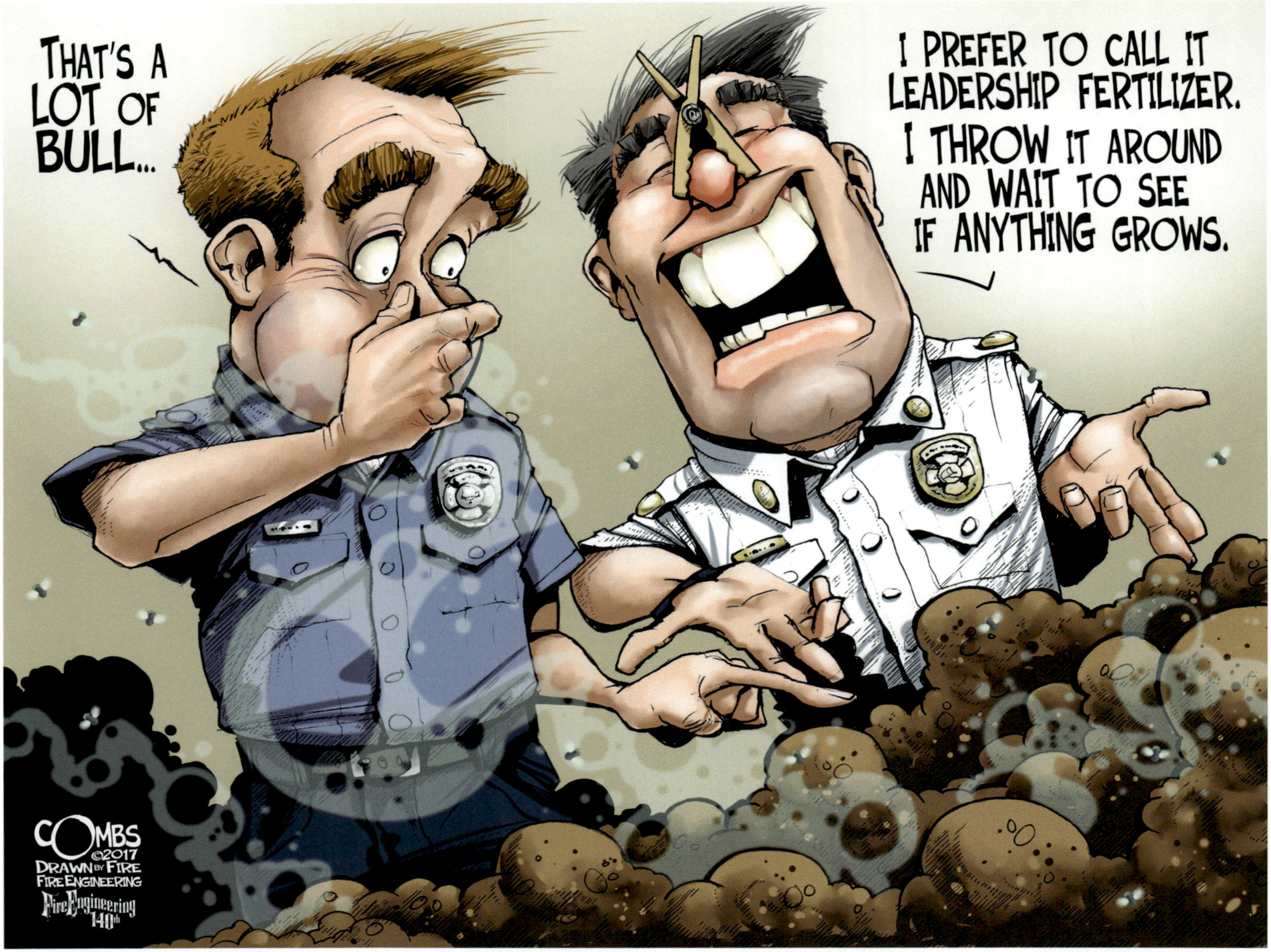

THE DEEPER THE PILE, THE SHALLOWER THE INTEGRITY!

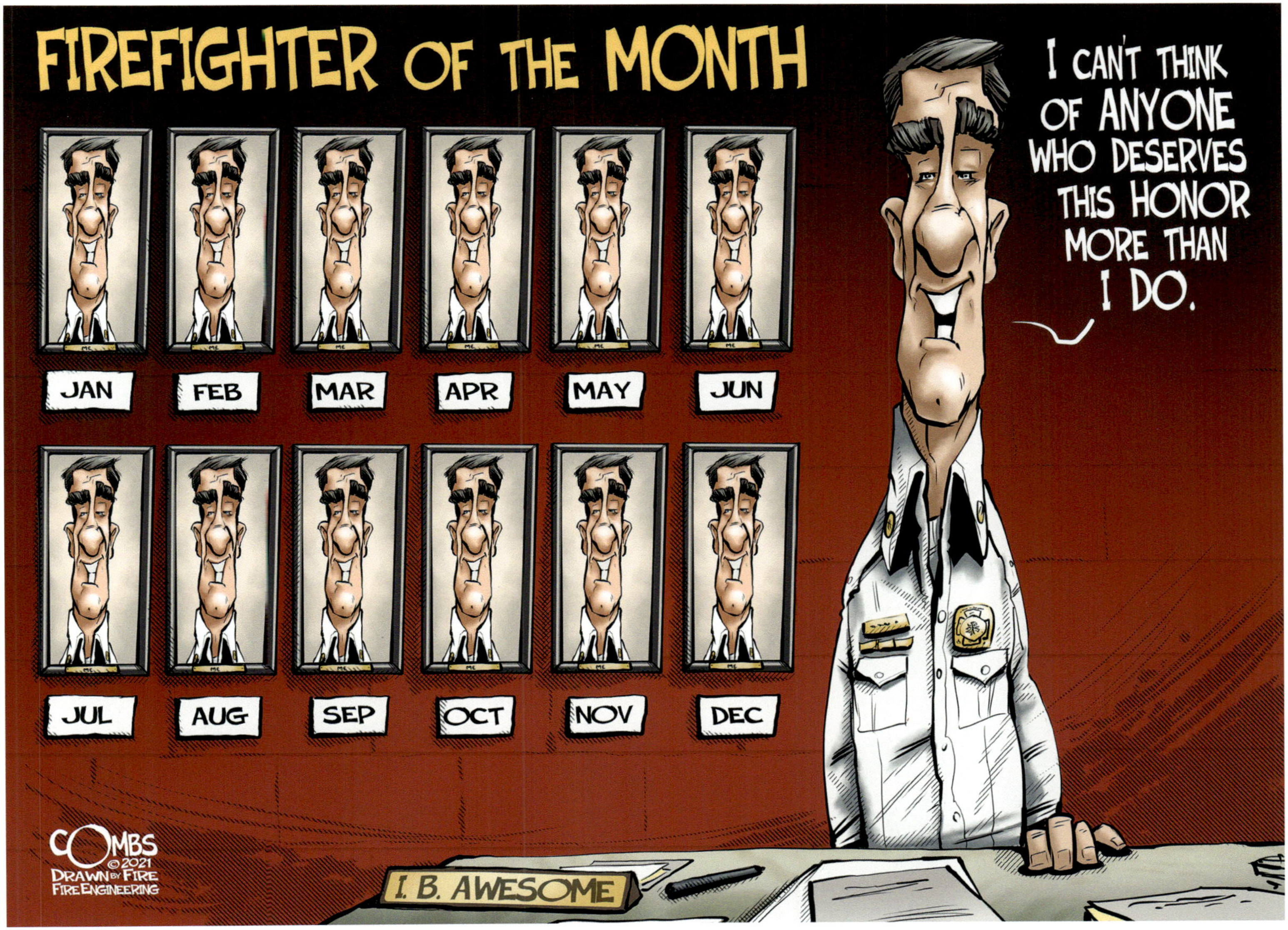

IT'S NEVER ABOUT YOU!

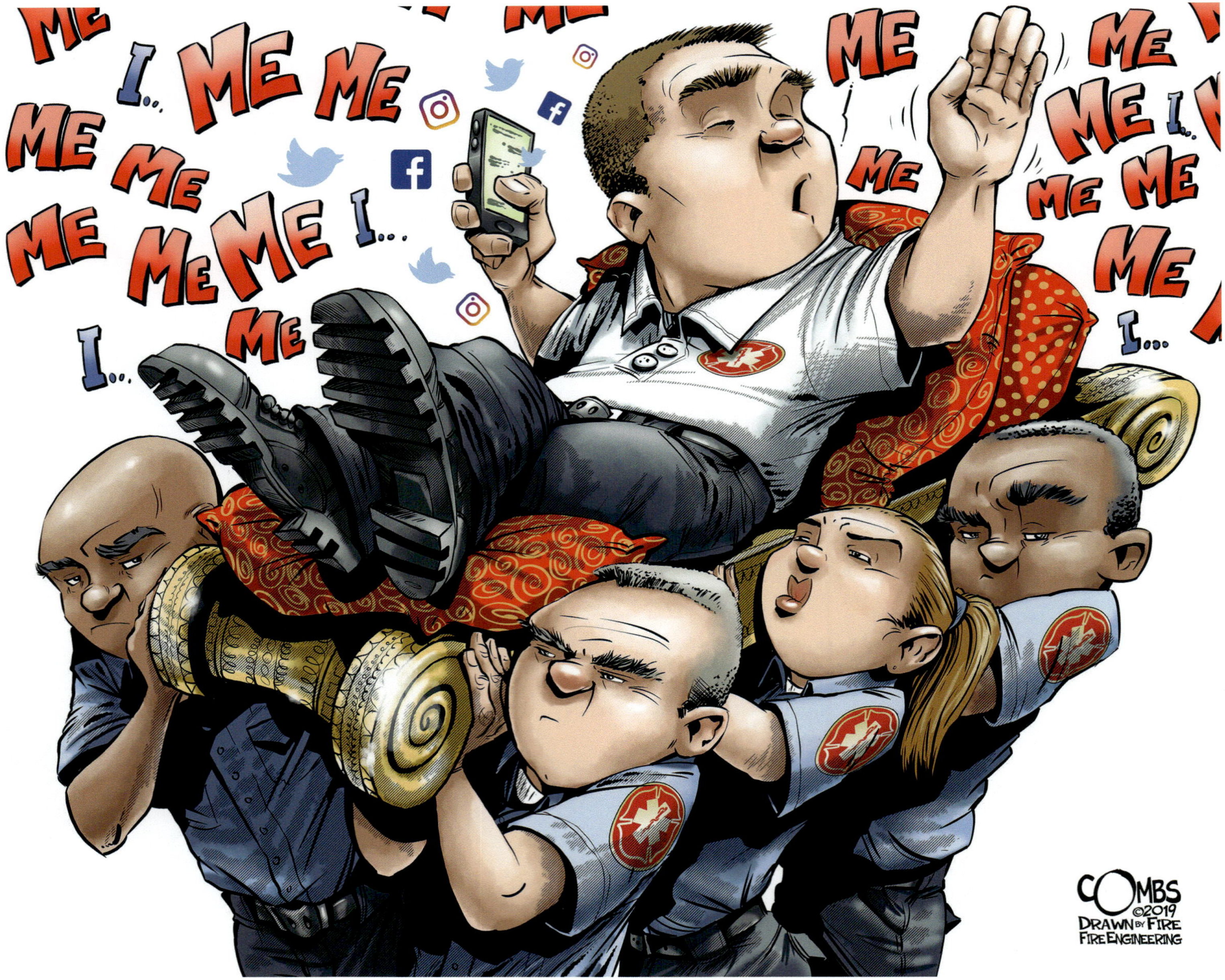
ME ME
I... ME ME
ME ME
ME ME
ME MEME I...
I... ME
ME
ME
ME ME I...
ME ME
ME
I...
COMBS
©2019
DRAWN BY FIRE
FIRE ENGINEERING

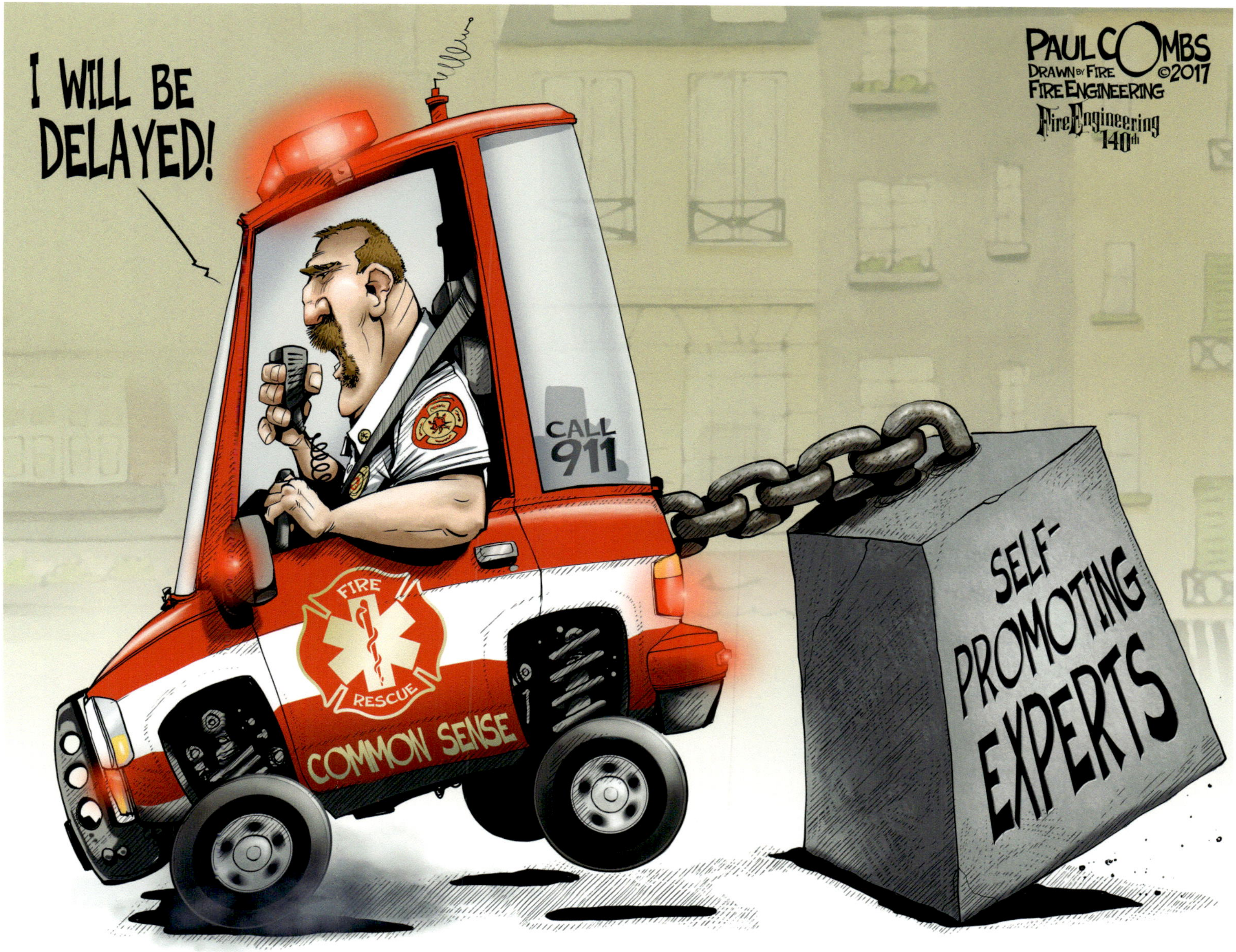
I WILL BE DELAYED!
PAUL COMBS
DRAWN BY FIRE ©2017
FIRE ENGINEERING
Fire Engineering 140th
CALL 911
FIRE
RESCUE
COMMON SENSE
SELF-PROMOTING EXPERTS

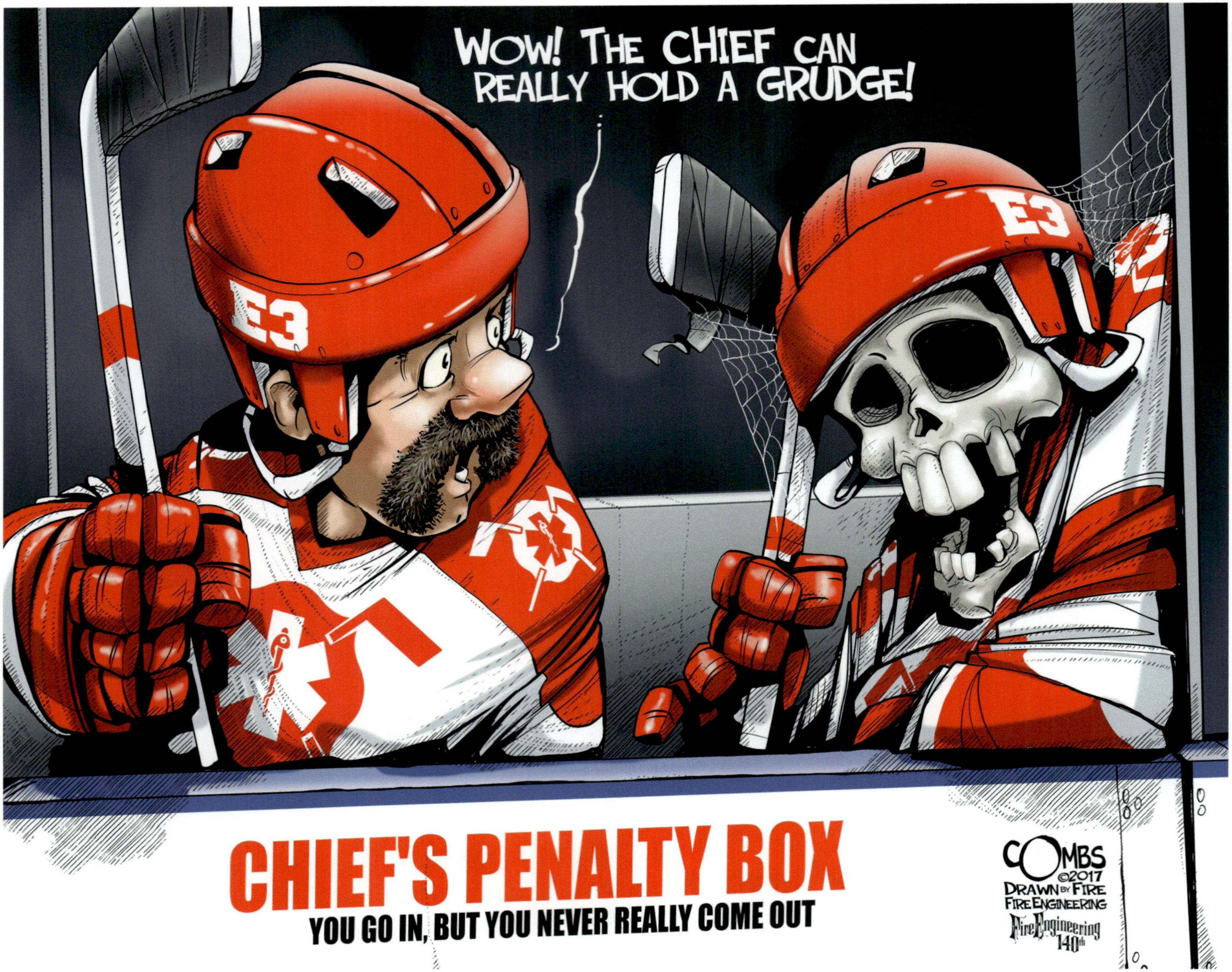
WOW! THE CHIEF CAN REALLY HOLD A GRUDGE!
E3
E3
CHIEF'S PENALTY BOX
YOU GO IN, BUT YOU NEVER REALLY COME OUT
COMBS ©2017 DRAWN BY FIRE FIRE ENGINEERING
Fire Engineering 140th

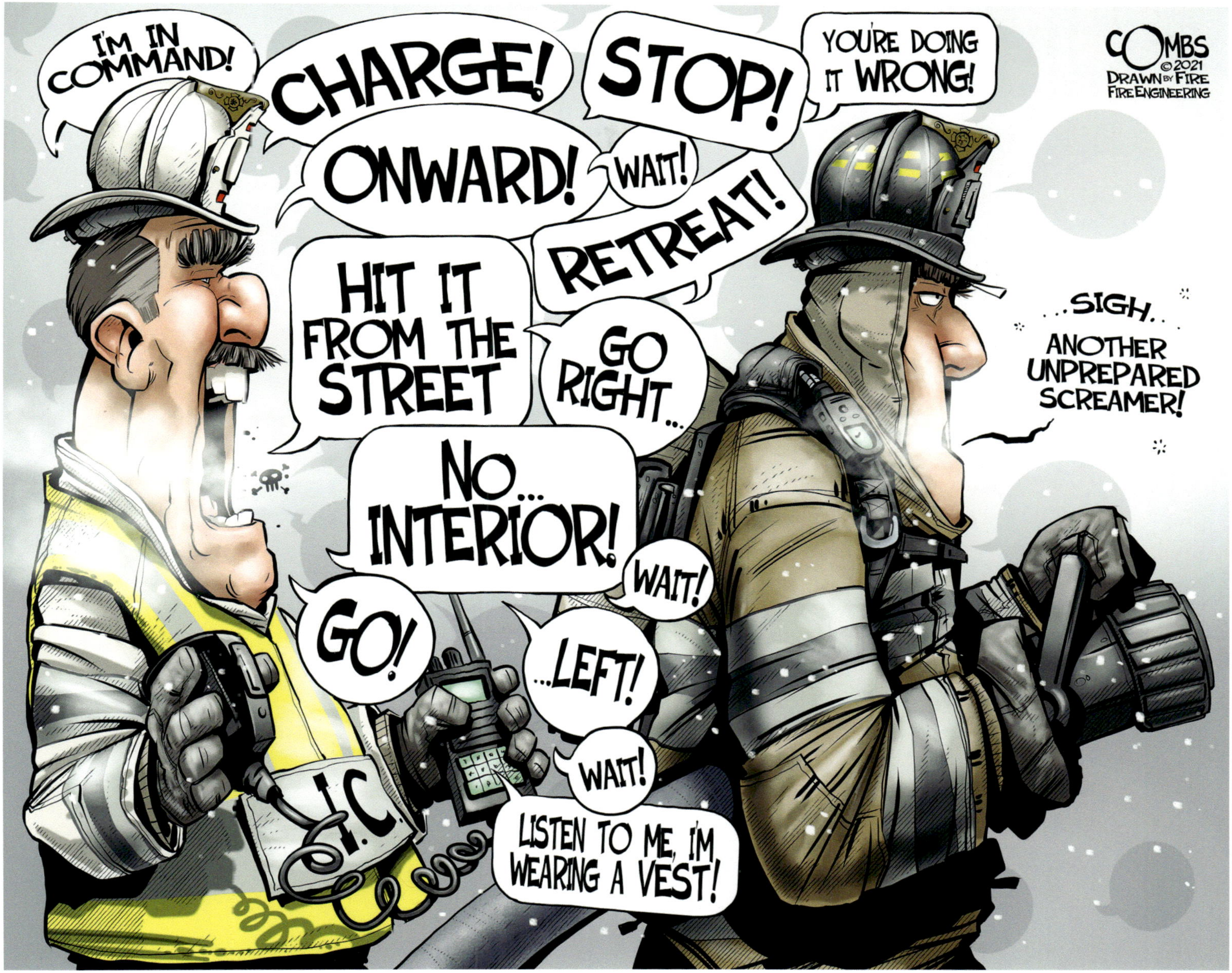
I'M IN COMMAND!
CHARGE!
STOP!
YOU'RE DOING IT WRONG!
ONWARD!
WAIT!
RETREAT!
HIT IT FROM THE STREET
GO RIGHT...
NO... INTERIOR!
WAIT!
GO!
...LEFT!
WAIT!
LISTEN TO ME, I'M WEARING A VEST!
...SIGH... ANOTHER UNPREPARED SCREAMER!
I.C.
COMBS ©2021 DRAWN BY FIRE FIRE ENGINEERING

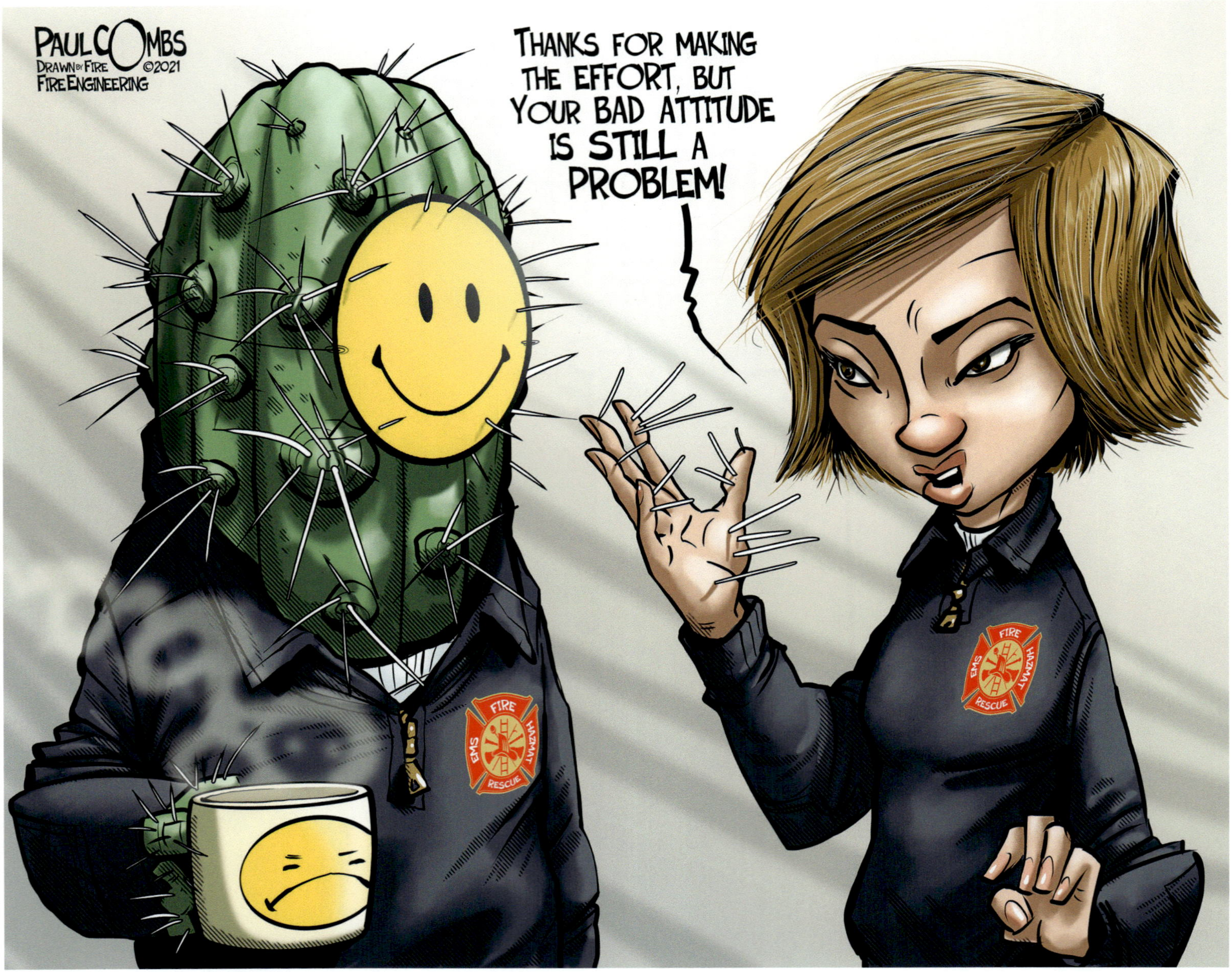
PAUL COMBS
DRAWN BY FIRE ©2021
FIRE ENGINEERING
THANKS FOR MAKING THE EFFORT, BUT YOUR BAD ATTITUDE IS STILL A PROBLEM!
FIRE
EMS
HAZMAT
RESCUE

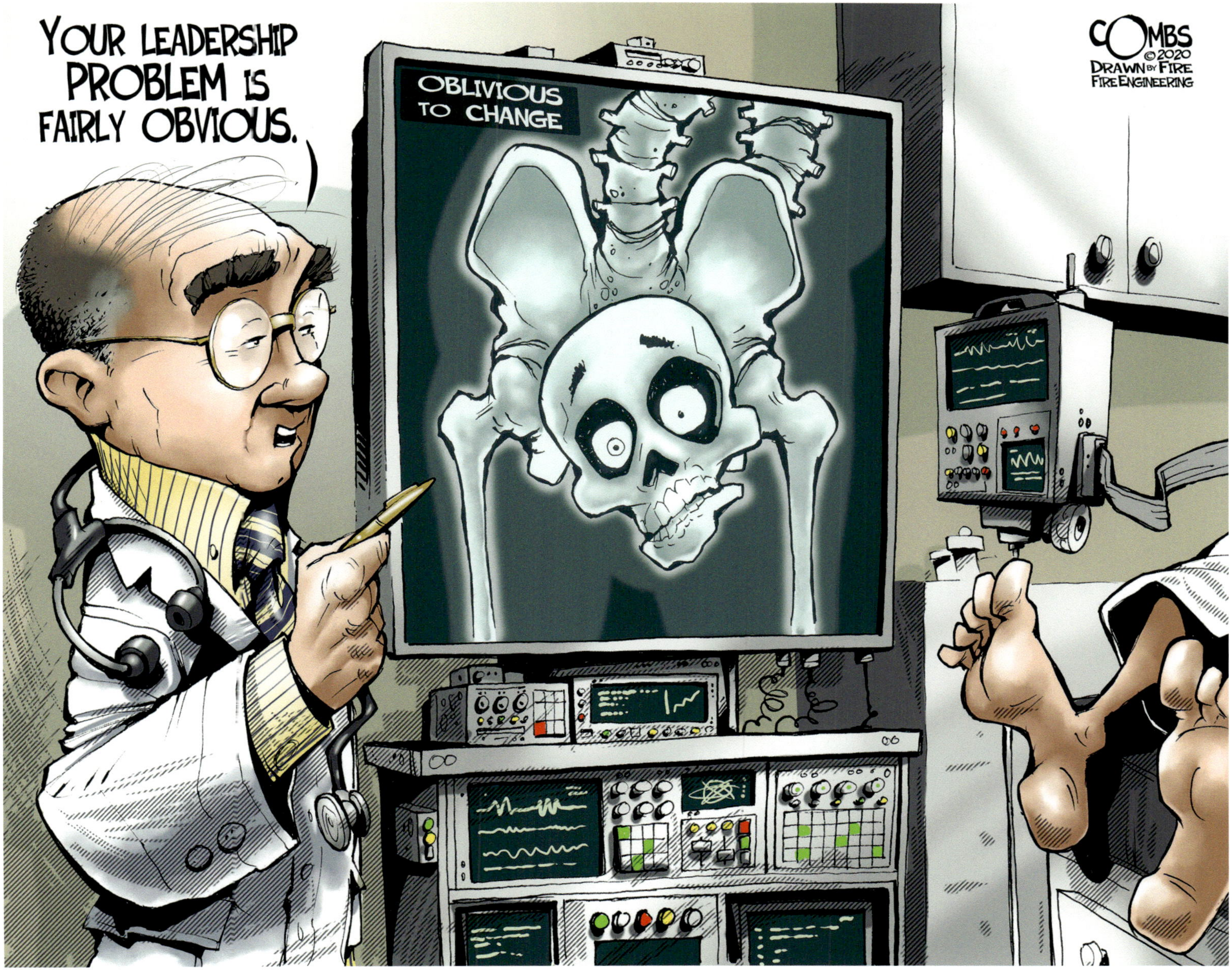
YOUR LEADERSHIP PROBLEM IS FAIRLY OBVIOUS.
OBLIVIOUS TO CHANGE
COMBS ©2020 DRAWN BY FIRE FIRE ENGINEERING

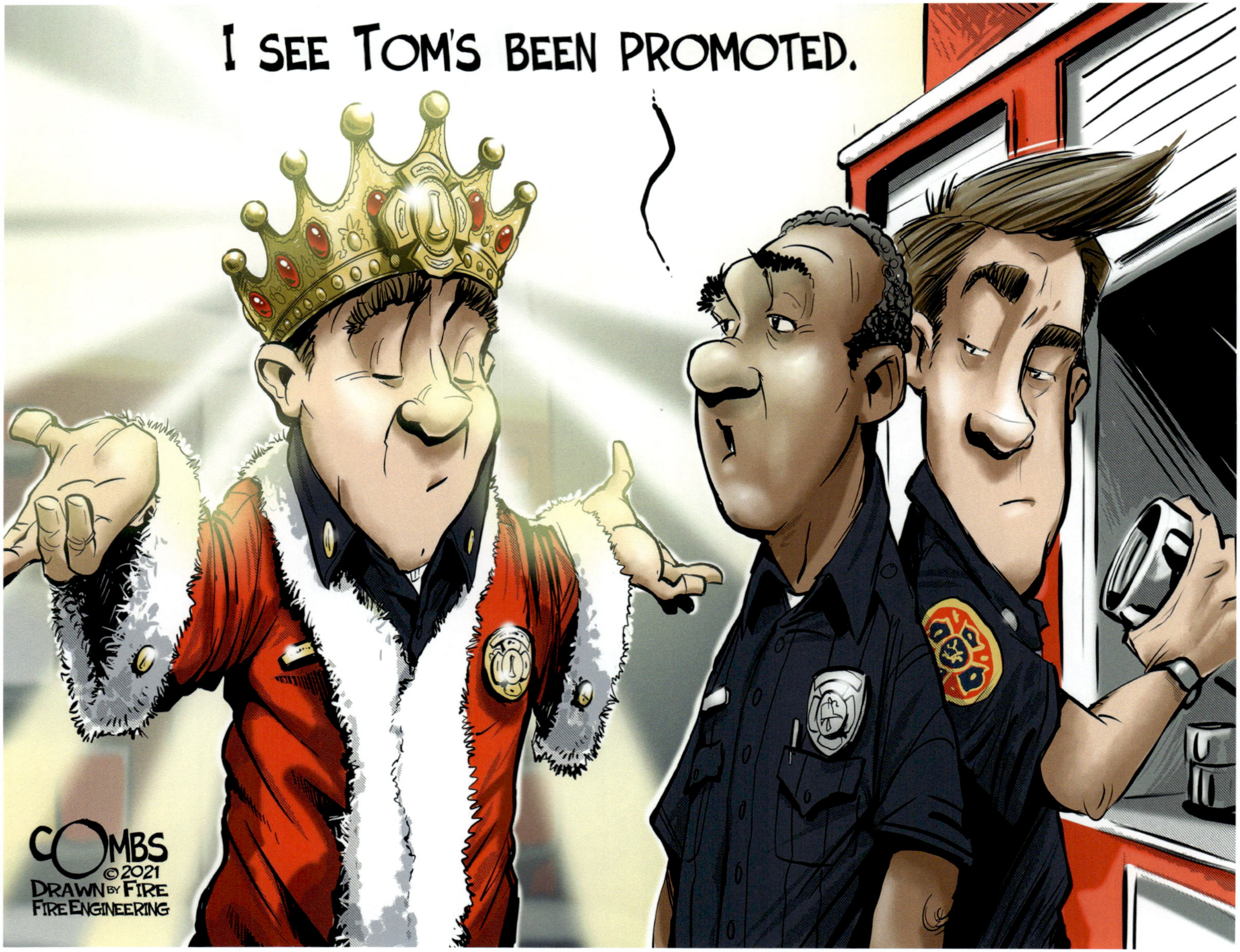
I SEE TOM'S BEEN PROMOTED.
COMBS
©2021
DRAWN BY FIRE
FIRE ENGINEERING

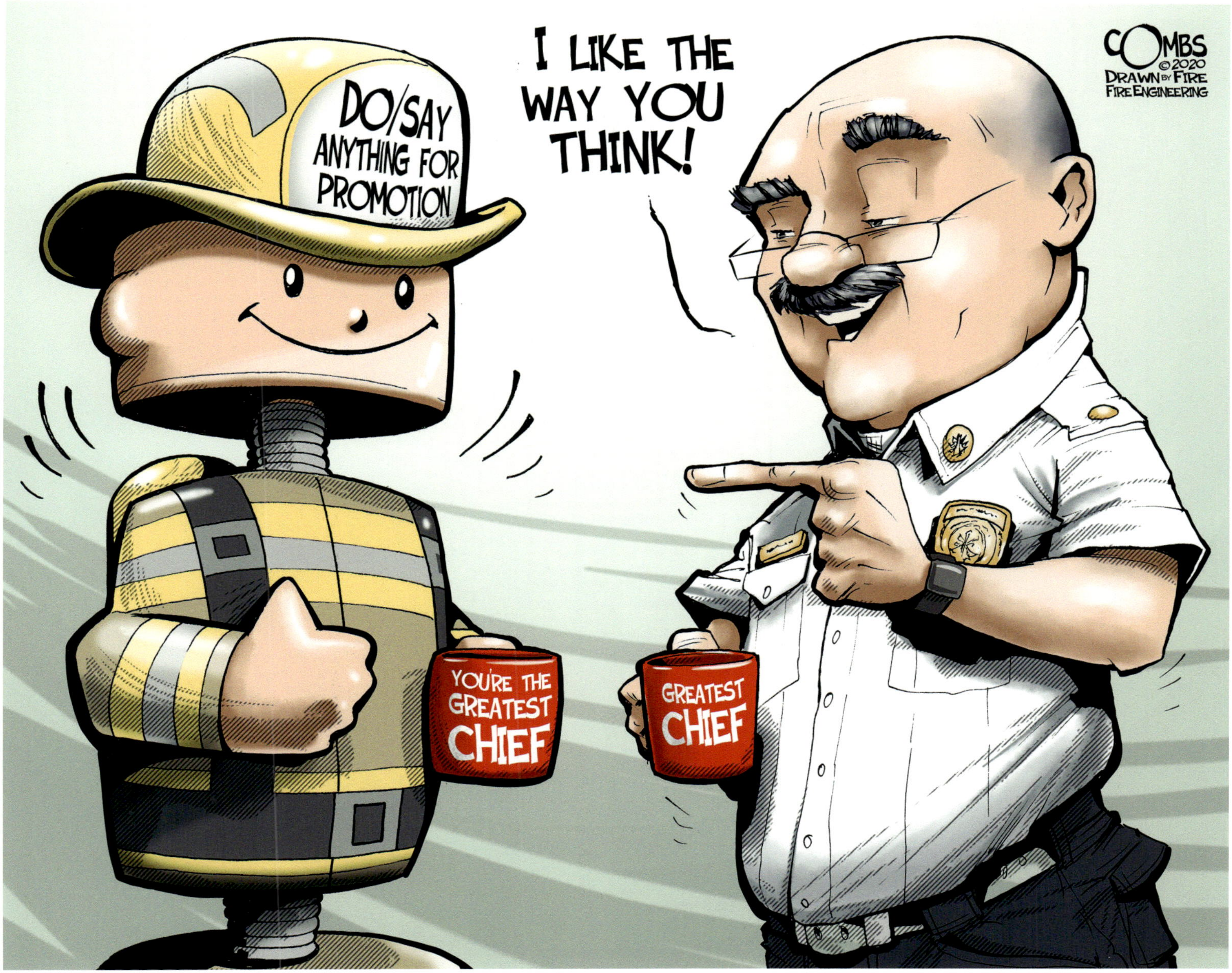
DO/SAY ANYTHING FOR PROMOTION
I LIKE THE WAY YOU THINK!
COMBS ©2020 DRAWN BY FIRE FIRE ENGINEERING
YOU'RE THE GREATEST CHIEF
GREATEST CHIEF

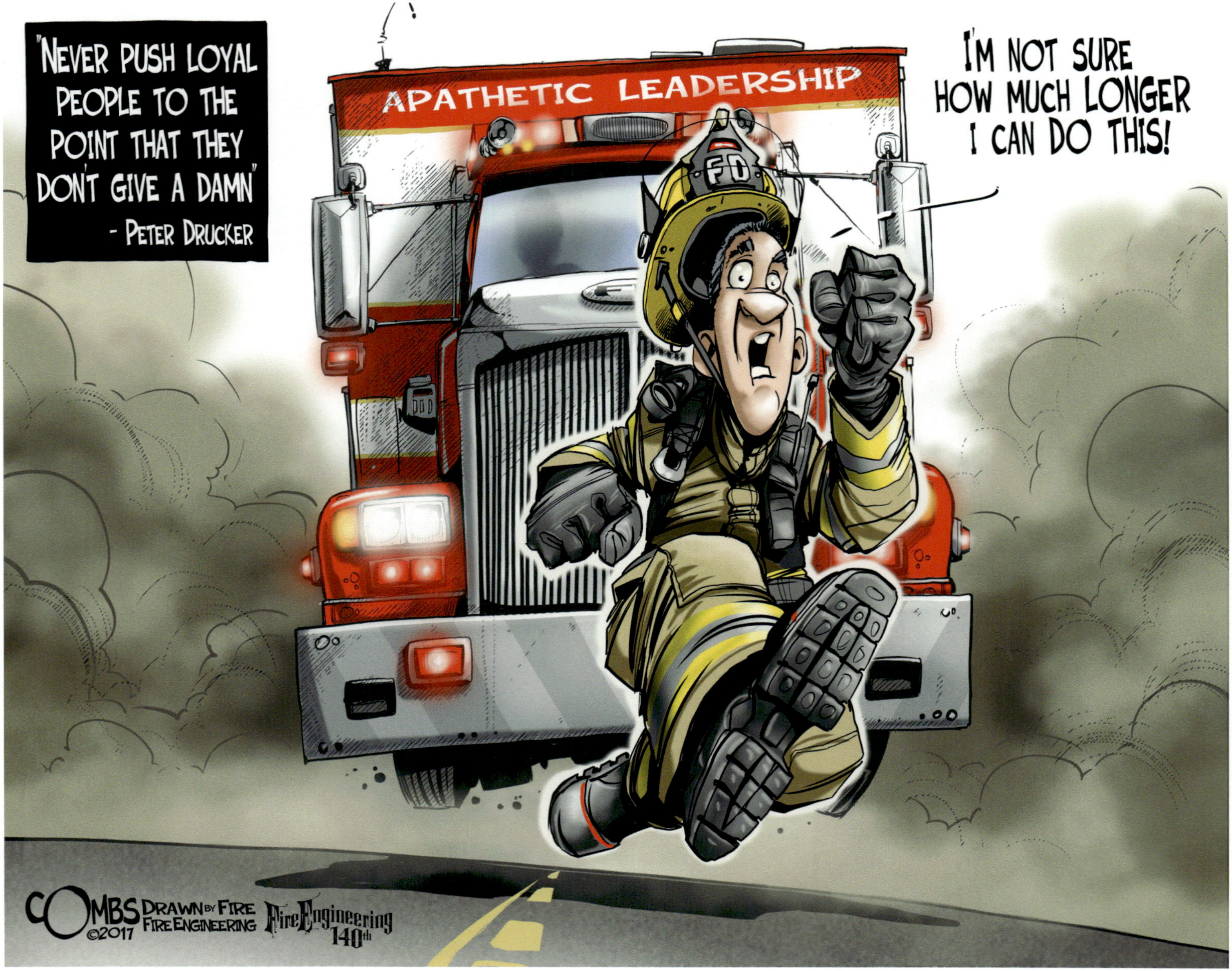
"NEVER PUSH LOYAL PEOPLE TO THE POINT THAT THEY DON'T GIVE A DAMN"
- PETER DRUCKER
APATHETIC LEADERSHIP
FD
I'M NOT SURE HOW MUCH LONGER I CAN DO THIS!
COMBS ©2017
DRAWN BY FIRE
FIRE ENGINEERING
Fire Engineering 140th

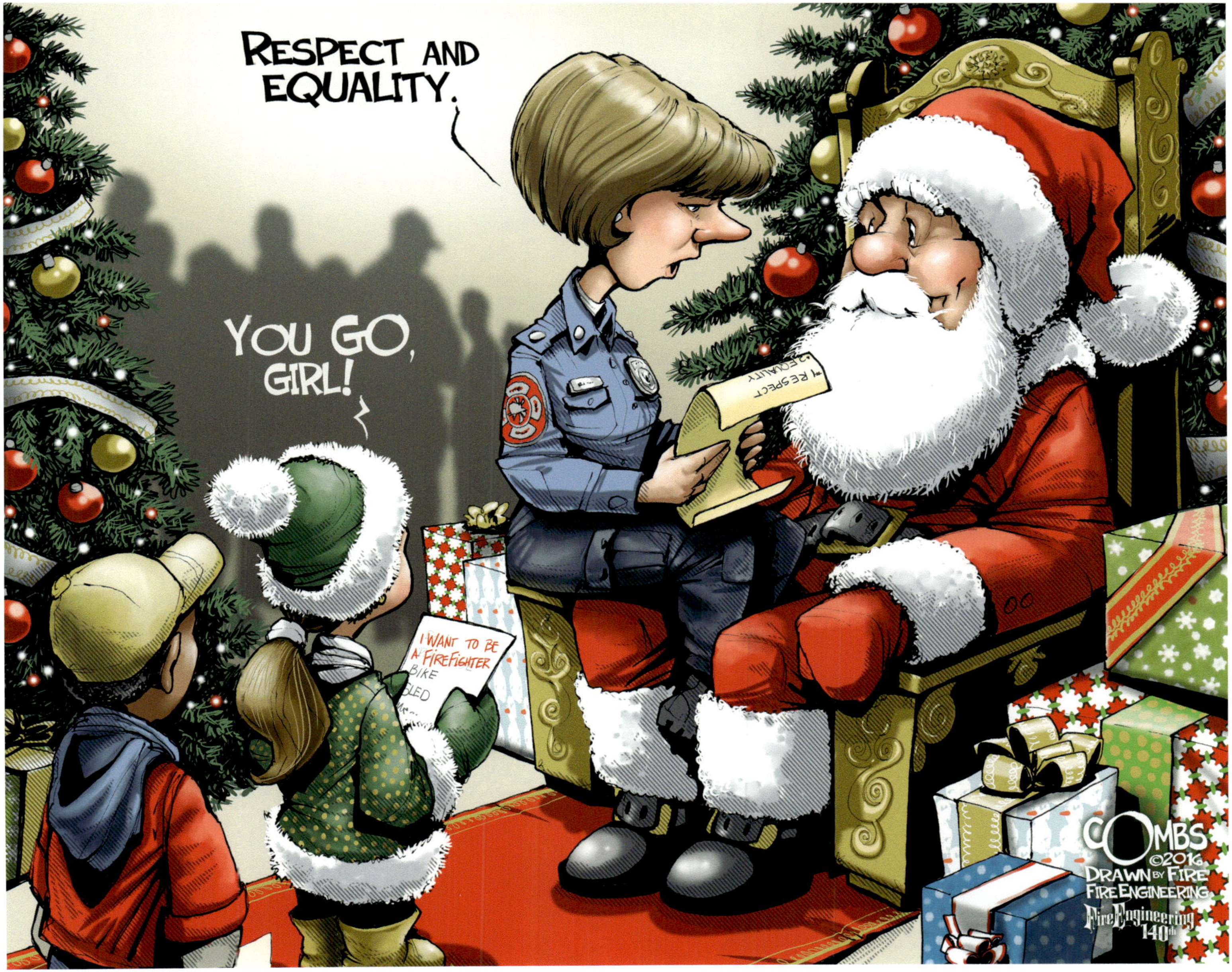
RESPECT AND EQUALITY.
YOU GO, GIRL!
I WANT TO BE A FIREFIGHTER
BIKE
SLED
RESPECT
EQUALITY
COMBS ©2016
DRAWN BY FIRE
FIRE ENGINEERING
Fire Engineering 140th

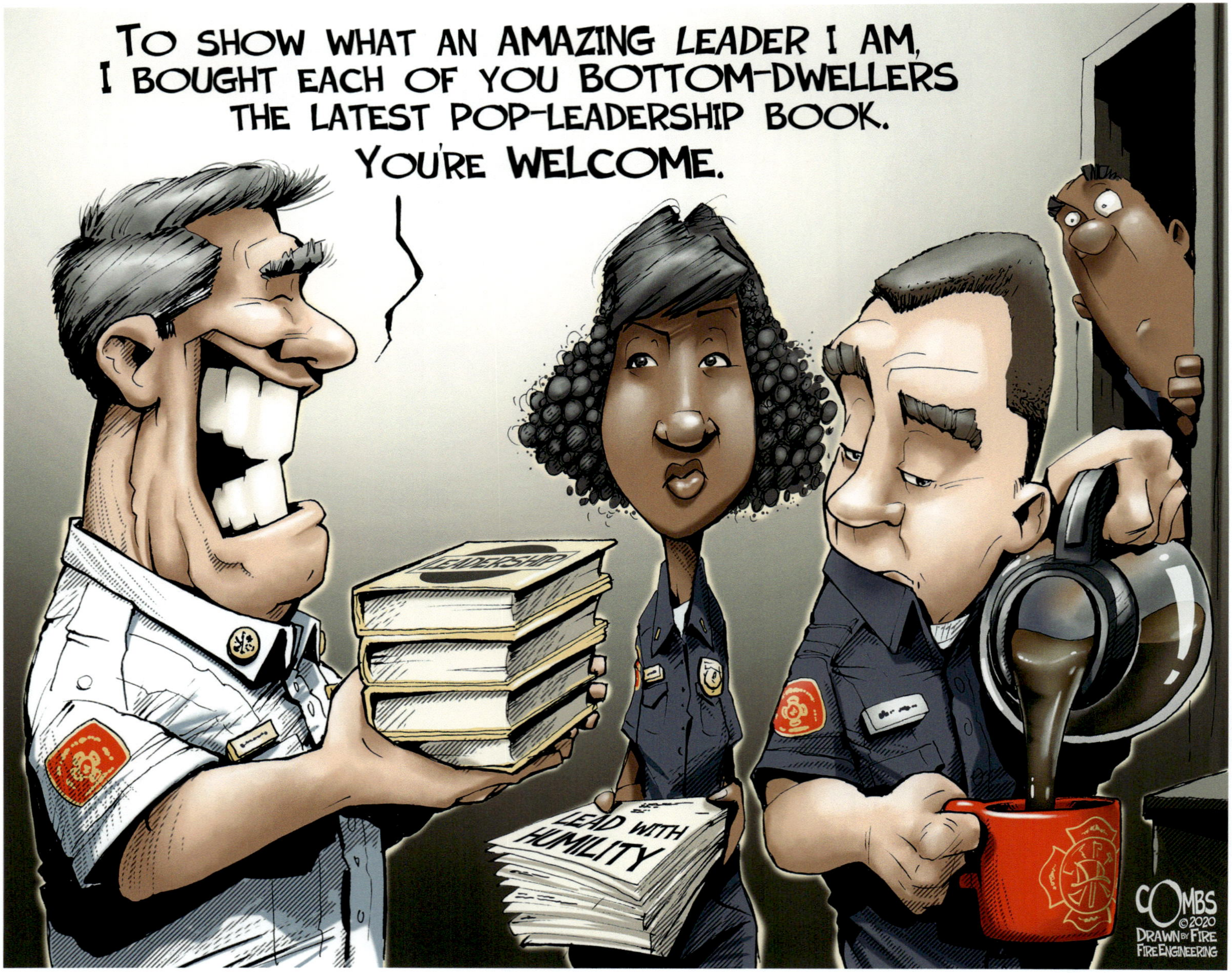
TO SHOW WHAT AN AMAZING LEADER I AM, I BOUGHT EACH OF YOU BOTTOM-DWELLERS THE LATEST POP-LEADERSHIP BOOK. YOU'RE WELCOME.
LEADERSHIP
LEAD WITH HUMILITY
COMBS ©2020 DRAWN BY FIRE FIRE ENGINEERING

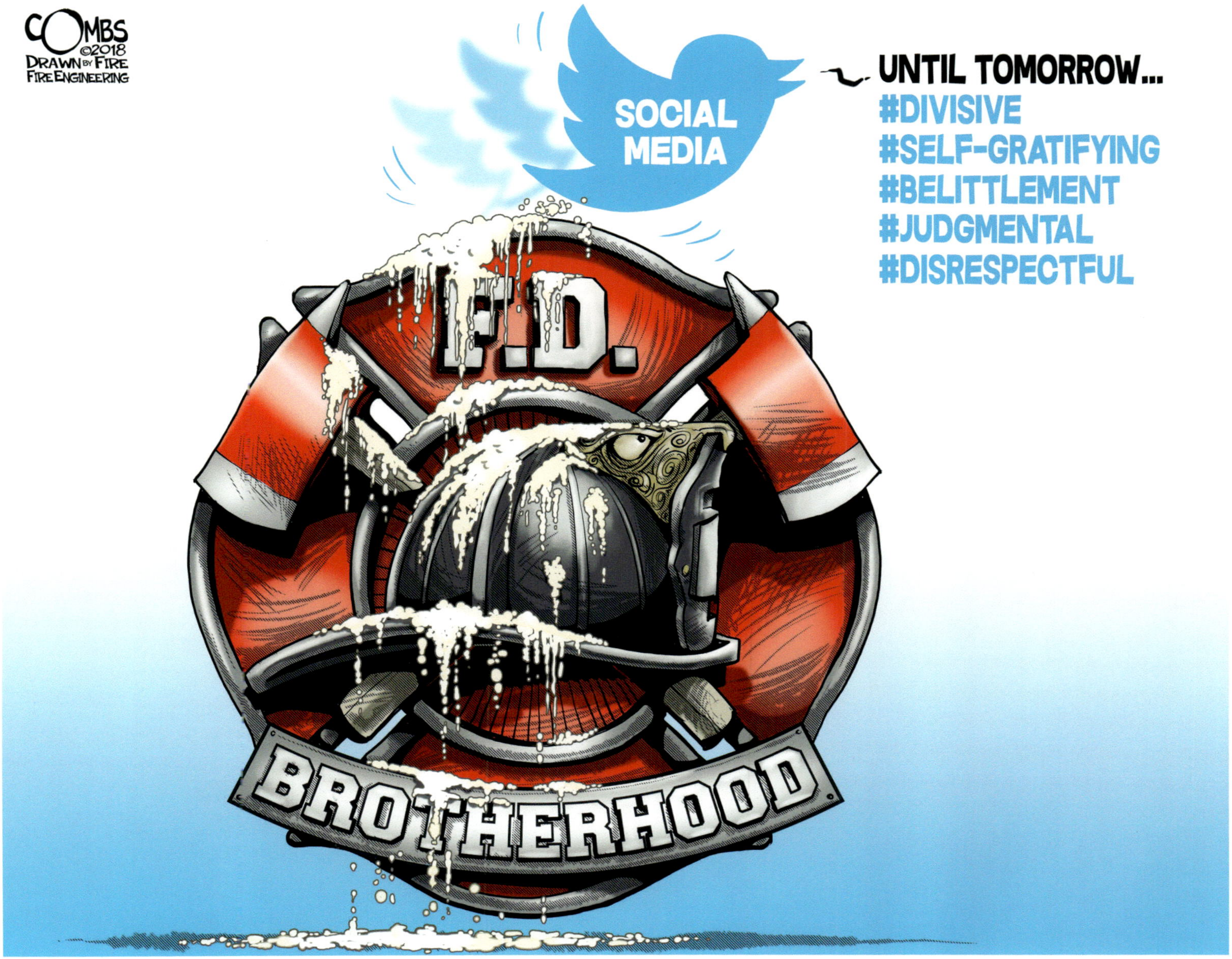
COMBS ©2018 DRAWN BY FIRE FIRE ENGINEERING
SOCIAL MEDIA
UNTIL TOMORROW...
#DIVISIVE
#SELF-GRATIFYING
#BELITTLEMENT
#JUDGMENTAL
#DISRESPECTFUL
F.D.
BROTHERHOOD

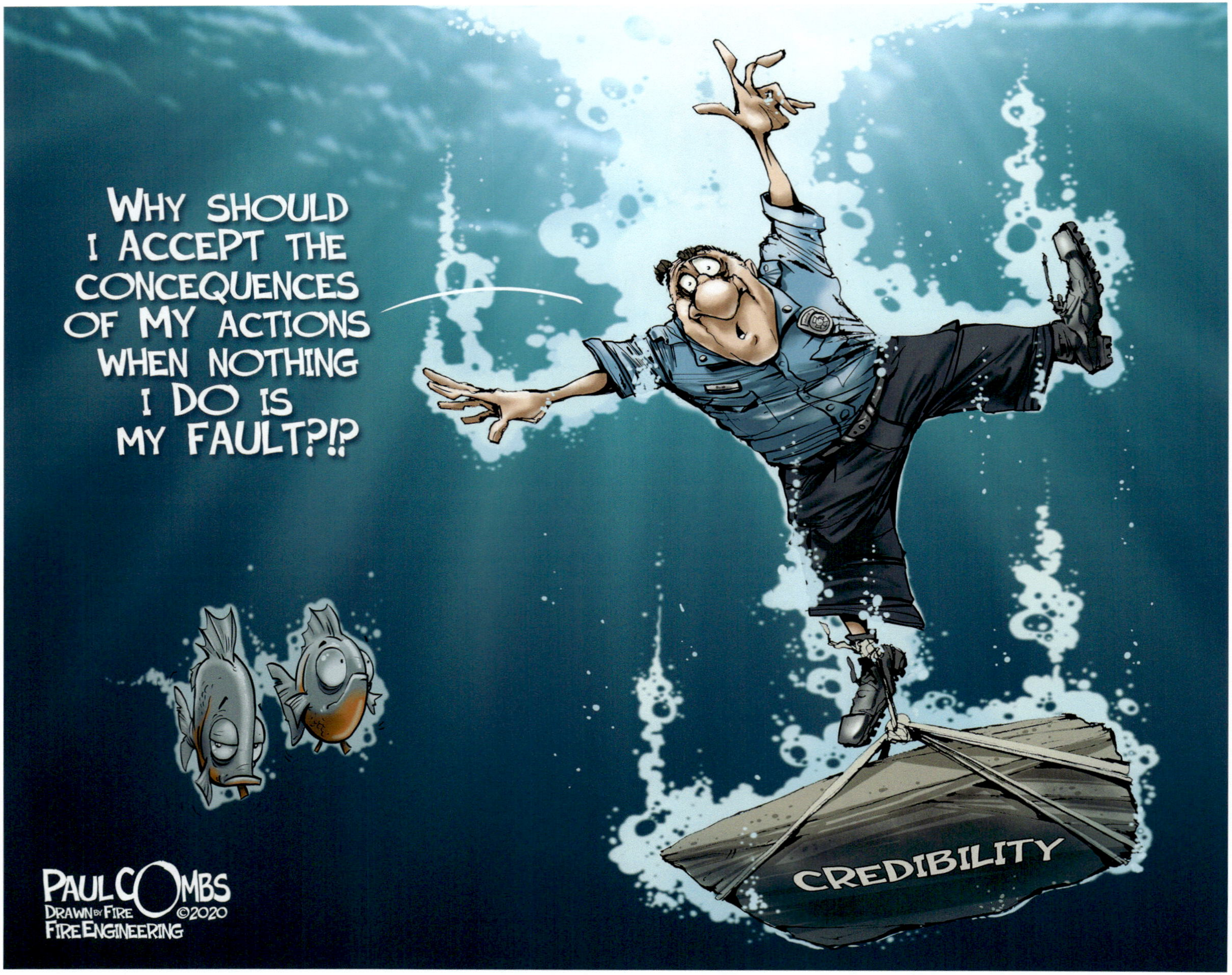
WHY SHOULD I ACCEPT THE CONCEQUENCES OF MY ACTIONS WHEN NOTHING I DO IS MY FAULT?!?
CREDIBILITY
PAUL COMBS
DRAWN BY FIRE ©2020
FIRE ENGINEERING

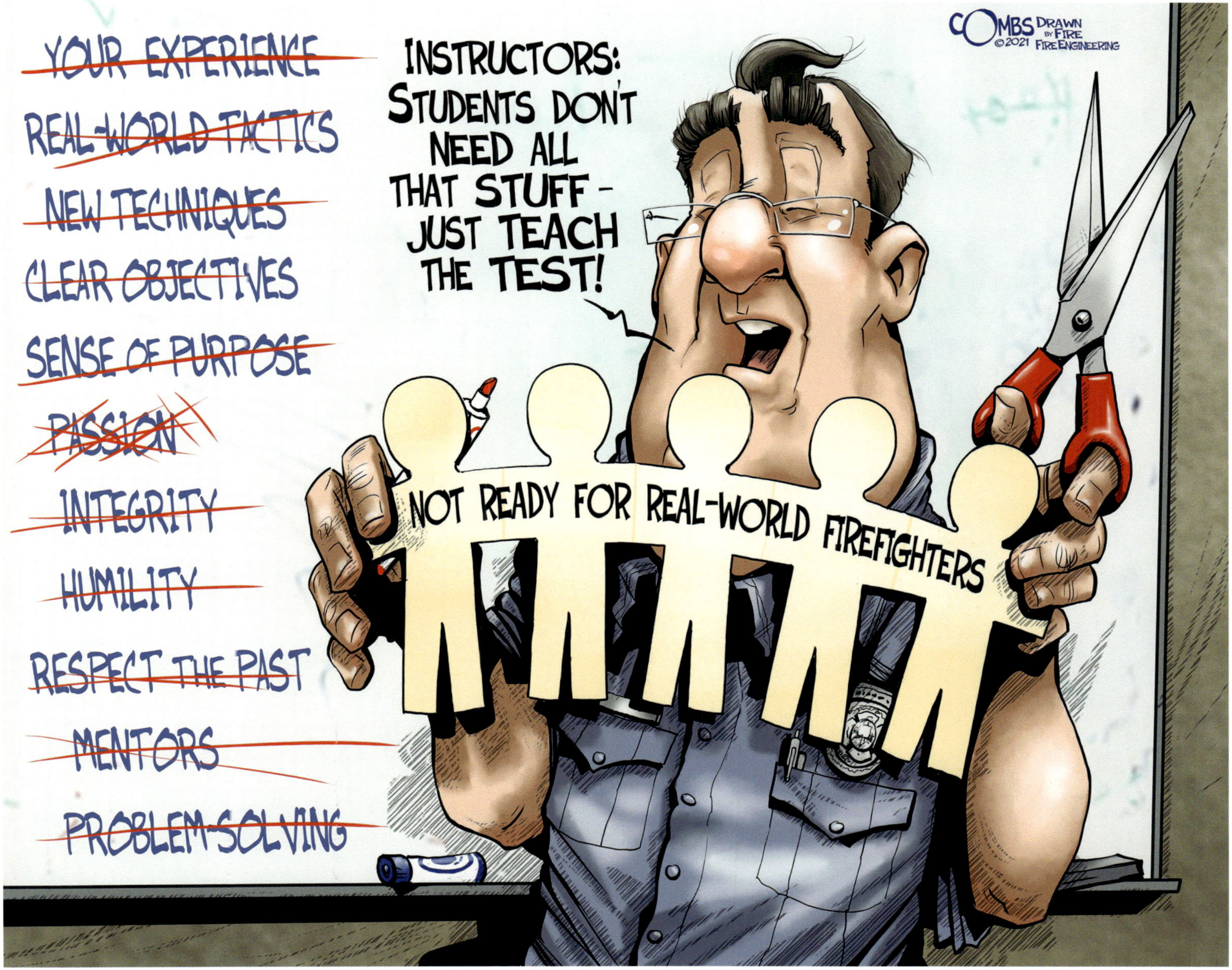
COMBS DRAWN BY FIRE ©2021 FIRE ENGINEERING
YOUR EXPERIENCE
REAL-WORLD TACTICS
NEW TECHNIQUES
CLEAR OBJECTIVES
SENSE OF PURPOSE
PASSION
INTEGRITY
HUMILITY
RESPECT THE PAST
MENTORS
PROBLEM-SOLVING
INSTRUCTORS: STUDENTS DON'T NEED ALL THAT STUFF - JUST TEACH THE TEST!
NOT READY FOR REAL-WORLD FIREFIGHTERS

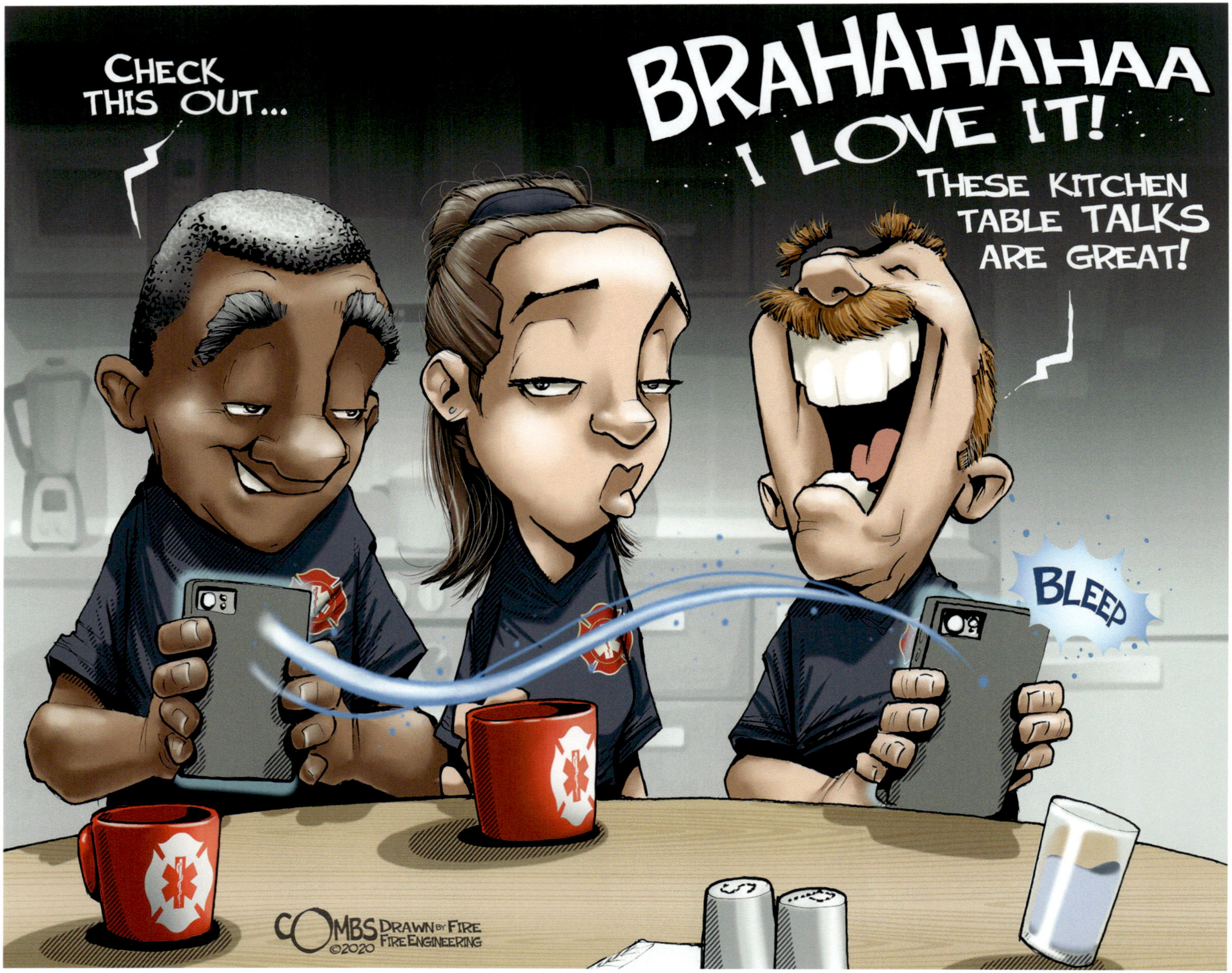
CHECK THIS OUT...
BRAHAHAHAHAA I LOVE IT!
THESE KITCHEN TABLE TALKS ARE GREAT!
BLEEP
COMBS ©2020 DRAWN BY FIRE FIRE ENGINEERING

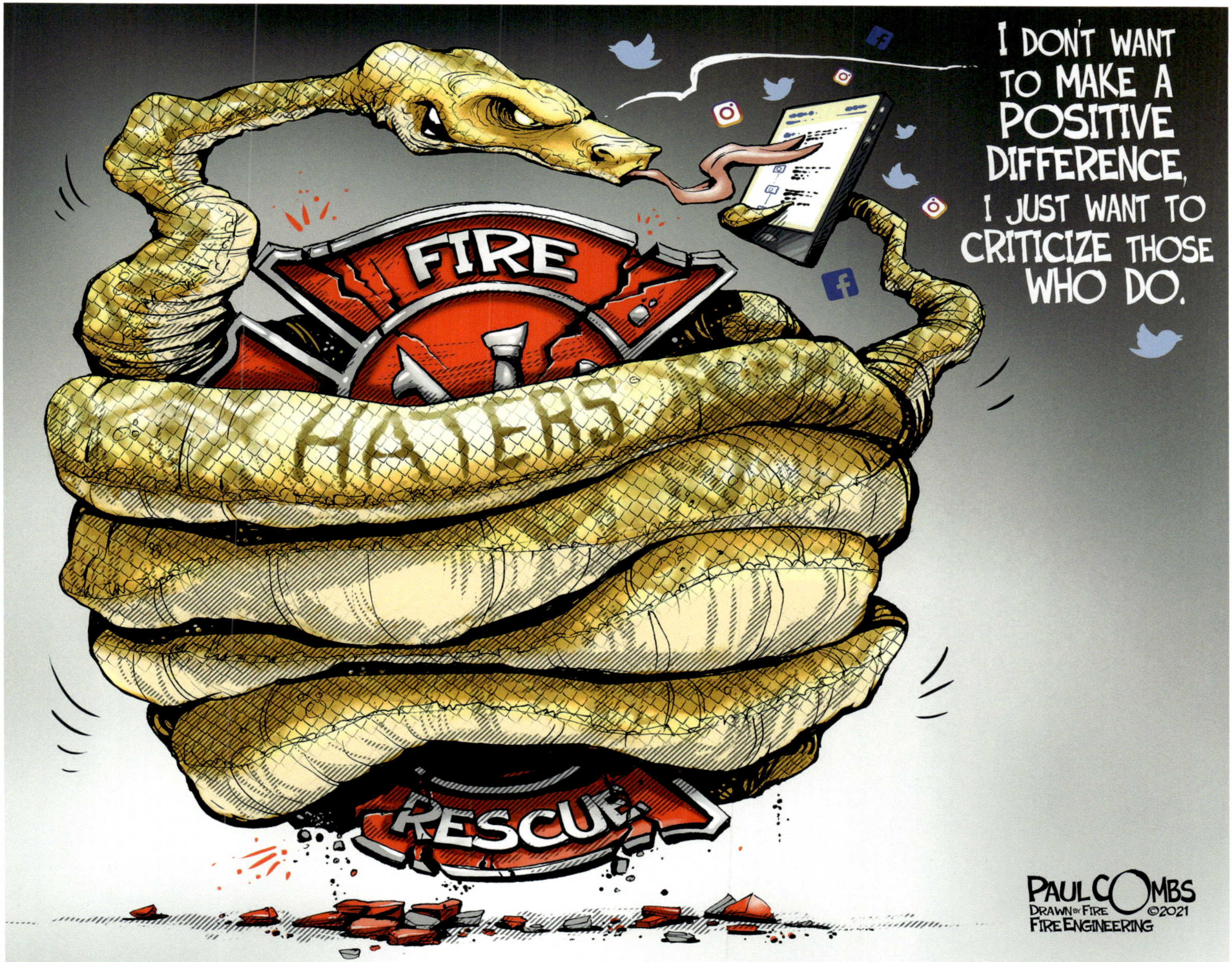
I DON'T WANT TO MAKE A POSITIVE DIFFERENCE, I JUST WANT TO CRITICIZE THOSE WHO DO.
FIRE
HATERS
RESCUE
PAUL COMBS
DRAWN BY FIRE
©2021
FIRE ENGINEERING

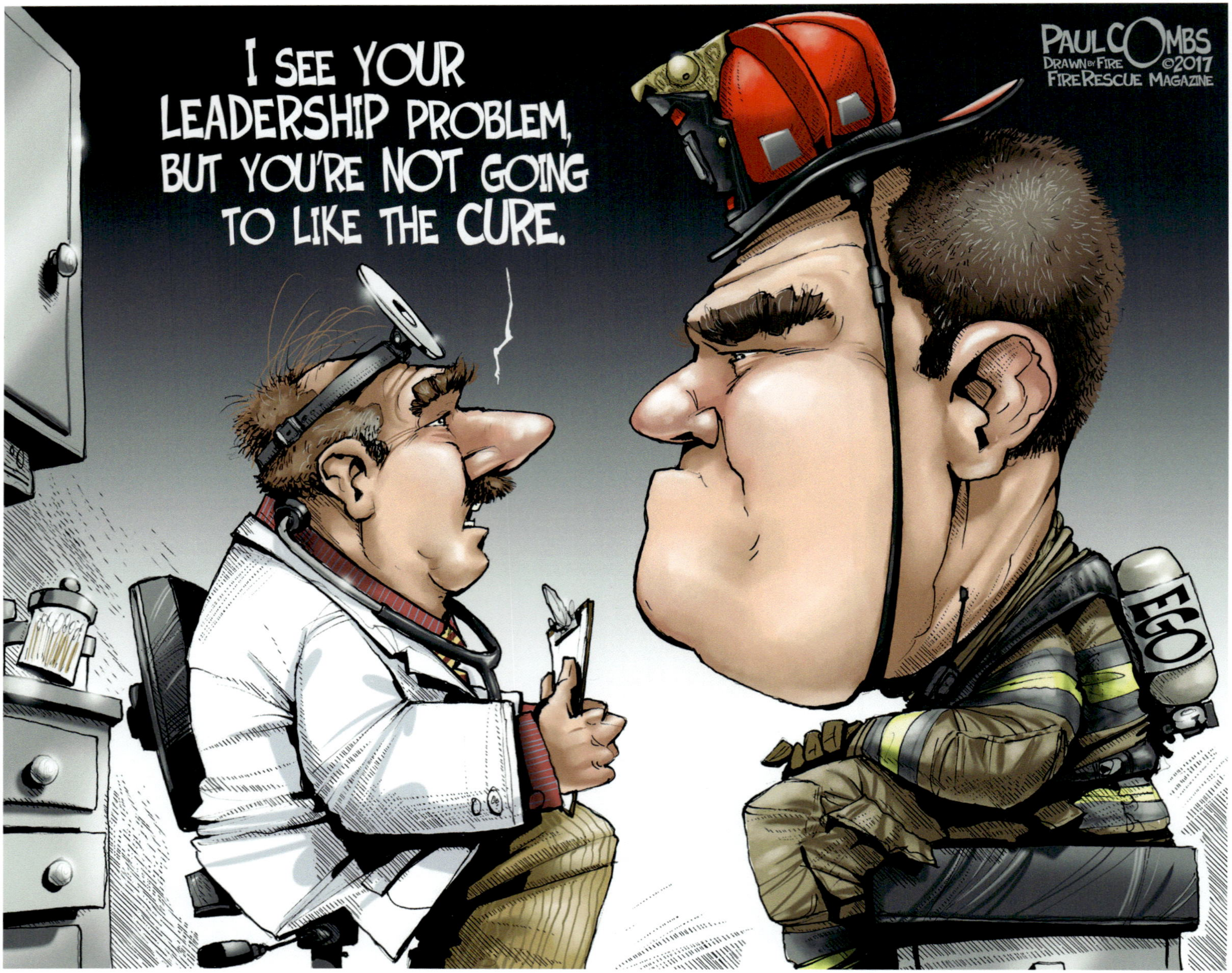
I SEE YOUR LEADERSHIP PROBLEM, BUT YOU'RE NOT GOING TO LIKE THE CURE.
PAUL COMBS
DRAWN BY FIRE ©2017
FIRERESCUE MAGAZINE
EGO

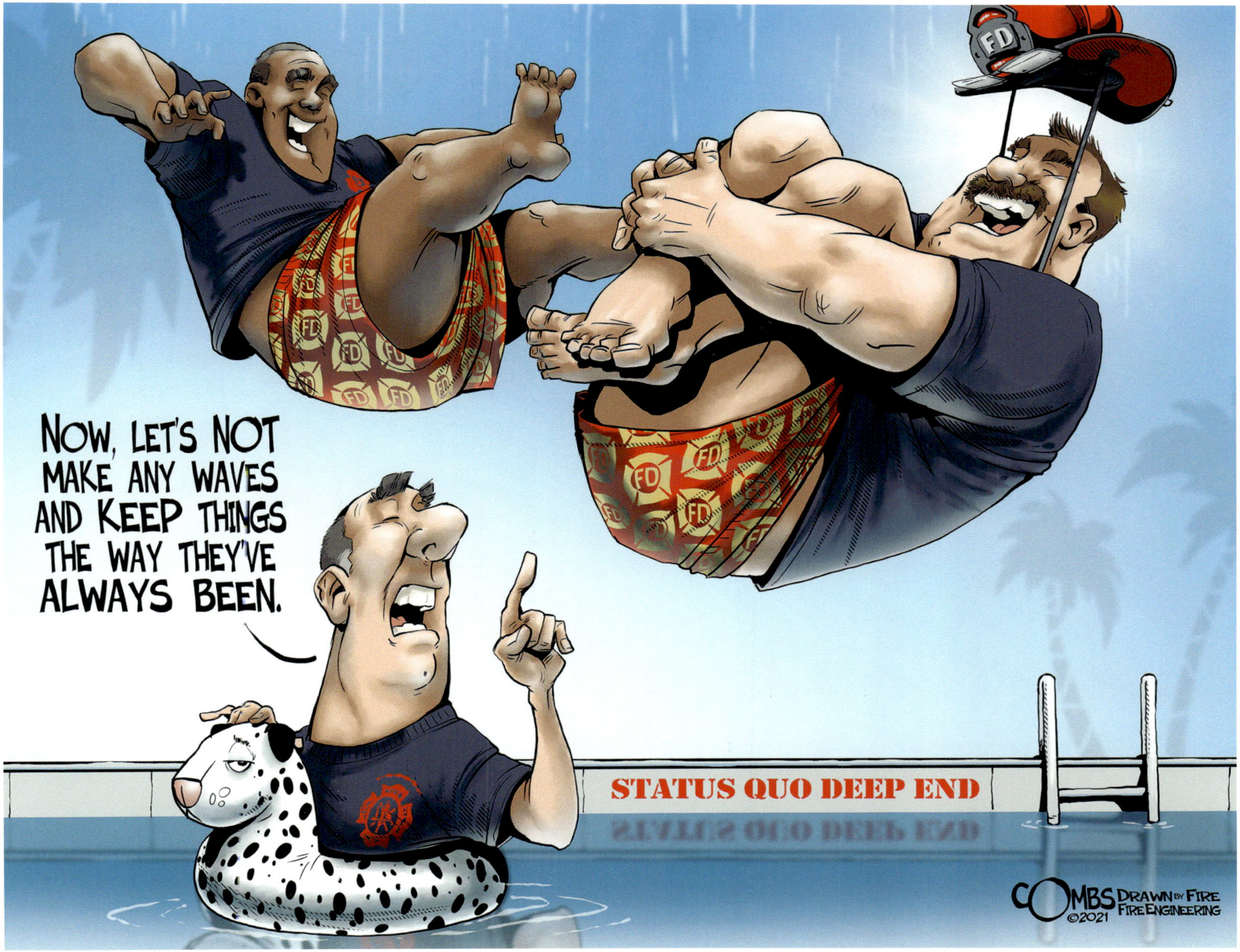
NOW, LET'S NOT MAKE ANY WAVES AND KEEP THINGS THE WAY THEY'VE ALWAYS BEEN.
STATUS QUO DEEP END
COMBS ©2021 DRAWN BY FIRE FIRE ENGINEERING

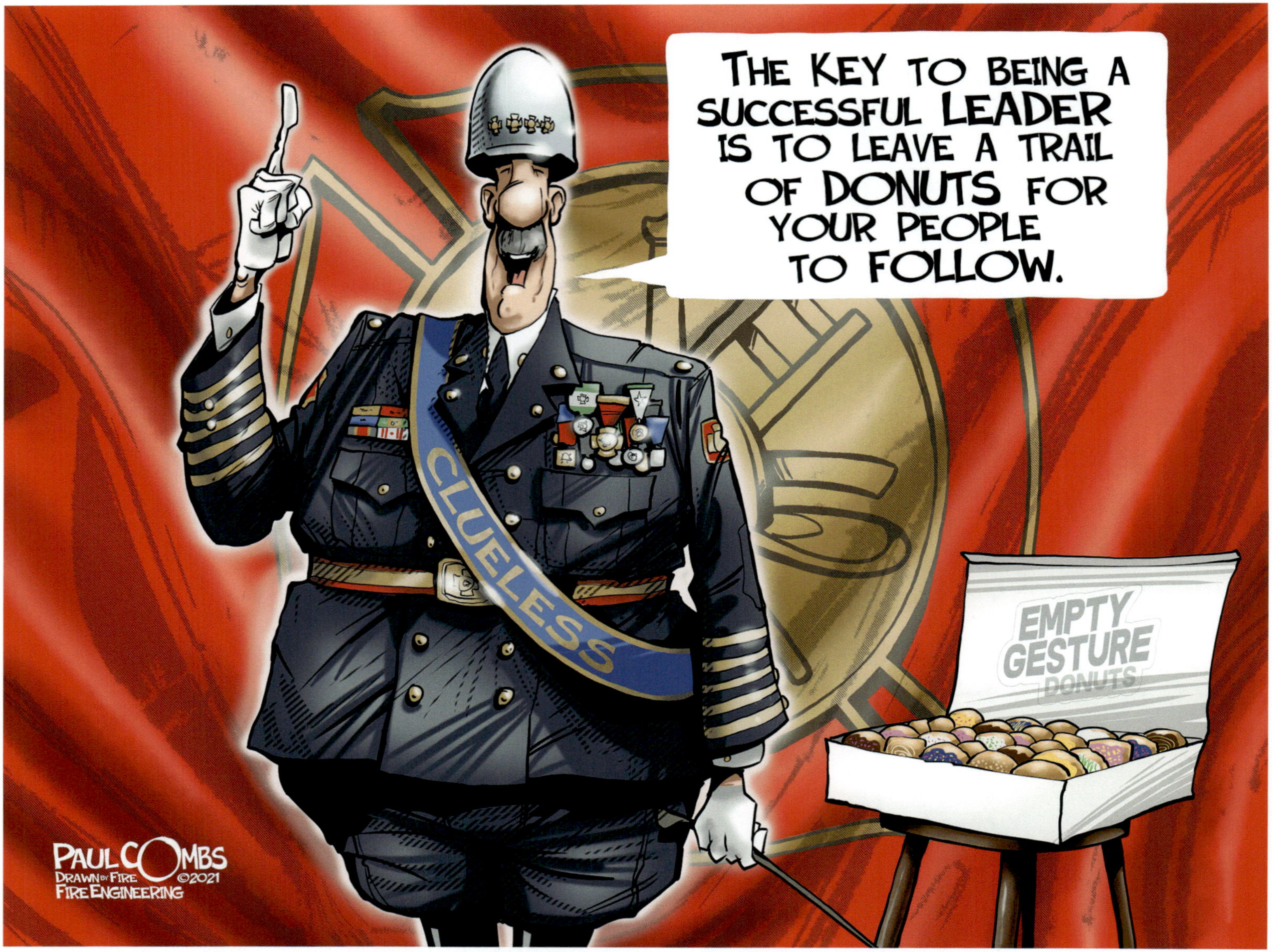

WHAT'S YOUR LEADERSHIP STRATEGY?

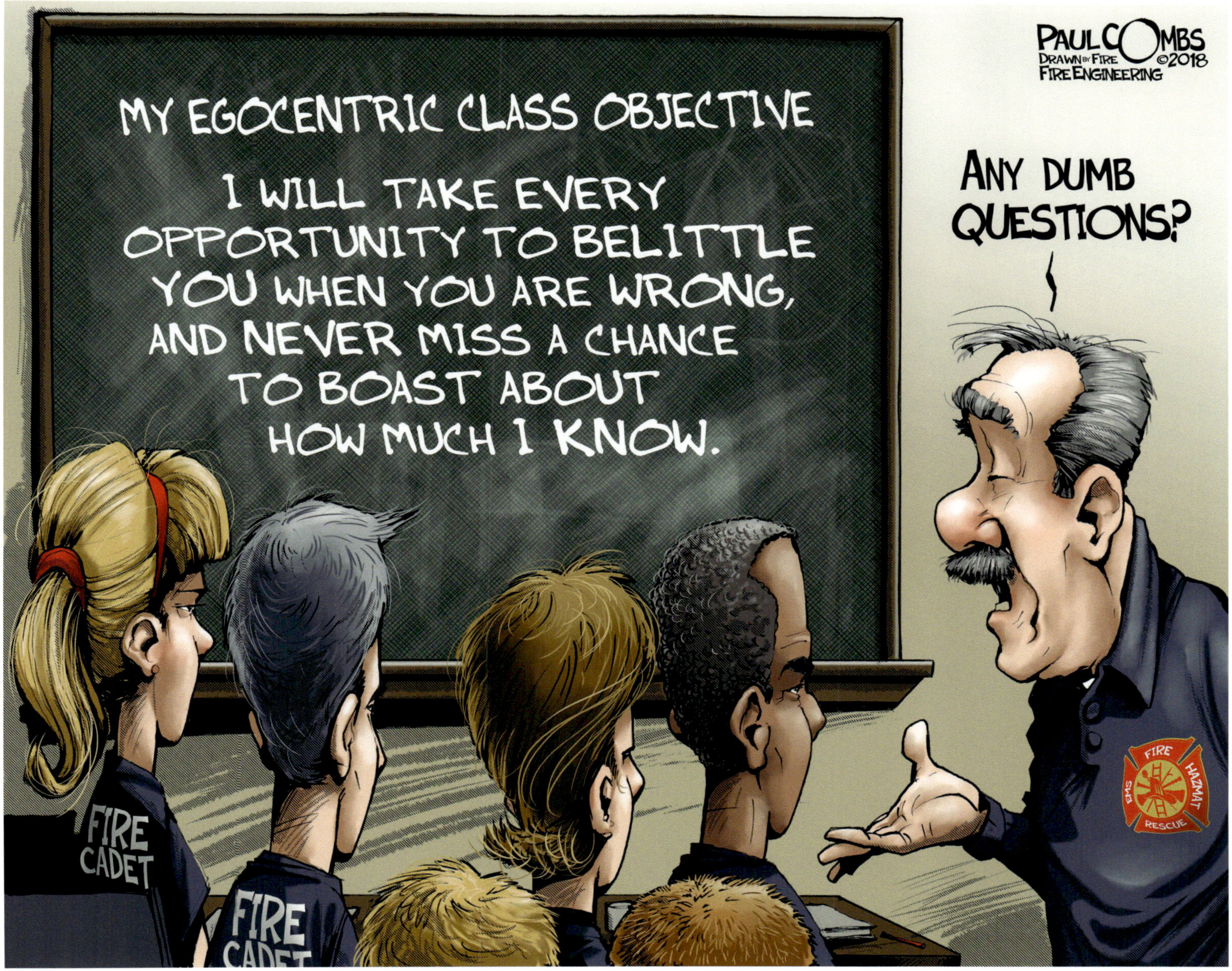
PAUL COMBS
DRAWN BY FIRE ©2018
FIRE ENGINEERING
MY EGOCENTRIC CLASS OBJECTIVE
I WILL TAKE EVERY OPPORTUNITY TO BELITTLE YOU WHEN YOU ARE WRONG, AND NEVER MISS A CHANCE TO BOAST ABOUT HOW MUCH I KNOW.
ANY DUMB QUESTIONS?
FIRE CADET
FIRE
HAZMAT
RESCUE
EMS

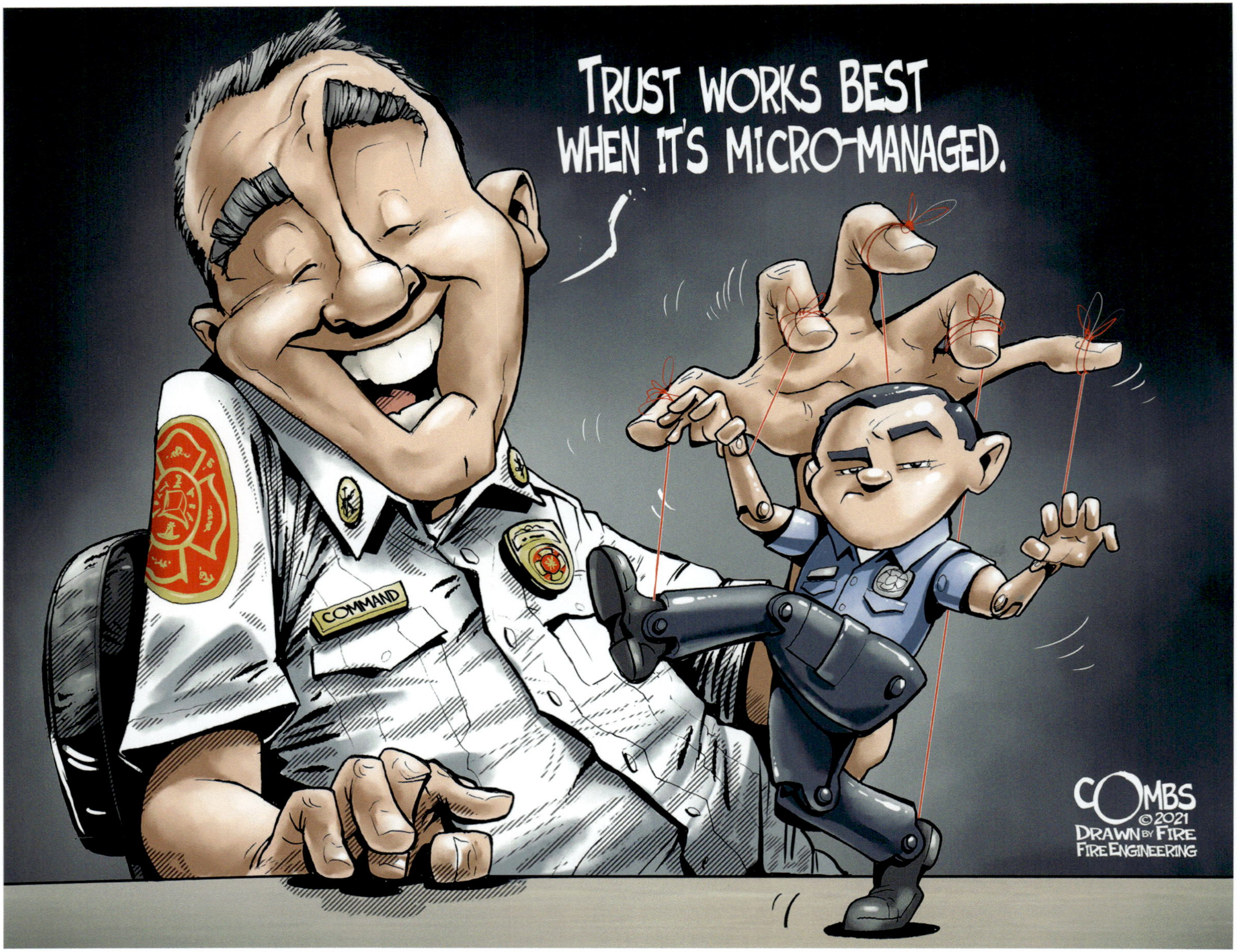
TRUST WORKS BEST
WHEN IT'S MICRO-MANAGED.
COMMAND
COMBS
©2021
DRAWN BY FIRE
FIRE ENGINEERING

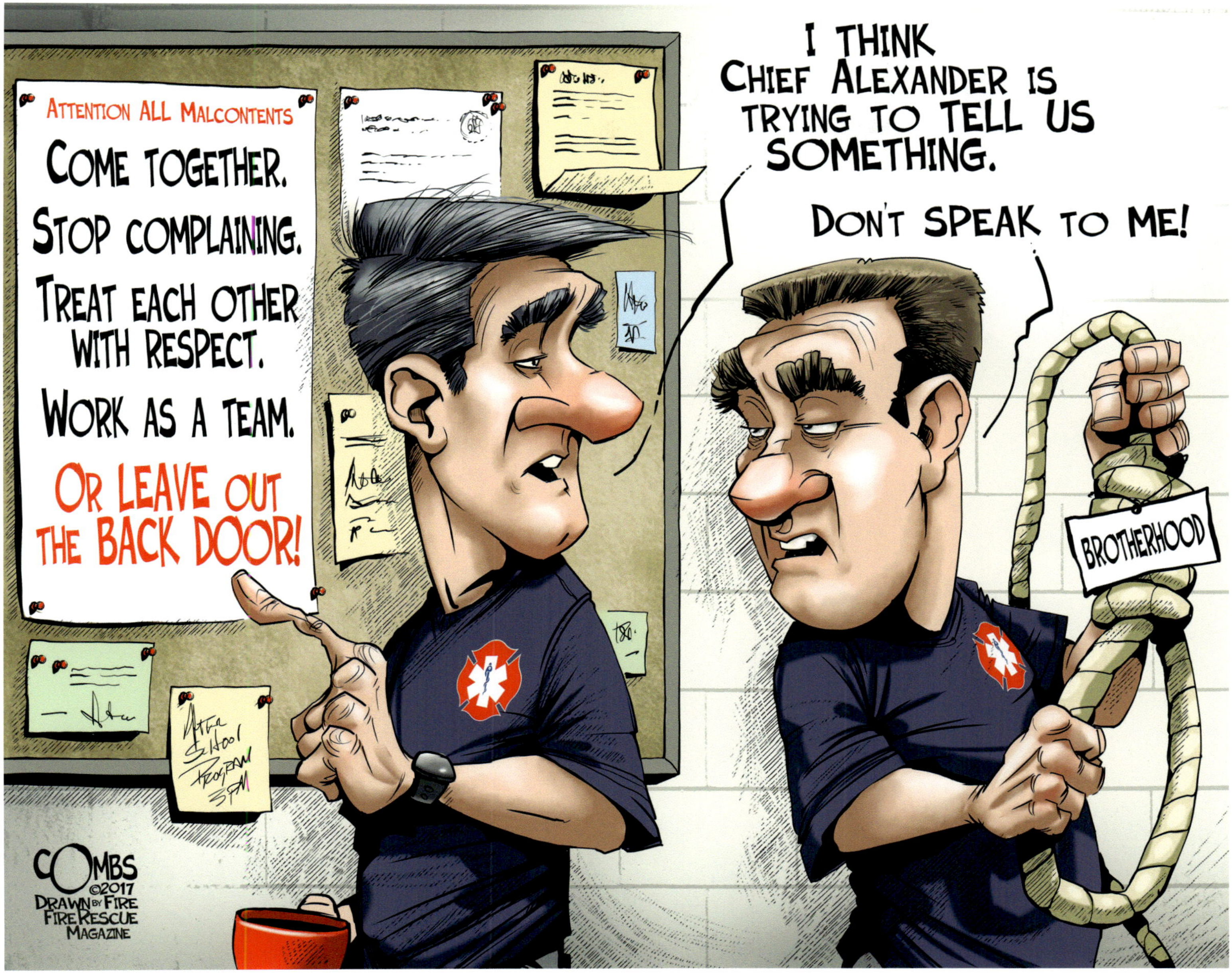
ATTENTION ALL MALCONTENTS
COME TOGETHER.
STOP COMPLAINING.
TREAT EACH OTHER WITH RESPECT.
WORK AS A TEAM.
OR LEAVE OUT THE BACK DOOR!
I THINK CHIEF ALEXANDER IS TRYING TO TELL US SOMETHING.
DON'T SPEAK TO ME!
BROTHERHOOD
COMBS ©2017 DRAWN BY FIRE FIRERESCUE MAGAZINE

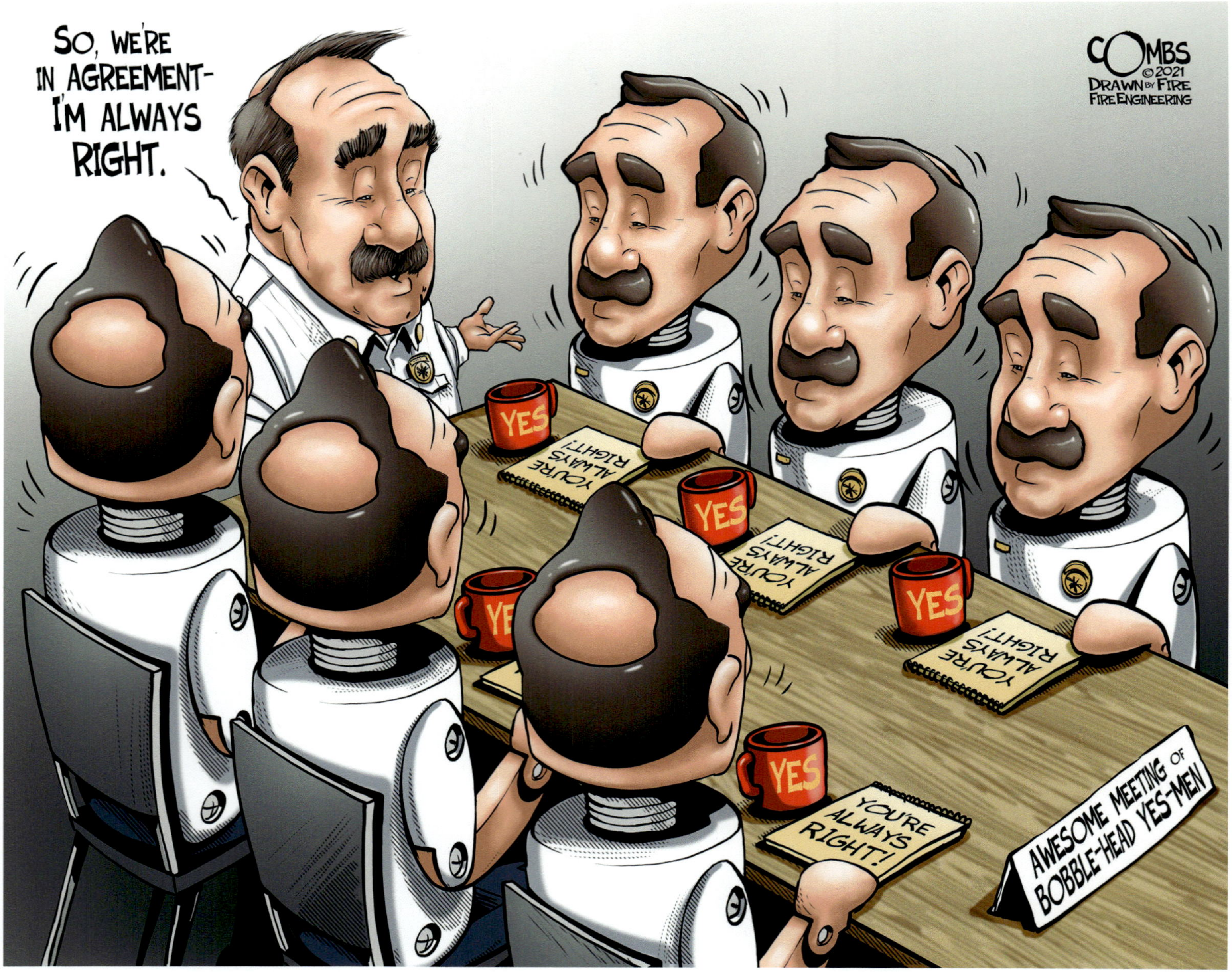
SO, WE'RE IN AGREEMENT- I'M ALWAYS RIGHT.
COMBS ©2021 DRAWN BY FIRE FIRE ENGINEERING
YES
YOU'RE ALWAYS RIGHT!
AWESOME MEETING OF BOBBLE-HEAD YES-MEN

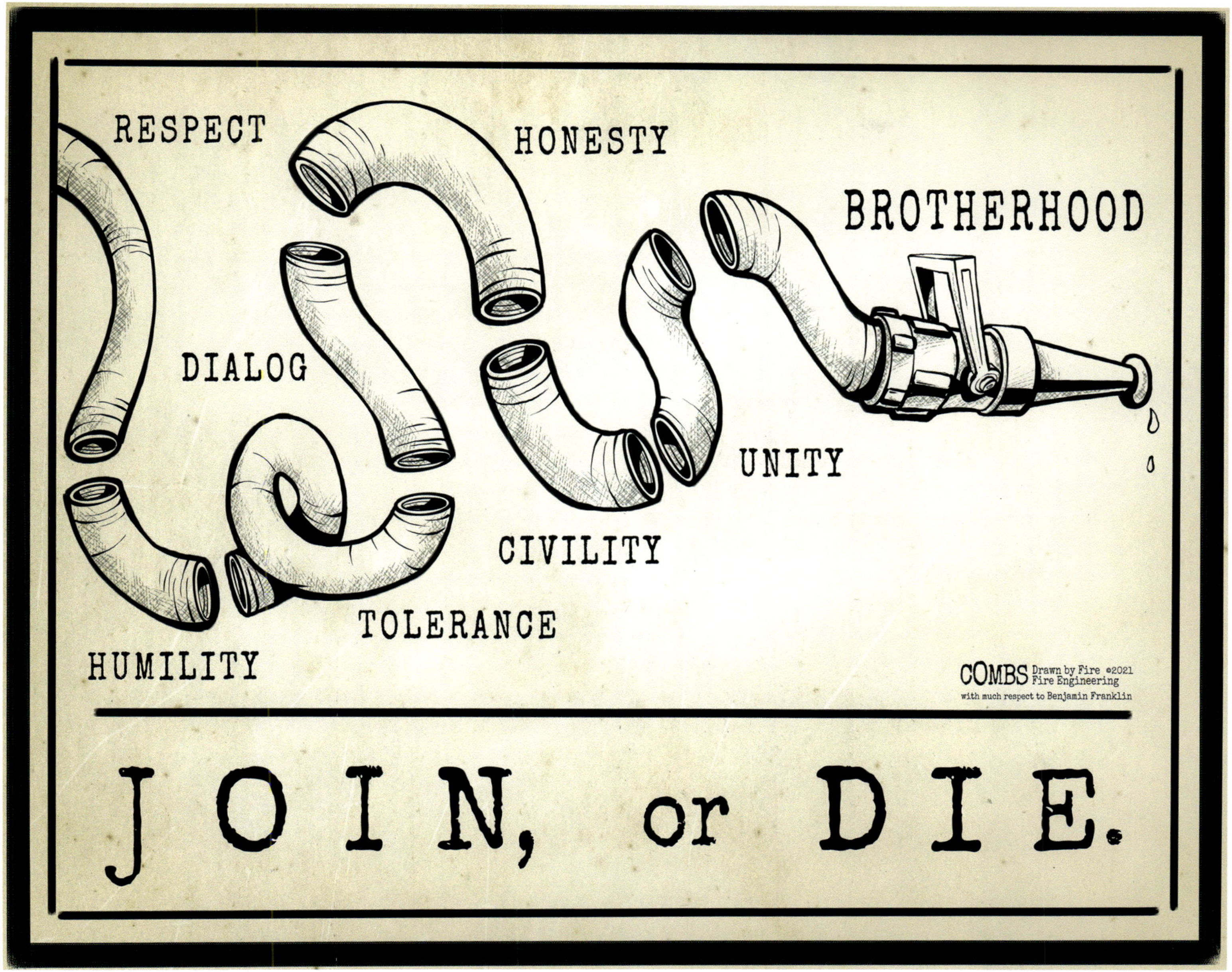
RESPECT
HONESTY
BROTHERHOOD
DIALOG
UNITY
CIVILITY
TOLERANCE
HUMILITY
COMBS Drawn by Fire ©2021 Fire Engineering
with much respect to Benjamin Franklin
JOIN, or DIE.

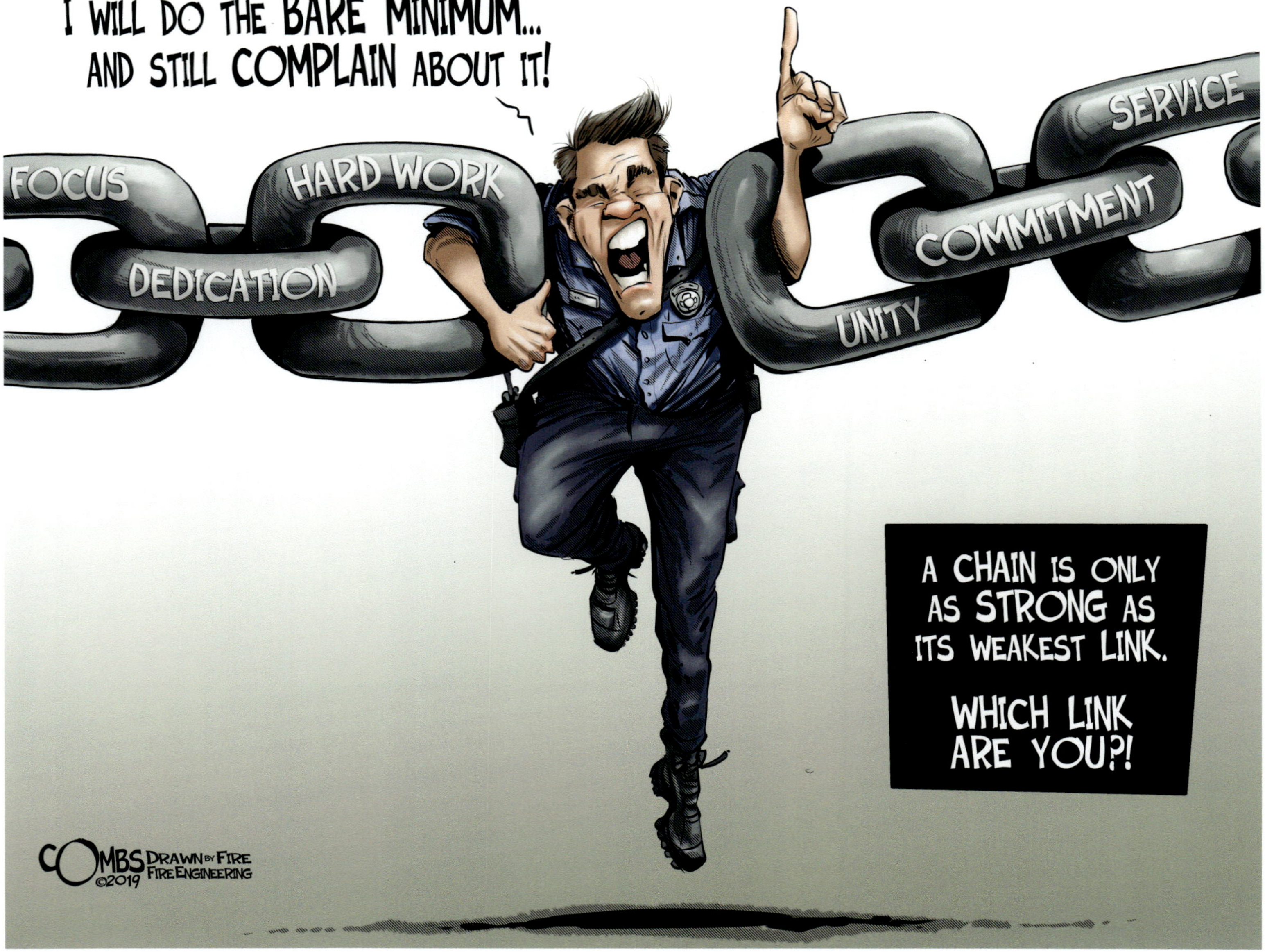
I WILL DO THE BARE MINIMUM...
AND STILL COMPLAIN ABOUT IT!
FOCUS
HARD WORK
DEDICATION
UNITY
COMMITMENT
SERVICE
A CHAIN IS ONLY AS STRONG AS ITS WEAKEST LINK.
WHICH LINK ARE YOU?!
COMBS ©2019 DRAWN BY FIRE FIRE ENGINEERING

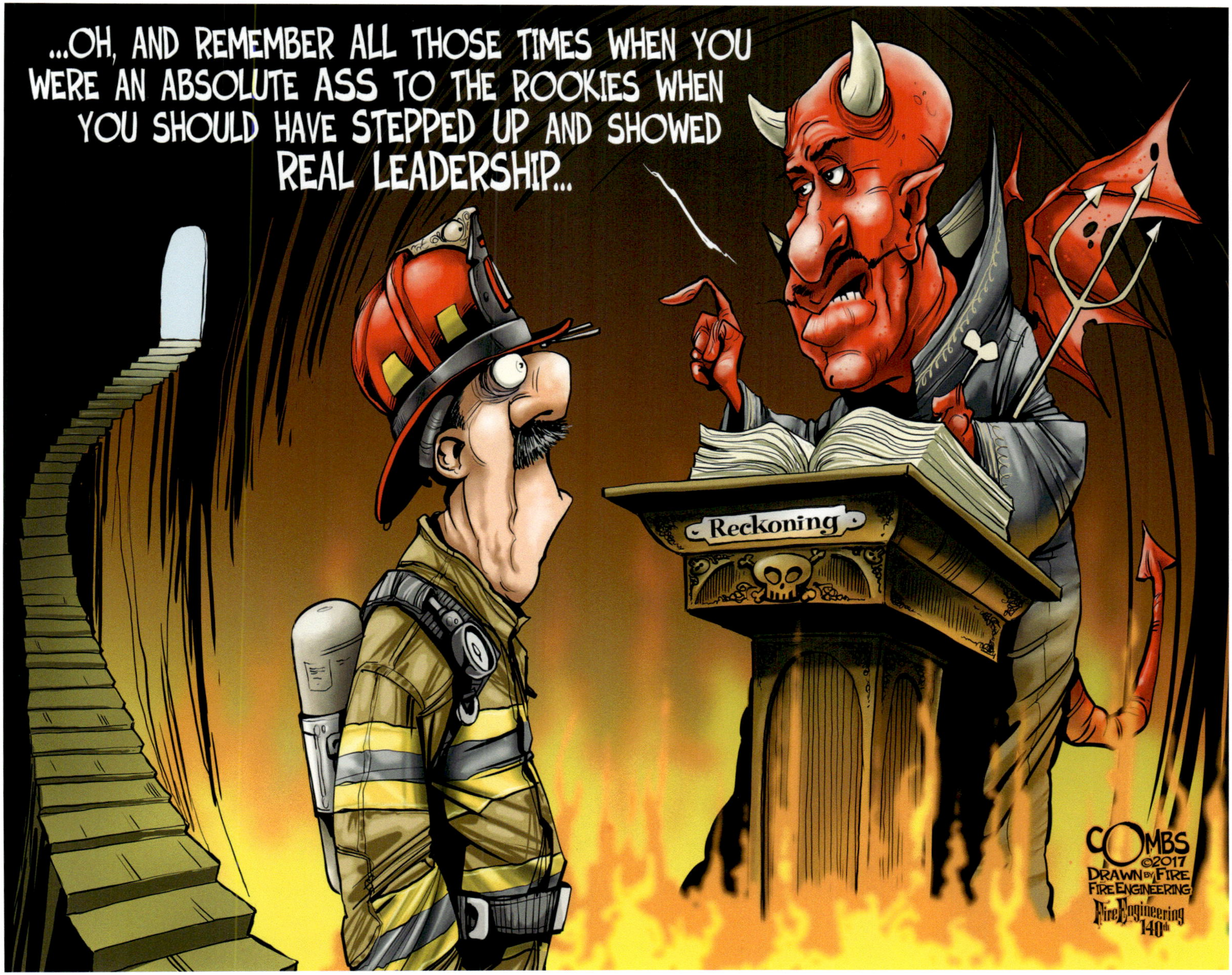
...OH, AND REMEMBER ALL THOSE TIMES WHEN YOU WERE AN ABSOLUTE ASS TO THE ROOKIES WHEN YOU SHOULD HAVE STEPPED UP AND SHOWED REAL LEADERSHIP...
Reckoning
COMBS ©2017 DRAWN BY FIRE FIRE ENGINEERING
Fire Engineering 140th

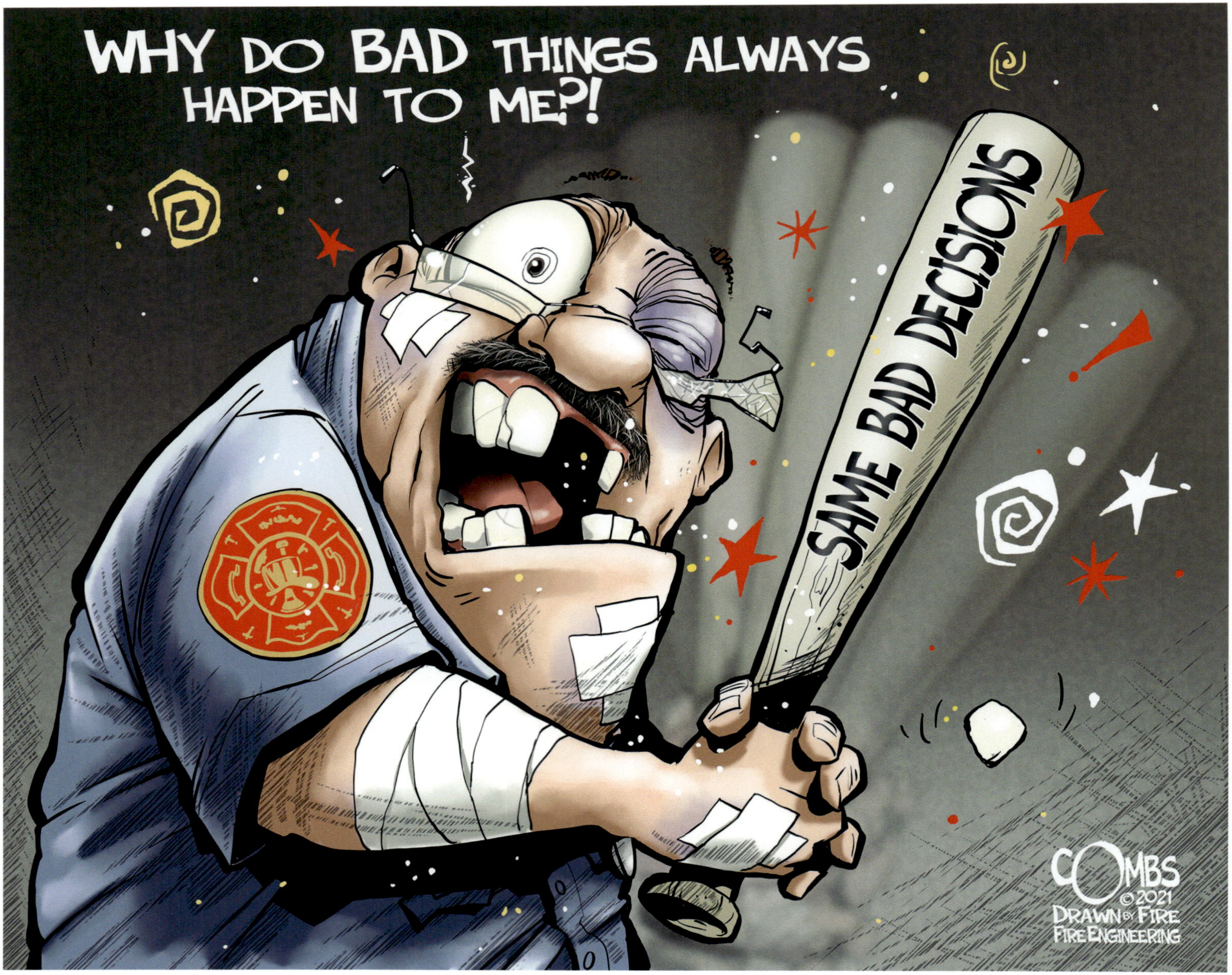
WHY DO BAD THINGS ALWAYS HAPPEN TO ME?!
SAME BAD DECISIONS
COMBS
©2021
DRAWN BY FIRE
FIRE ENGINEERING

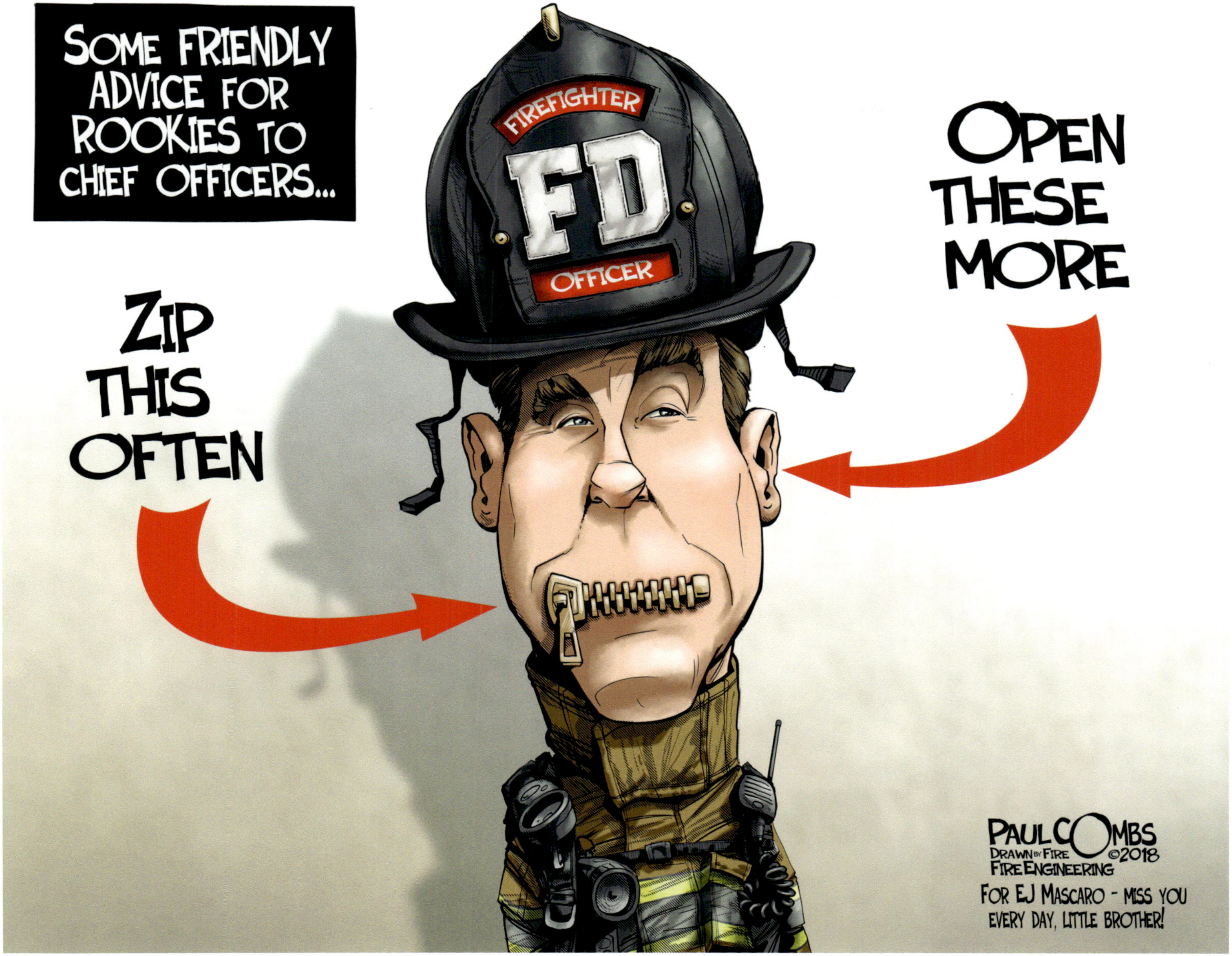
SOME FRIENDLY ADVICE FOR ROOKIES TO CHIEF OFFICERS...
FIREFIGHTER
FD
OFFICER
ZIP THIS OFTEN
OPEN THESE MORE
PAUL COMBS
DRAWN BY FIRE ©2018
FIRE ENGINEERING
FOR EJ MASCARO - MISS YOU EVERY DAY, LITTLE BROTHER!

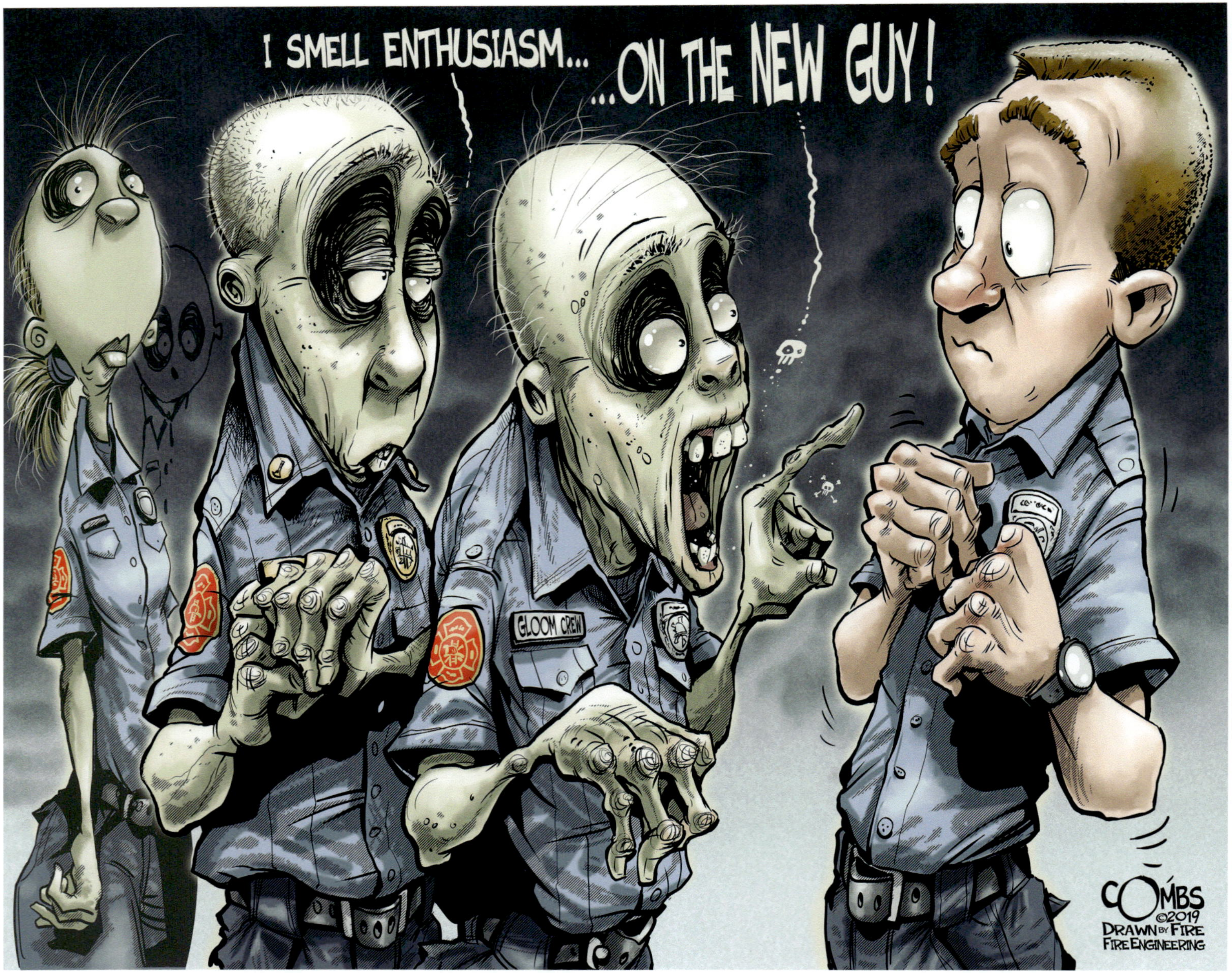
I SMELL ENTHUSIASM...
...ON THE NEW GUY!
GLOOM CREW
COMBS
©2019
DRAWN BY FIRE
FIRE ENGINEERING

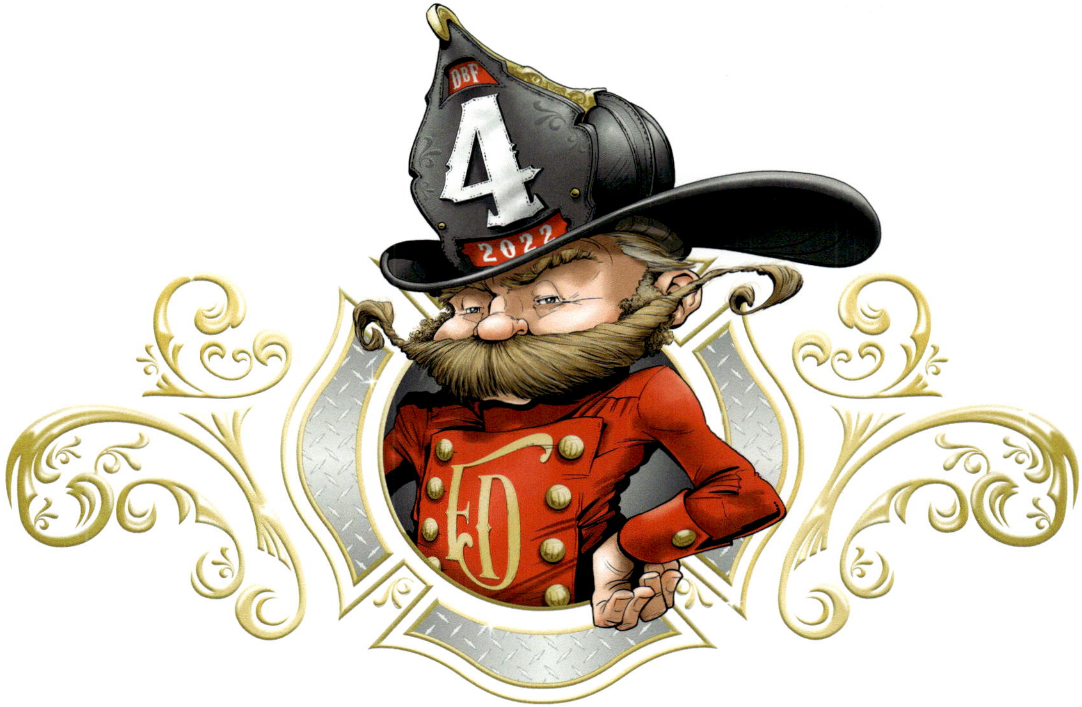

Let's Get Serious . . . Sort of

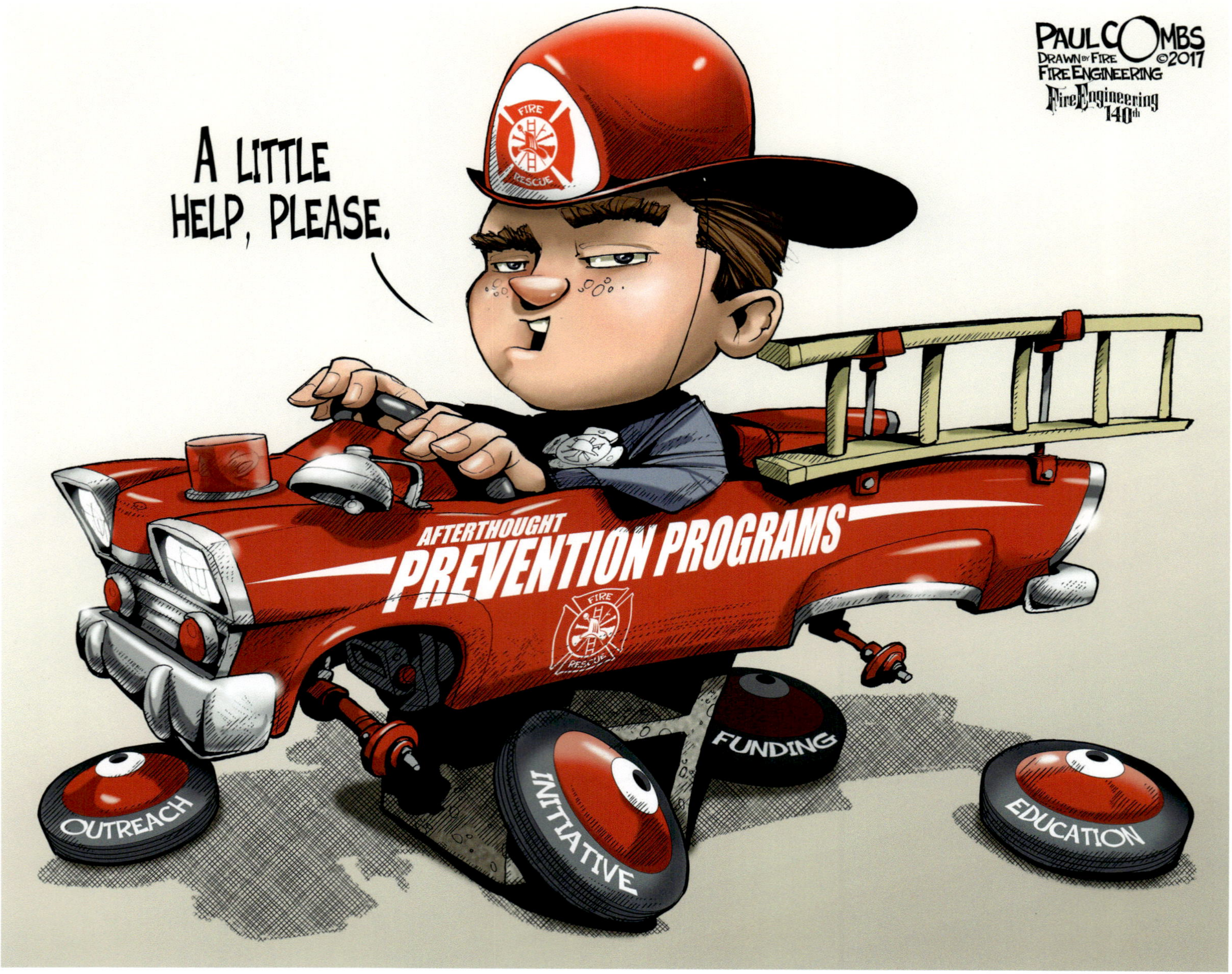
A LITTLE HELP, PLEASE.
PAUL COMBS
DRAWN BY FIRE ©2017
FIRE ENGINEERING
FireEngineering 140th
AFTERTHOUGHT
PREVENTION PROGRAMS
FIRE
RESCUE
OUTREACH
INITIATIVE
FUNDING
EDUCATION

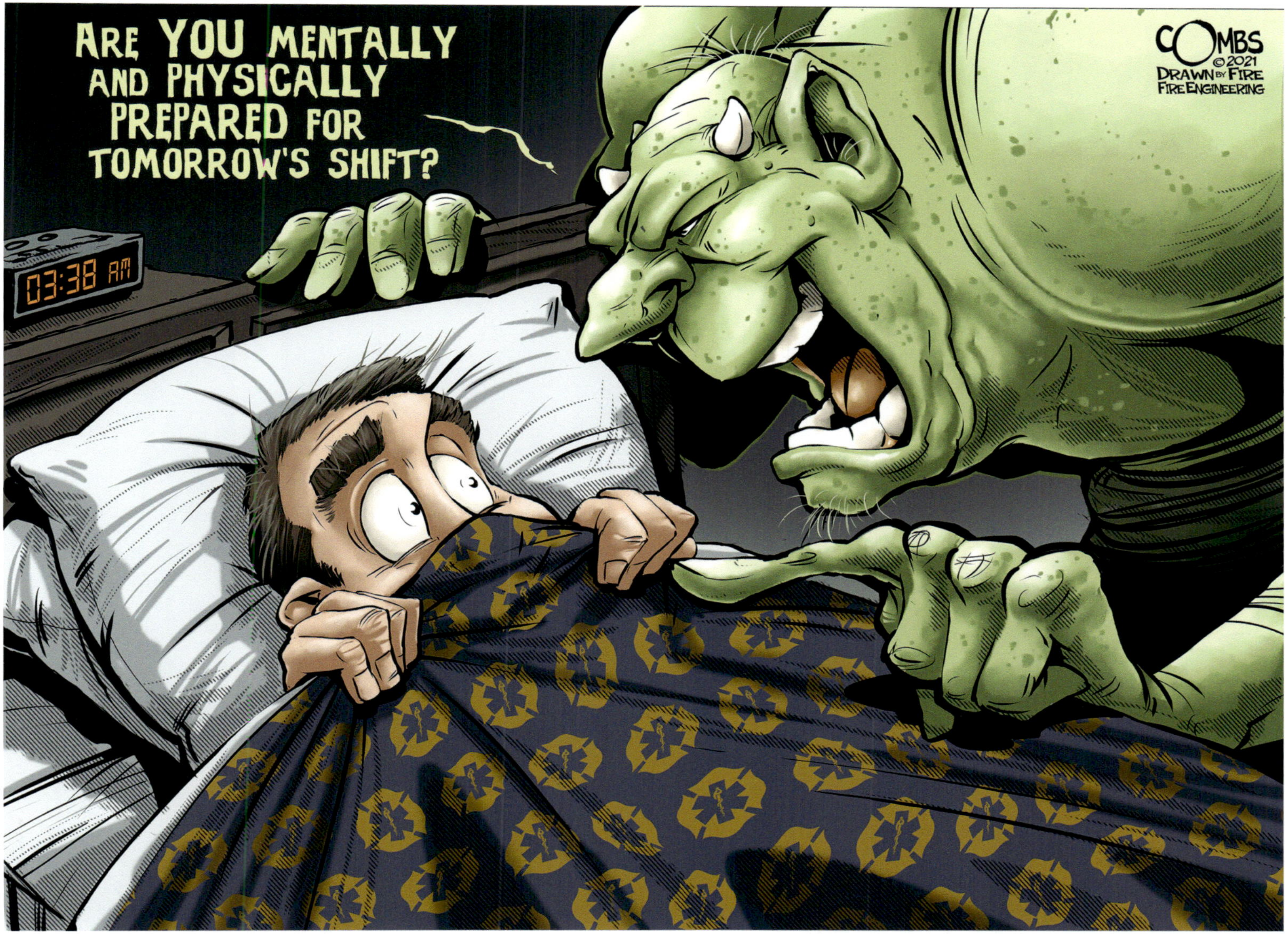

DON'T BE TORMENTED BY THE ACCOUNTABILITY MONSTER – TRAIN!

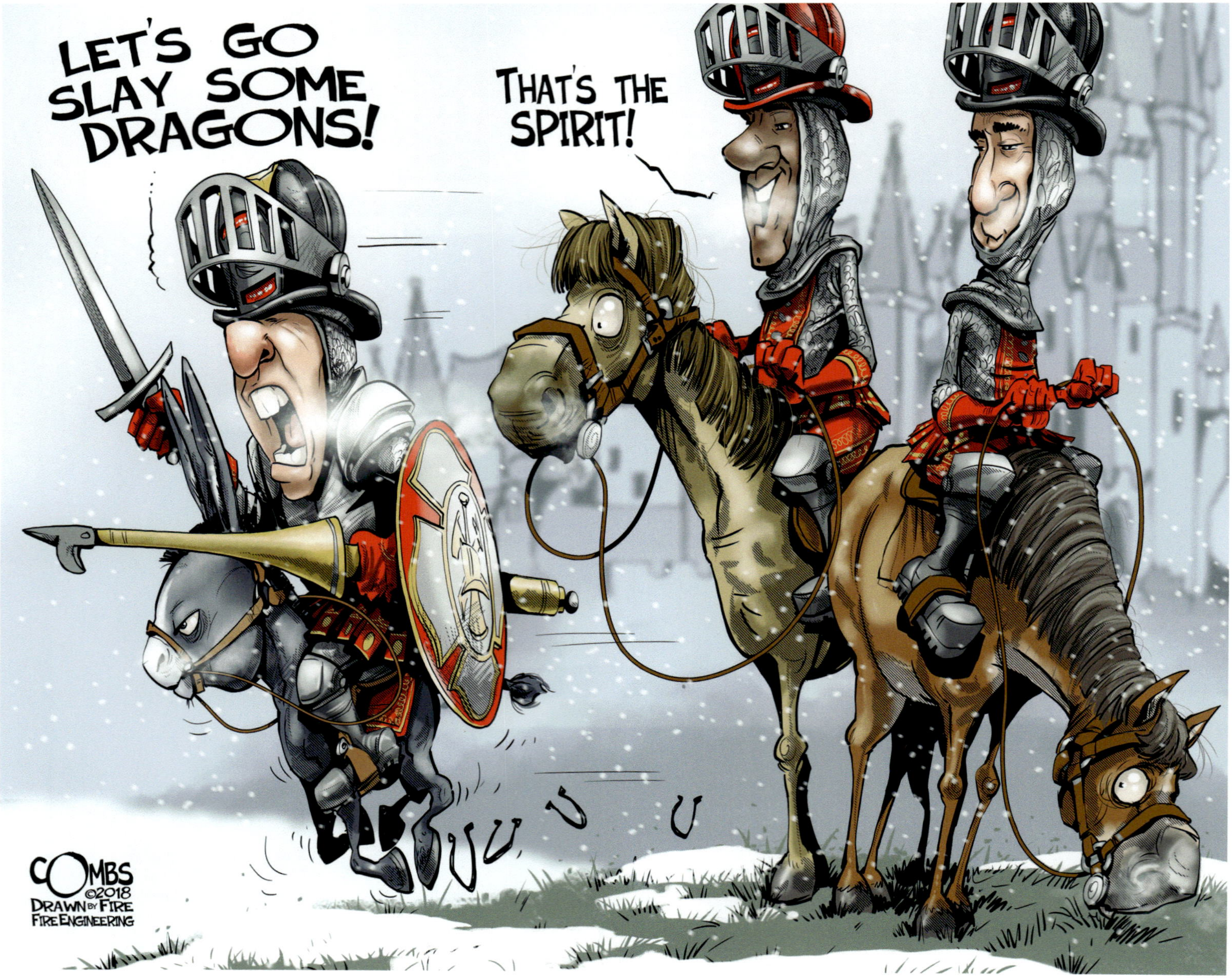
LET'S GO SLAY SOME DRAGONS!
THAT'S THE SPIRIT!
COMBS ©2018 DRAWN BY FIRE FIRE ENGINEERING

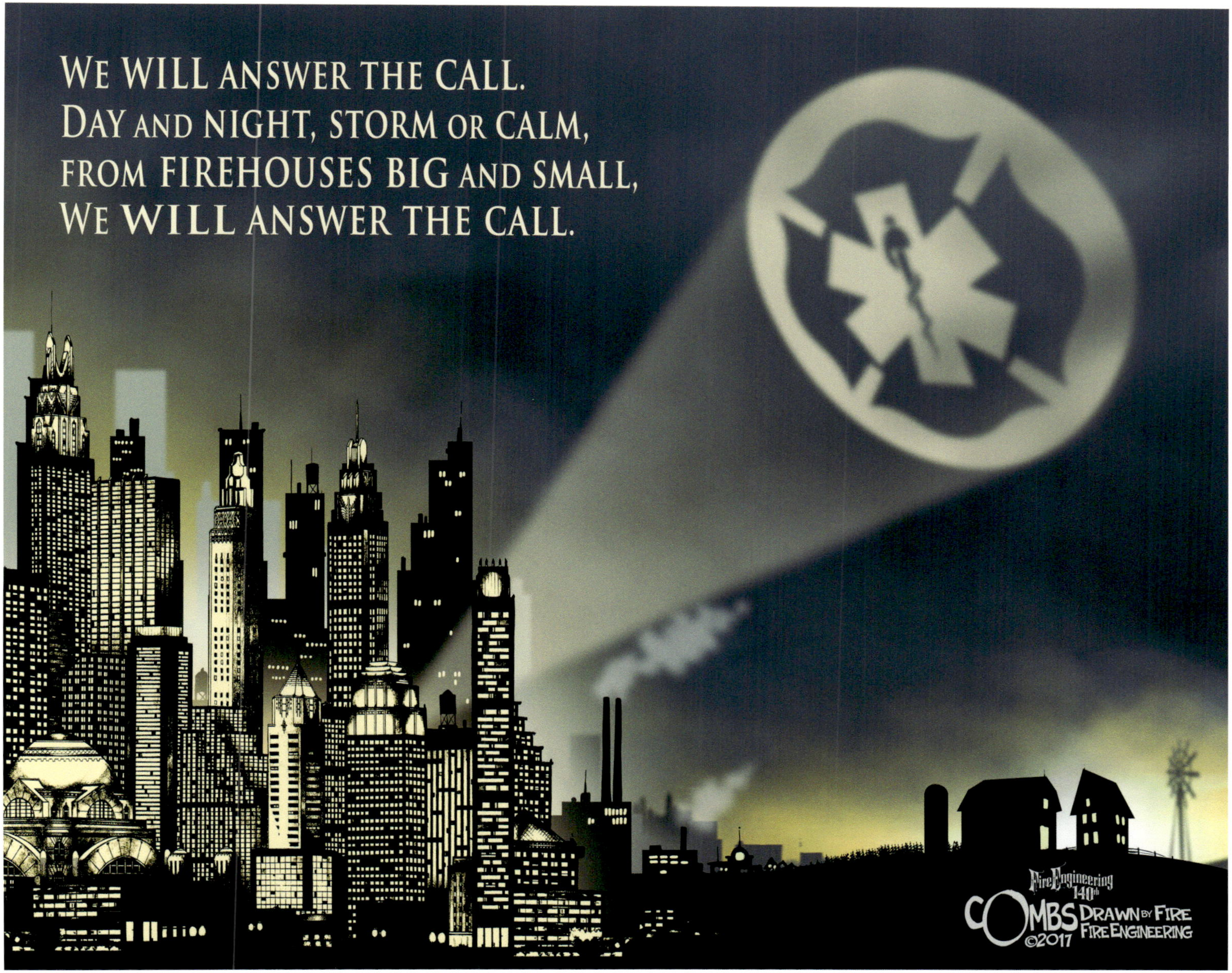
WE WILL ANSWER THE CALL.
DAY AND NIGHT, STORM OR CALM,
FROM FIREHOUSES BIG AND SMALL,
WE WILL ANSWER THE CALL.
Fire Engineering 140th
COMBS ©2017
DRAWN BY FIRE
FIRE ENGINEERING

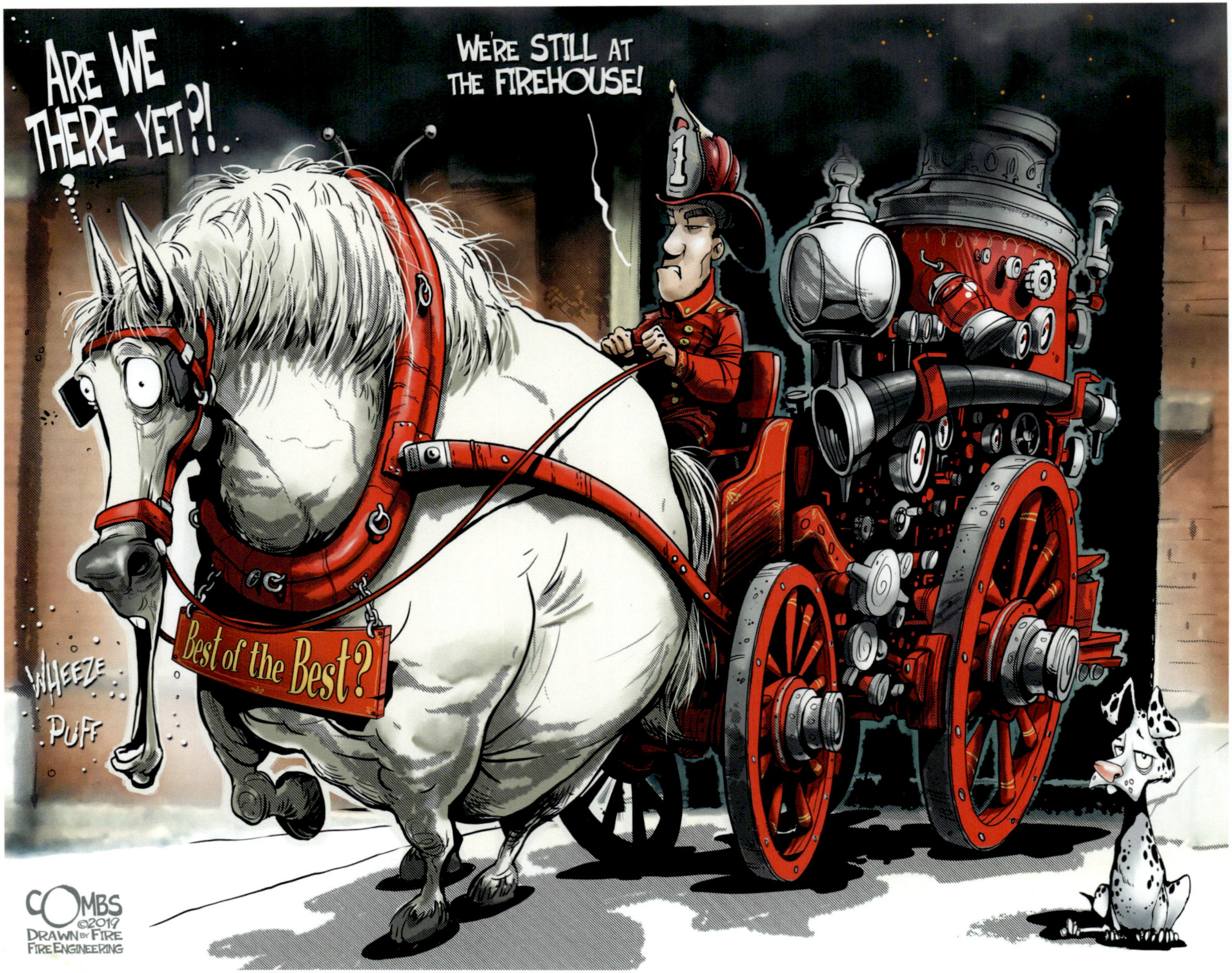
ARE WE THERE YET?!
WE'RE STILL AT THE FIREHOUSE!
1
Best of the Best?
WHEEZE
PUFF
COMBS ©2019 DRAWN BY FIRE FIRE ENGINEERING

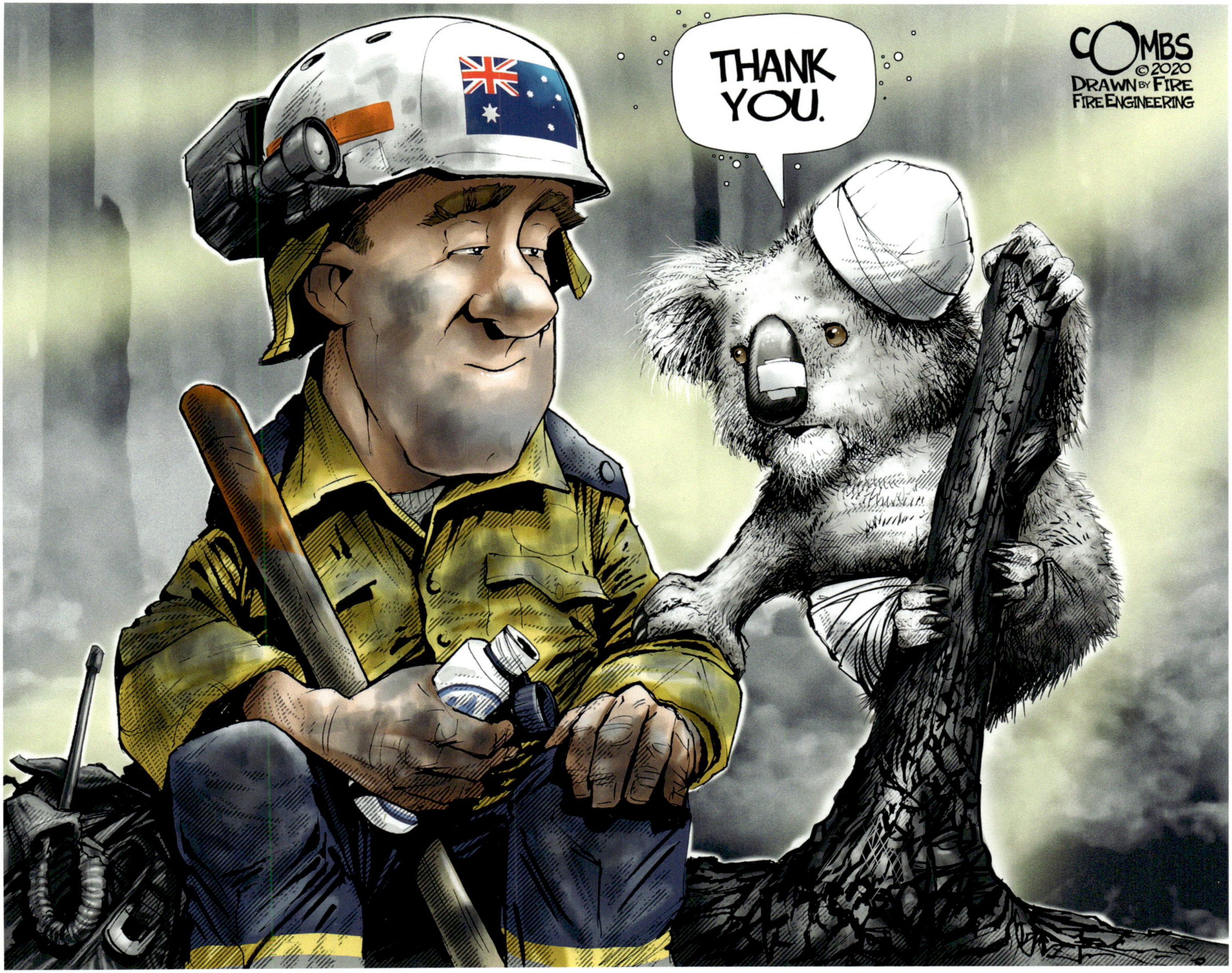
THANK YOU.
COMBS ©2020 DRAWN BY FIRE FIRE ENGINEERING

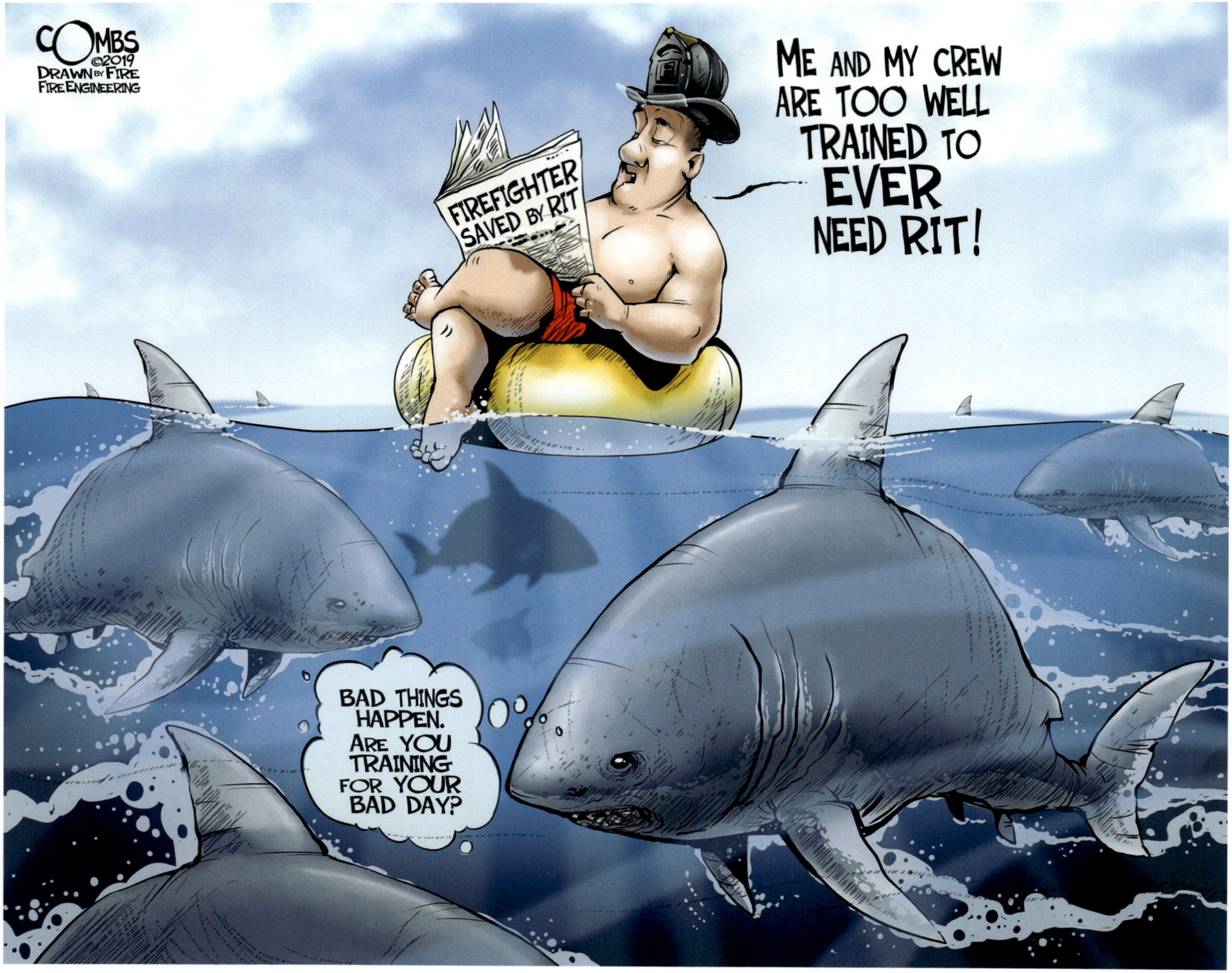
COMBS ©2019 DRAWN BY FIRE FIRE ENGINEERING
FIREFIGHTER SAVED BY RIT
ME AND MY CREW ARE TOO WELL TRAINED TO EVER NEED RIT!
BAD THINGS HAPPEN. ARE YOU TRAINING FOR YOUR BAD DAY?

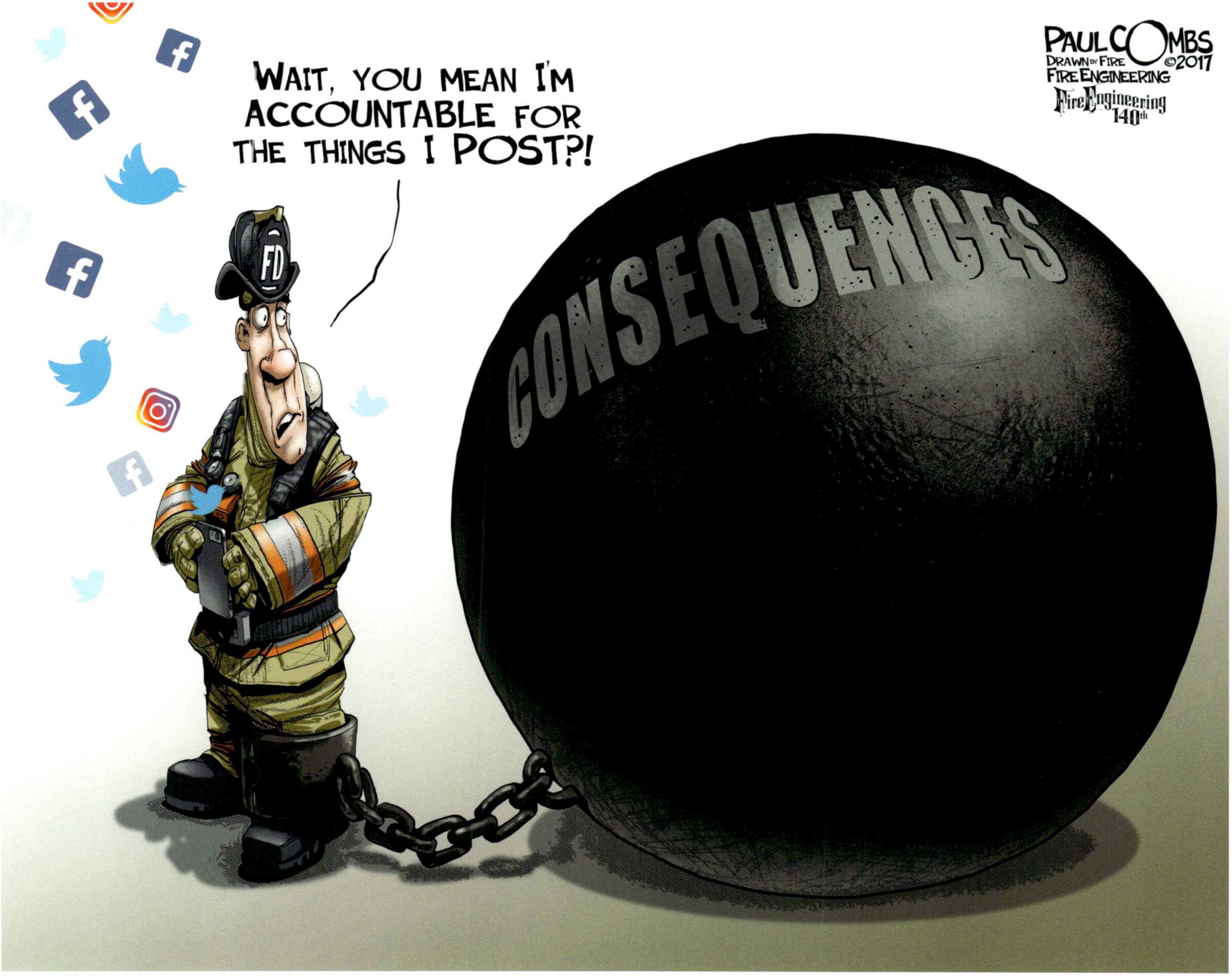
WAIT, YOU MEAN I'M ACCOUNTABLE FOR THE THINGS I POST?!
FD
CONSEQUENCES
PAUL COMBS
DRAWN BY FIRE ©2017
FIRE ENGINEERING
Fire Engineering 140th

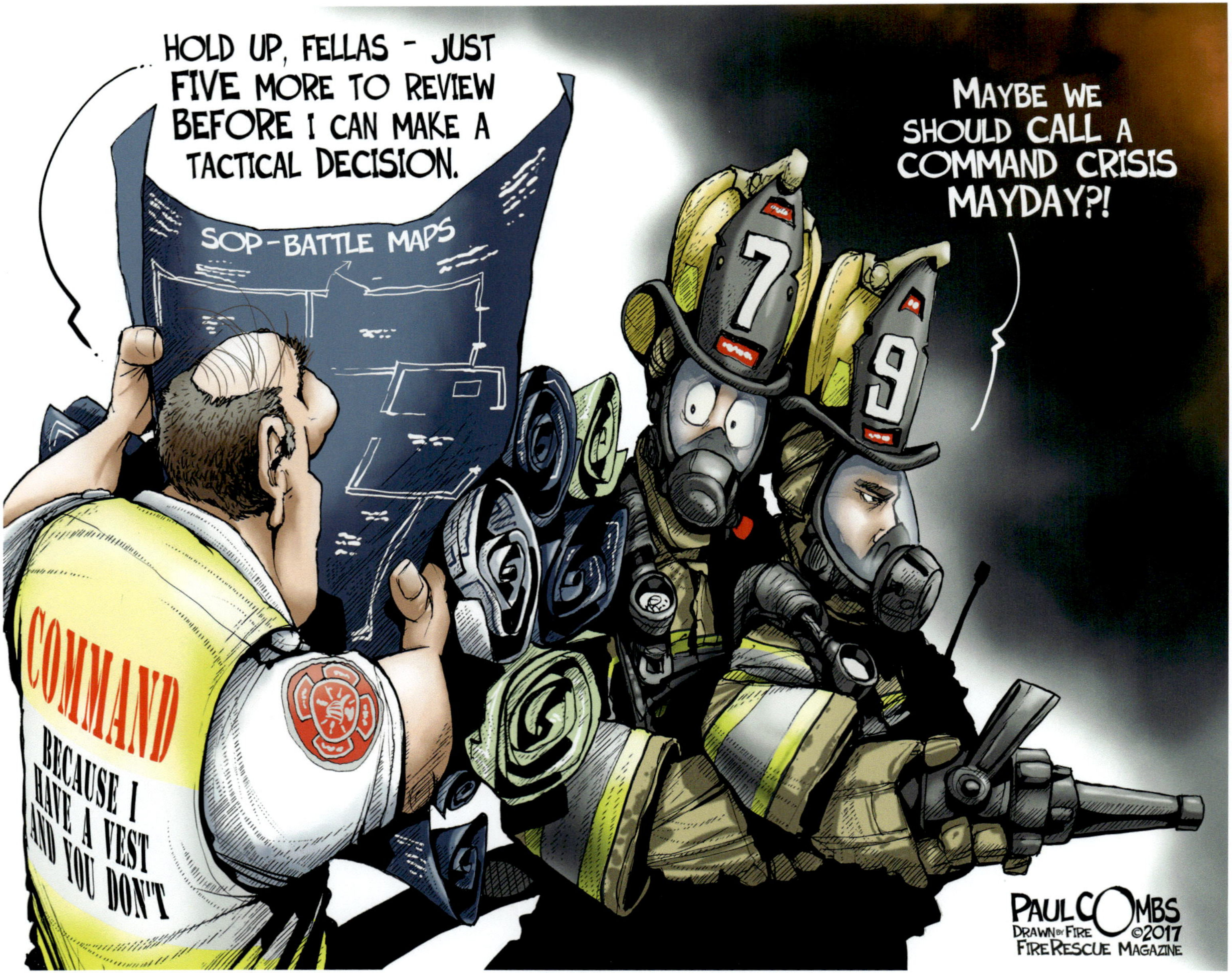
HOLD UP, FELLAS - JUST FIVE MORE TO REVIEW BEFORE I CAN MAKE A TACTICAL DECISION.
MAYBE WE SHOULD CALL A COMMAND CRISIS MAYDAY?!
SOP-BATTLE MAPS
COMMAND
BECAUSE I HAVE A VEST AND YOU DON'T
7
9
PAUL COMBS
DRAWN BY FIRE ©2017
FIRERESCUE MAGAZINE

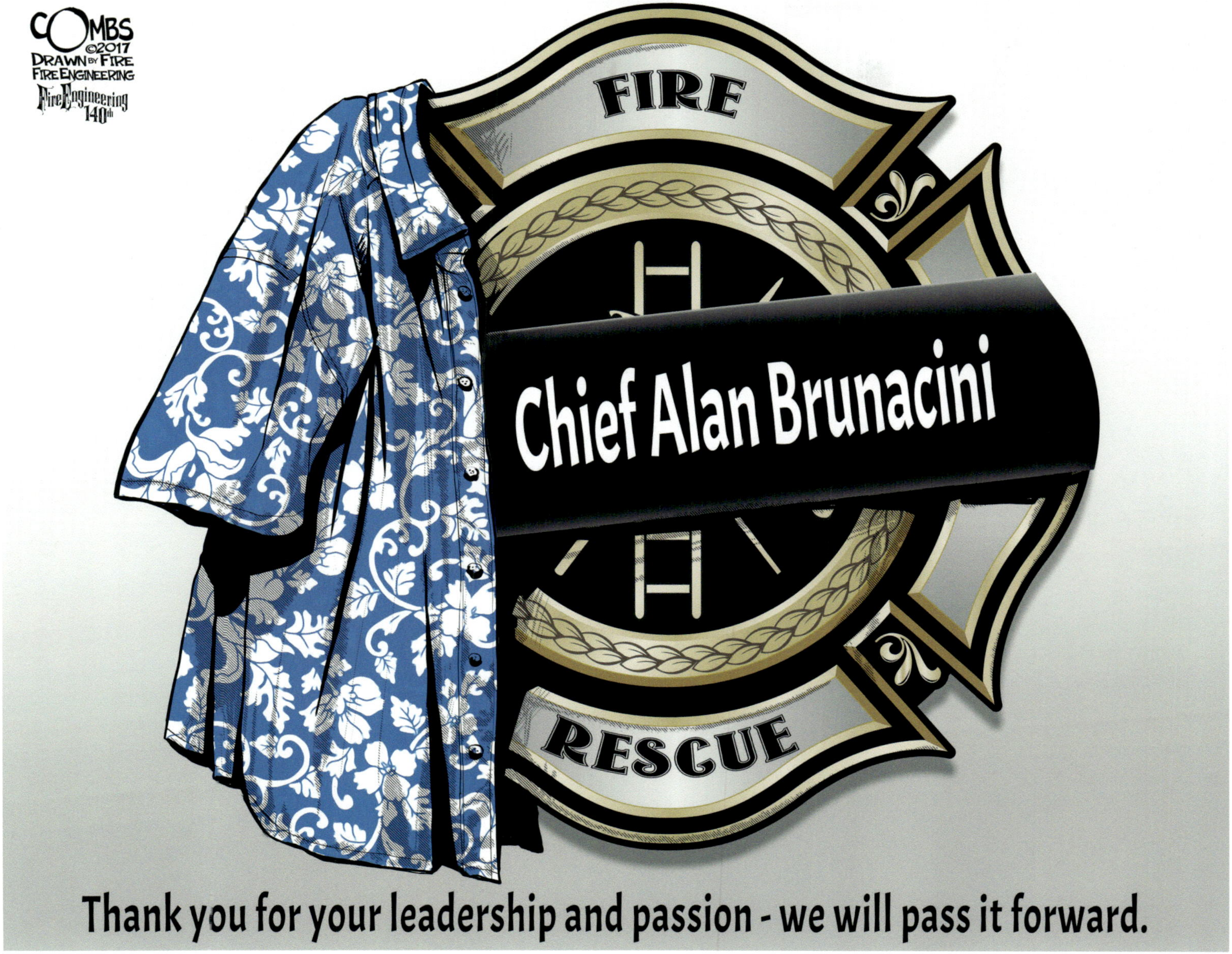
COMBS
©2017
DRAWN BY FIRE
FIRE ENGINEERING
Fire Engineering
140th
FIRE
Chief Alan Brunacini
RESCUE
Thank you for your leadership and passion - we will pass it forward.

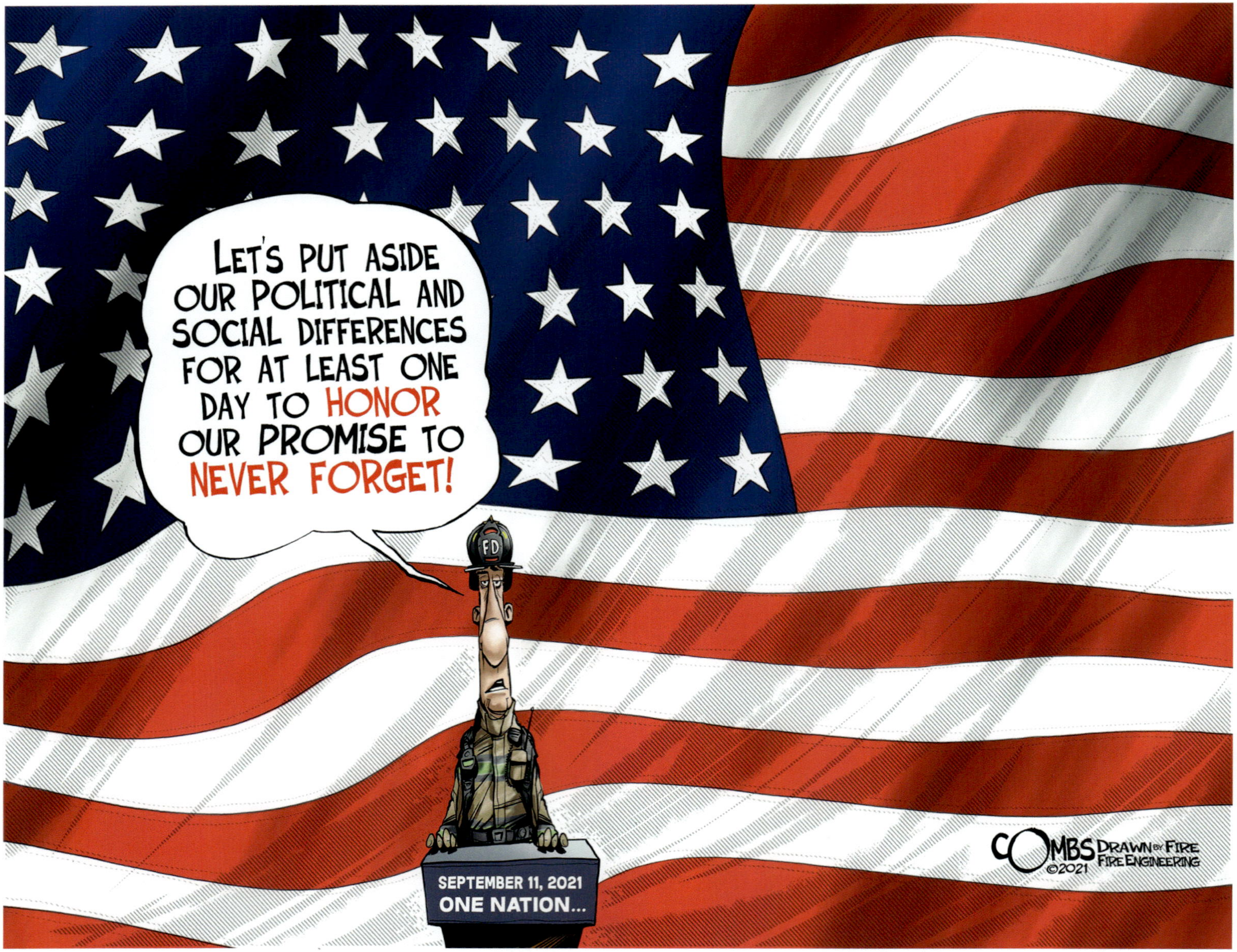
LET'S PUT ASIDE OUR POLITICAL AND SOCIAL DIFFERENCES FOR AT LEAST ONE DAY TO HONOR OUR PROMISE TO NEVER FORGET!
FD
SEPTEMBER 11, 2021
ONE NATION...
COMBS DRAWN BY FIRE FIRE ENGINEERING ©2021

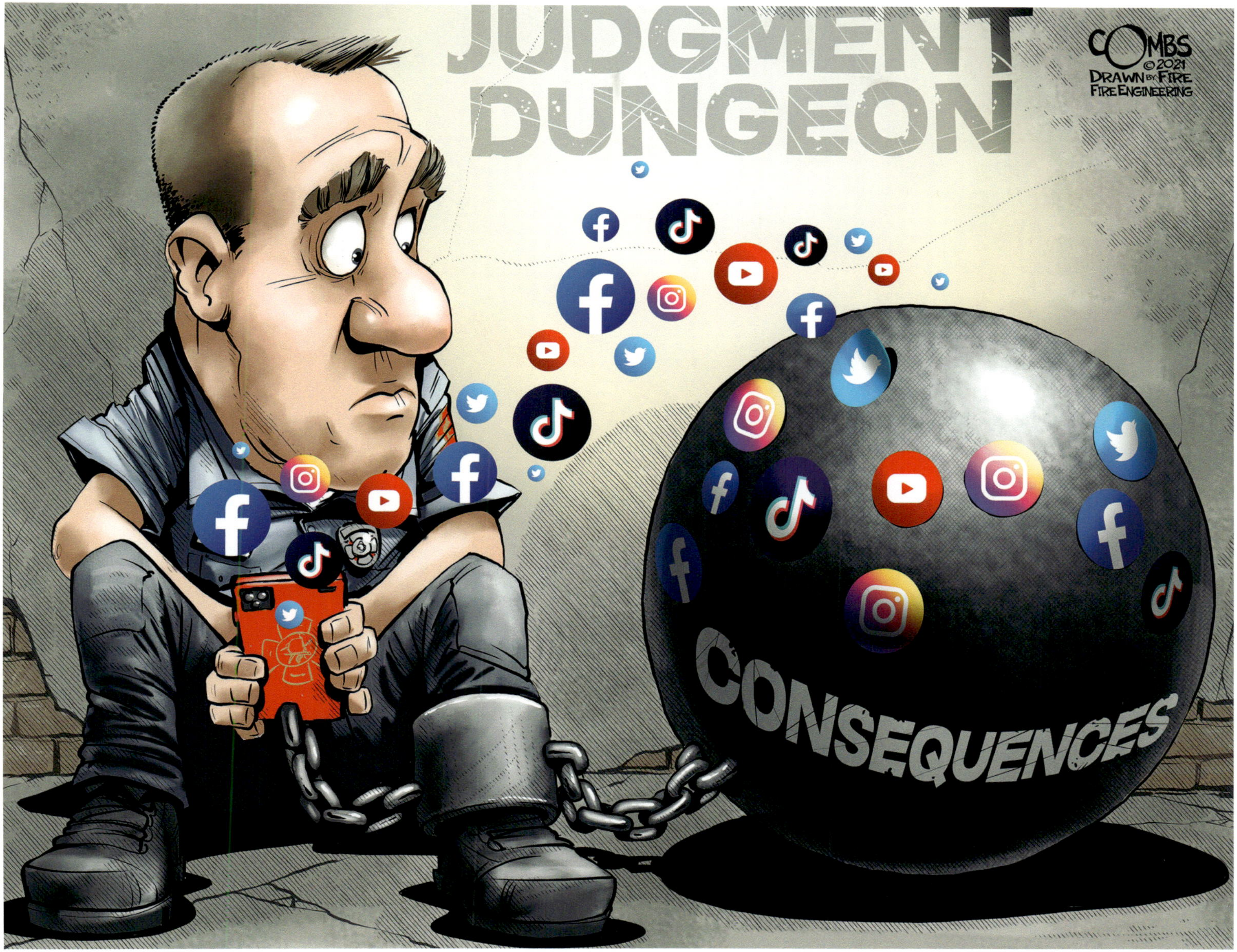
JUDGMENT
DUNGEON
COMBS
©2021
DRAWN BY FIRE
FIRE ENGINEERING
CONSEQUENCES

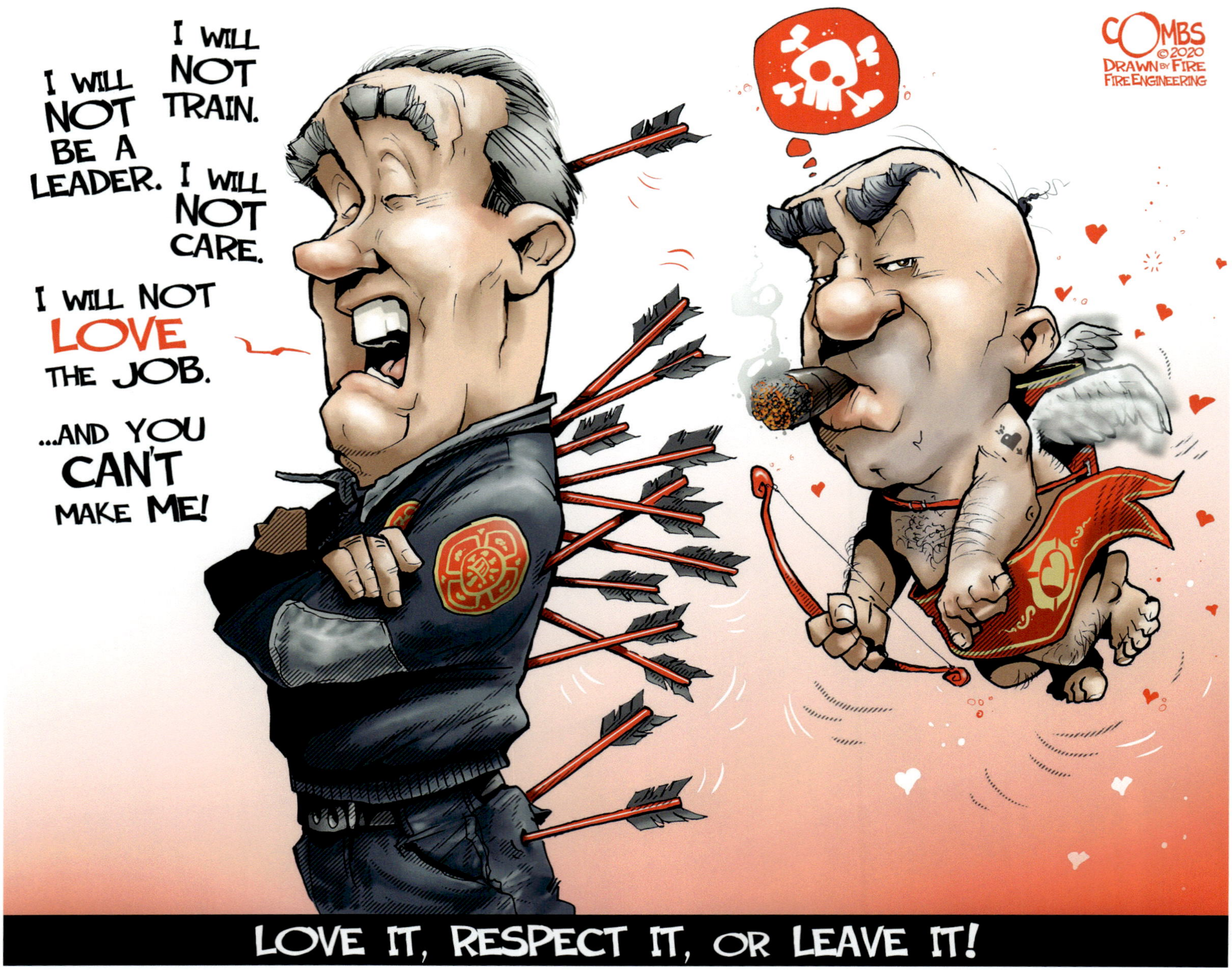
I WILL NOT BE A LEADER.
I WILL NOT TRAIN.
I WILL NOT CARE.
I WILL NOT LOVE THE JOB.
...AND YOU CAN'T MAKE ME!
COMBS ©2020 DRAWN BY FIRE FIRE ENGINEERING
LOVE IT, RESPECT IT, OR LEAVE IT!

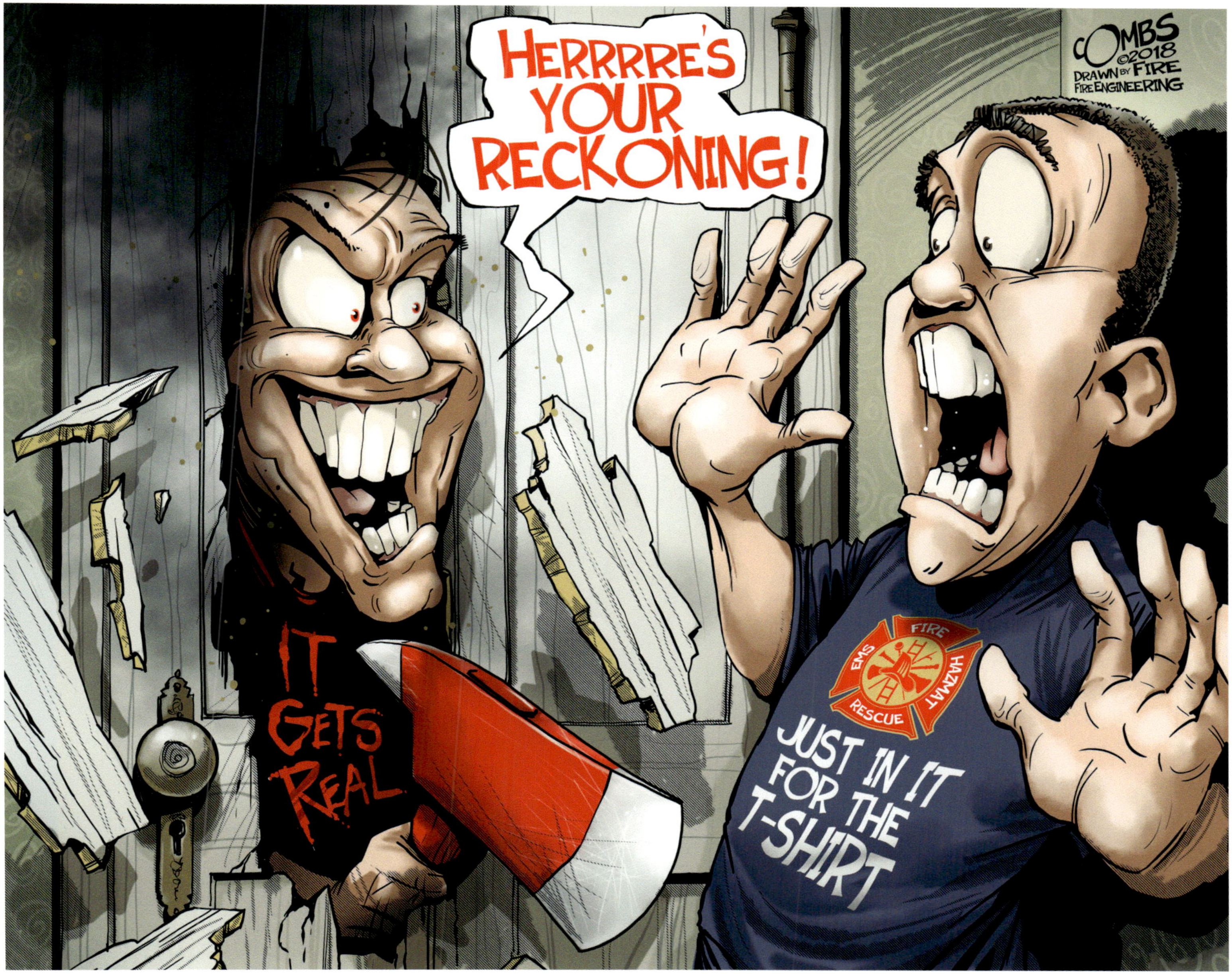
HERRRRE'S YOUR RECKONING!
COMBS ©2018 DRAWN BY FIRE FIRE ENGINEERING
IT GETS REAL
FIRE
EMS
HAZMAT
RESCUE
JUST IN IT FOR THE T-SHIRT

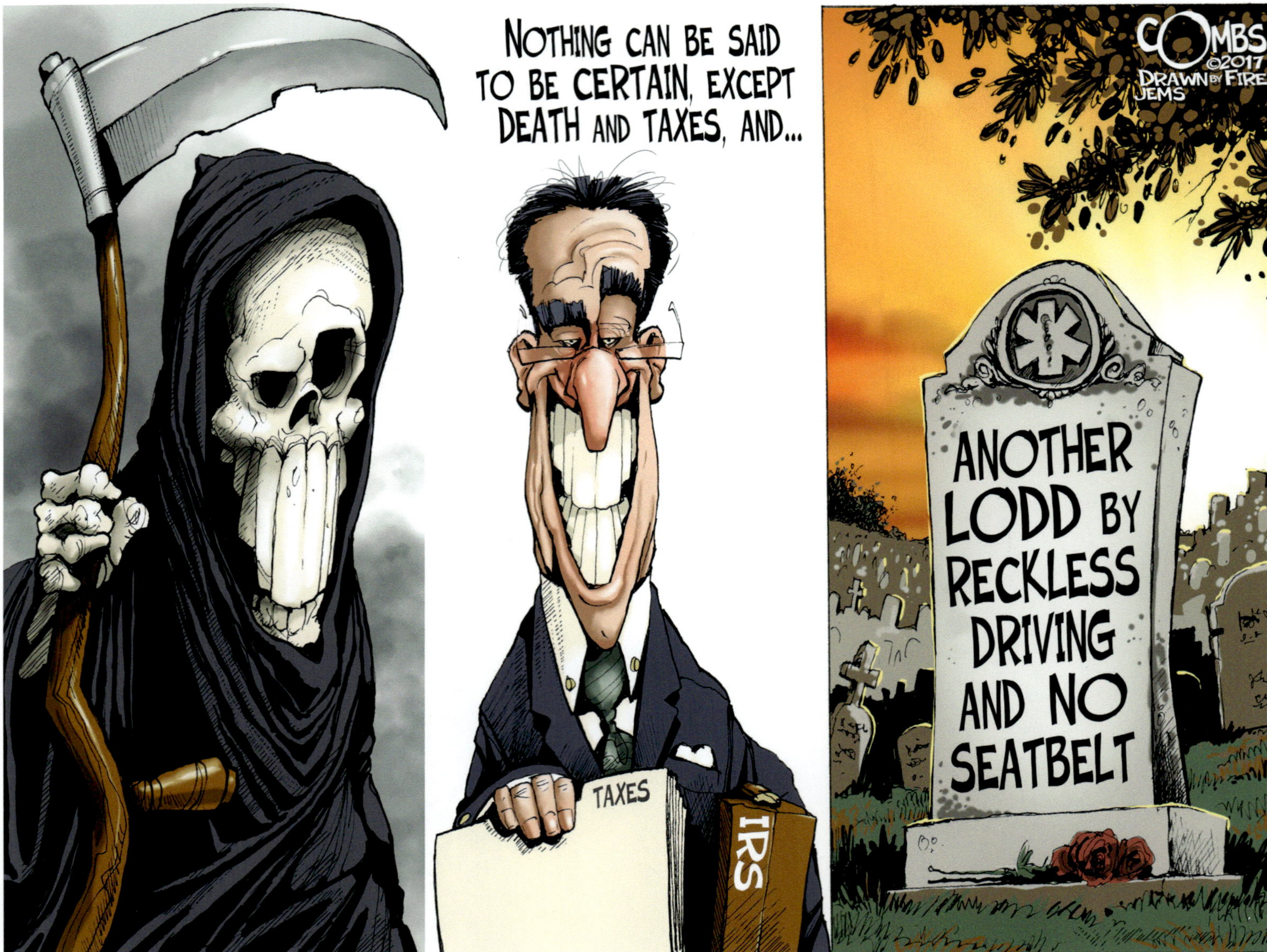

DON'T BE A CERTAINTY - SLOW DOWN AND BUCKLE UP!

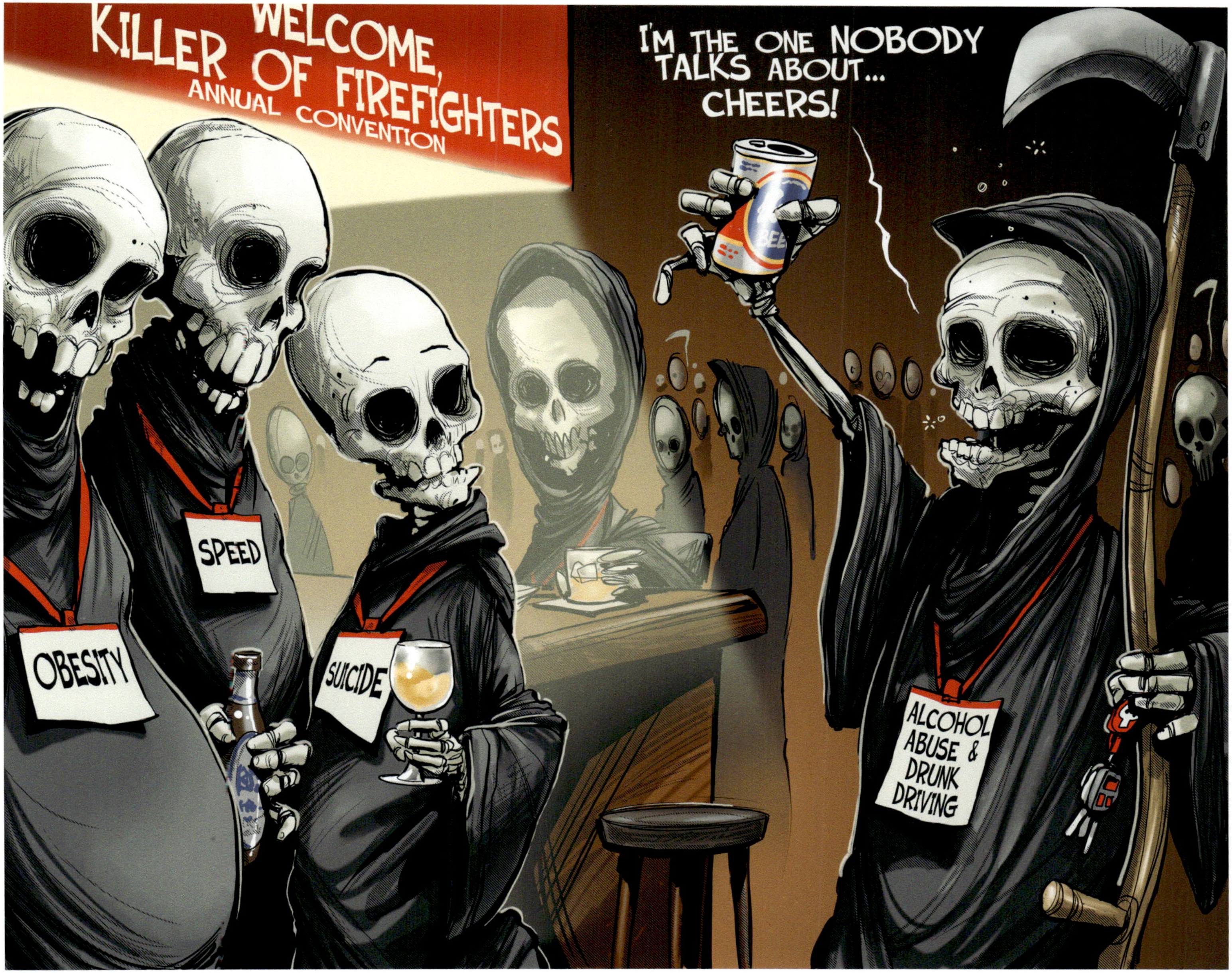
WELCOME, KILLER OF FIREFIGHTERS ANNUAL CONVENTION
I'M THE ONE NOBODY TALKS ABOUT... CHEERS!
OBESITY
SPEED
SUICIDE
ALCOHOL ABUSE & DRUNK DRIVING

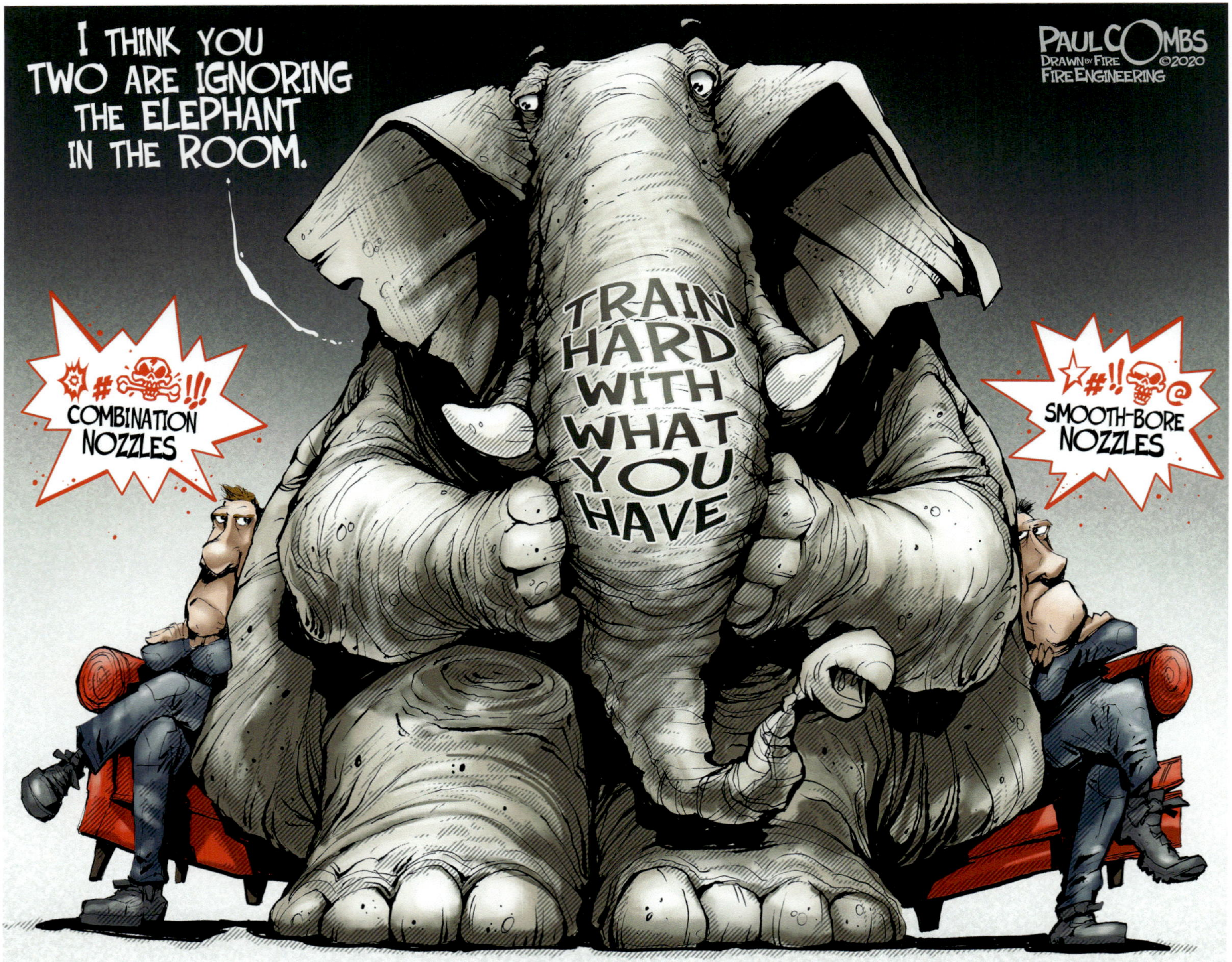
I THINK YOU TWO ARE IGNORING THE ELEPHANT IN THE ROOM.
PAUL COMBS
DRAWN BY FIRE ©2020
FIRE ENGINEERING
TRAIN HARD WITH WHAT YOU HAVE
COMBINATION NOZZLES
SMOOTH-BORE NOZZLES

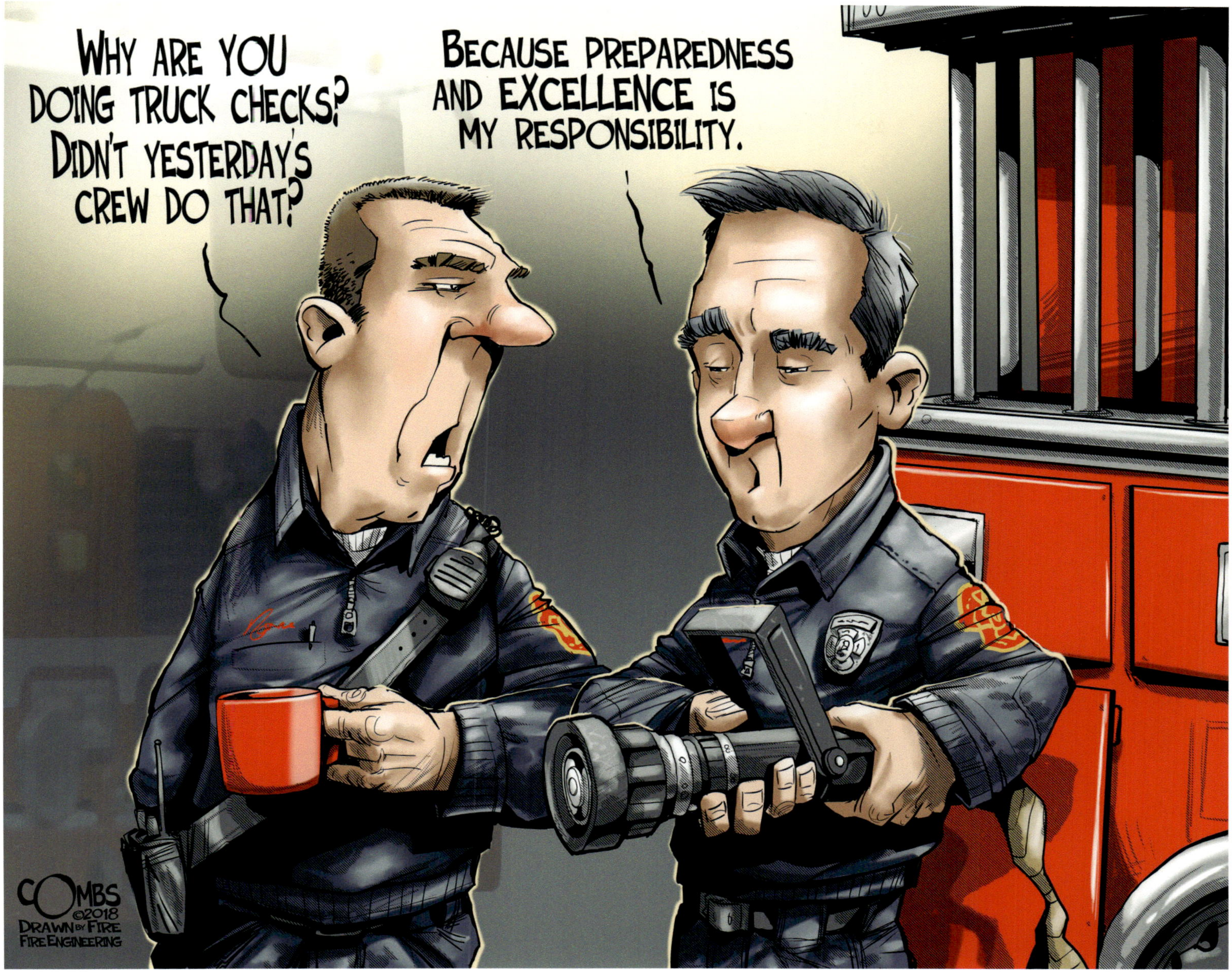
WHY ARE YOU DOING TRUCK CHECKS? DIDN'T YESTERDAY'S CREW DO THAT?
BECAUSE PREPAREDNESS AND EXCELLENCE IS MY RESPONSIBILITY.
COMBS ©2018 DRAWN BY FIRE FIRE ENGINEERING

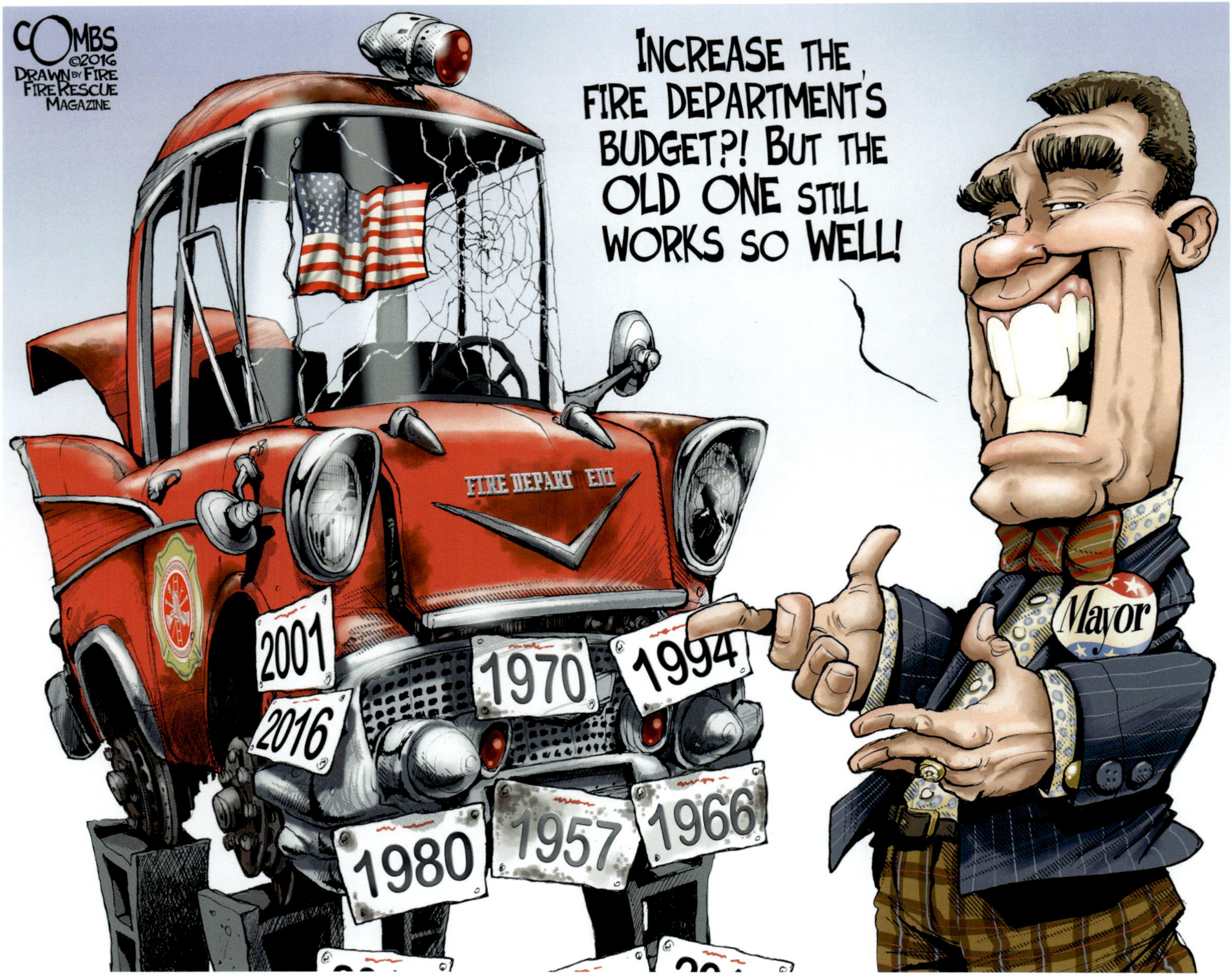
COMBS ©2016 DRAWN BY FIRE FIRERESCUE MAGAZINE
INCREASE THE FIRE DEPARTMENT'S BUDGET?! BUT THE OLD ONE STILL WORKS SO WELL!
FIRE DEPART ENT
2001
2016
1970
1994
1980
1957
1966
Mayor

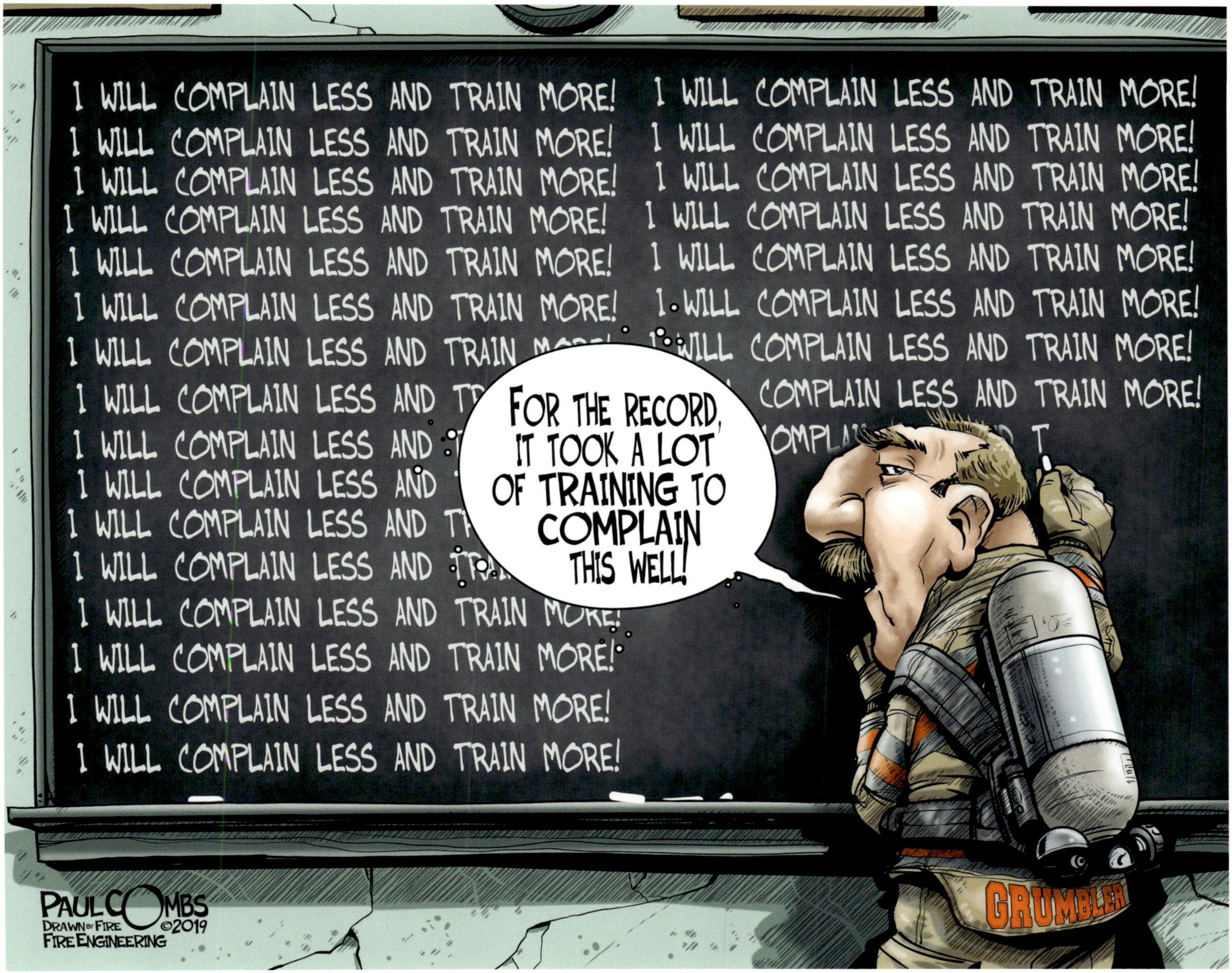
I WILL COMPLAIN LESS AND TRAIN MORE!
FOR THE RECORD, IT TOOK A LOT OF TRAINING TO COMPLAIN THIS WELL!
GRUMBLER
PAUL COMBS
DRAWN BY FIRE ©2019
FIRE ENGINEERING

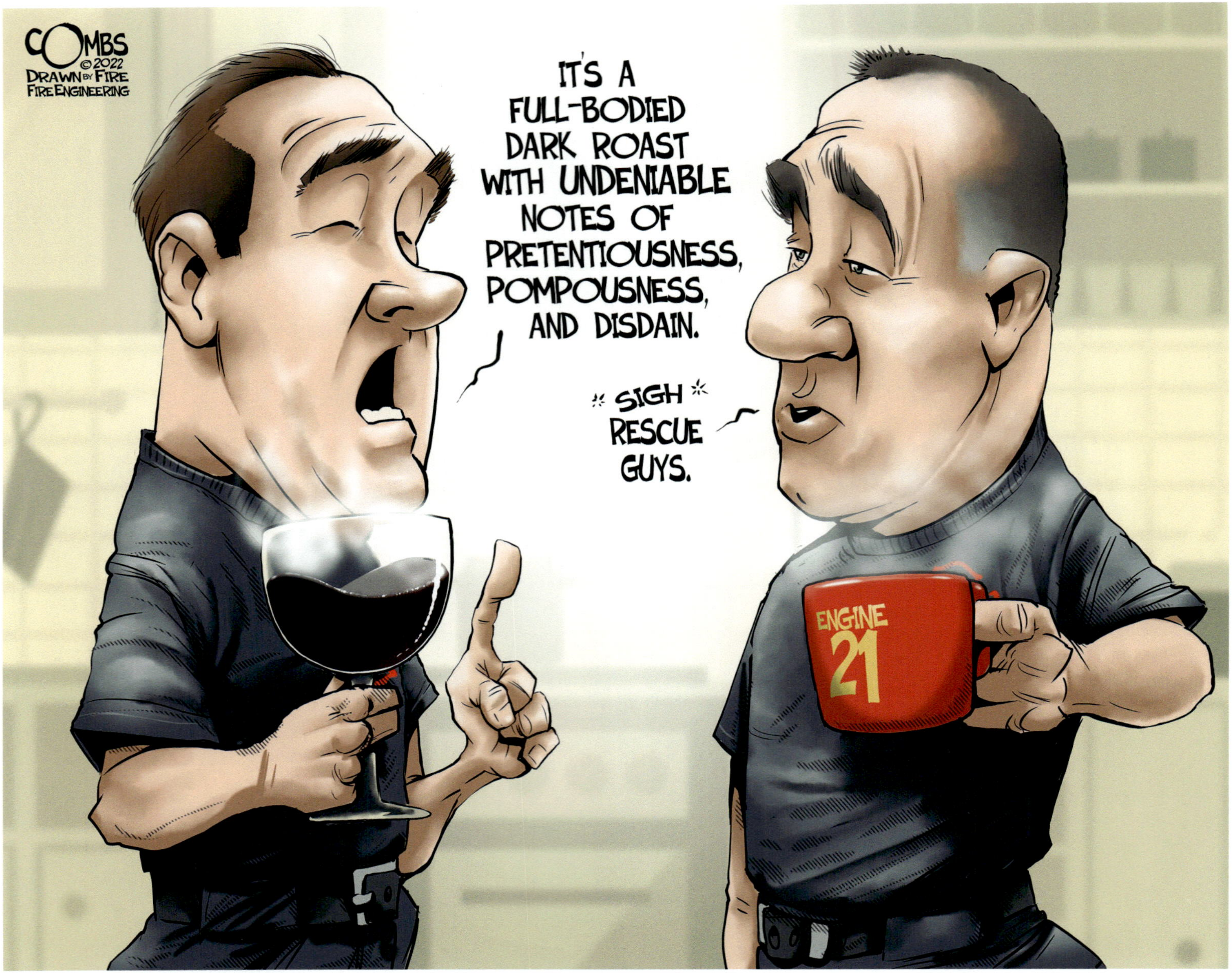
COMBS
©2022
DRAWN BY FIRE
FIRE ENGINEERING
IT'S A FULL-BODIED DARK ROAST WITH UNDENIABLE NOTES OF PRETENTIOUSNESS, POMPOUSNESS, AND DISDAIN.
SIGH RESCUE GUYS.
ENGINE 21

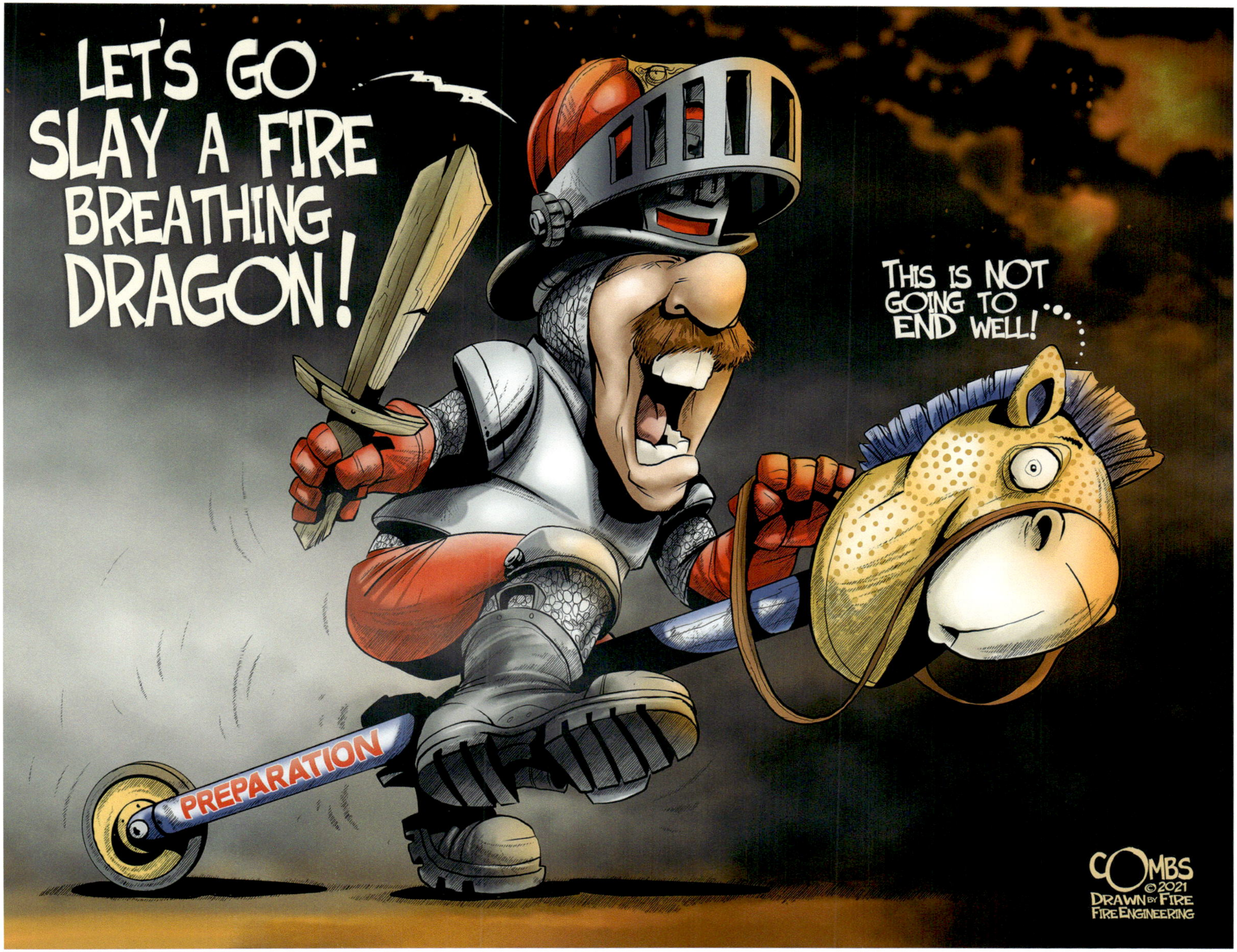
LET'S GO SLAY A FIRE BREATHING DRAGON!
THIS IS NOT GOING TO END WELL!
PREPARATION
COMBS ©2021 DRAWN BY FIRE FIRE ENGINEERING

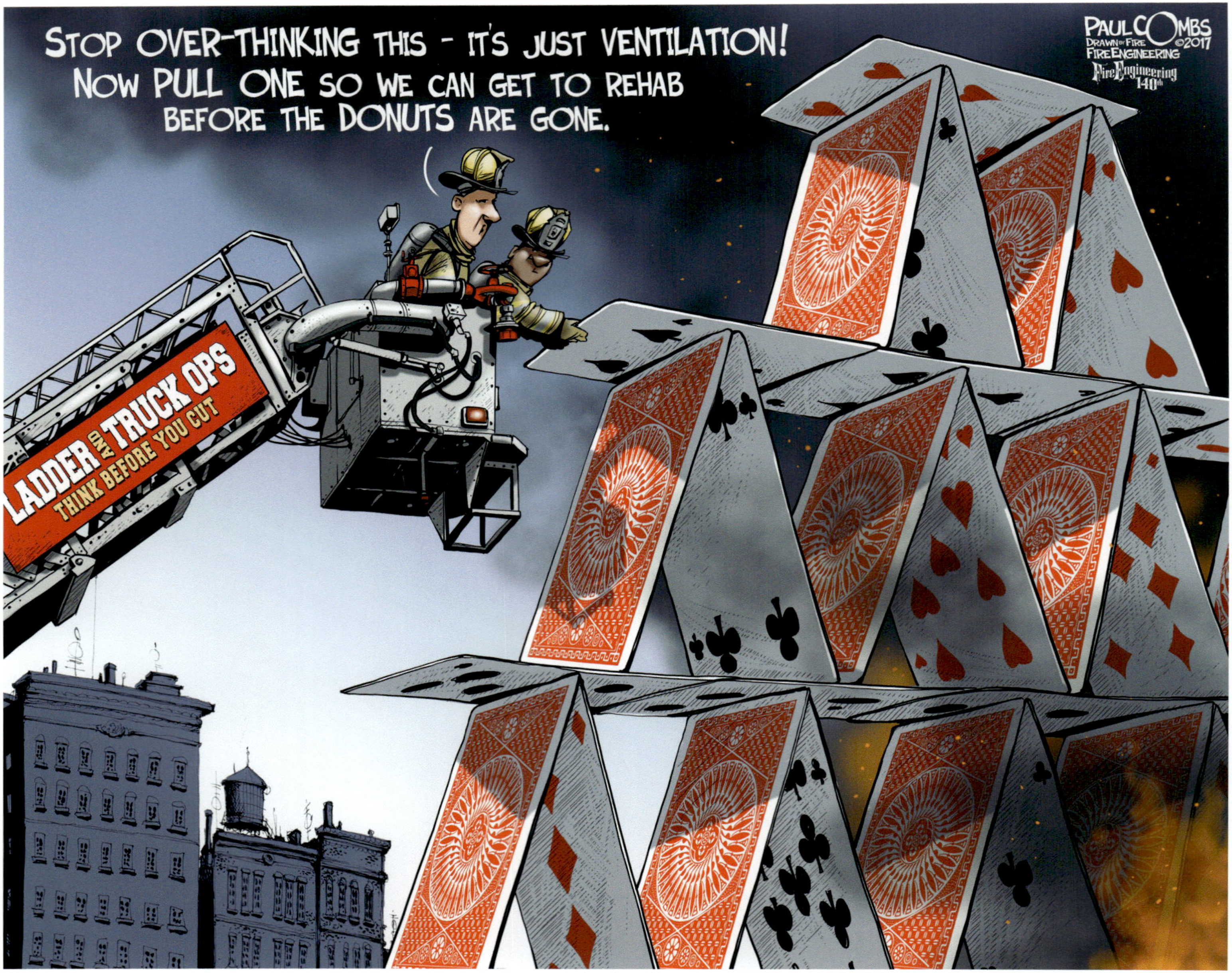
STOP OVER-THINKING THIS - IT'S JUST VENTILATION! NOW PULL ONE SO WE CAN GET TO REHAB BEFORE THE DONUTS ARE GONE.
LADDER AND TRUCK OPS
THINK BEFORE YOU CUT
PAUL COMBS
DRAWN BY FIRE ©2017
FIRE ENGINEERING
Fire Engineering 140th

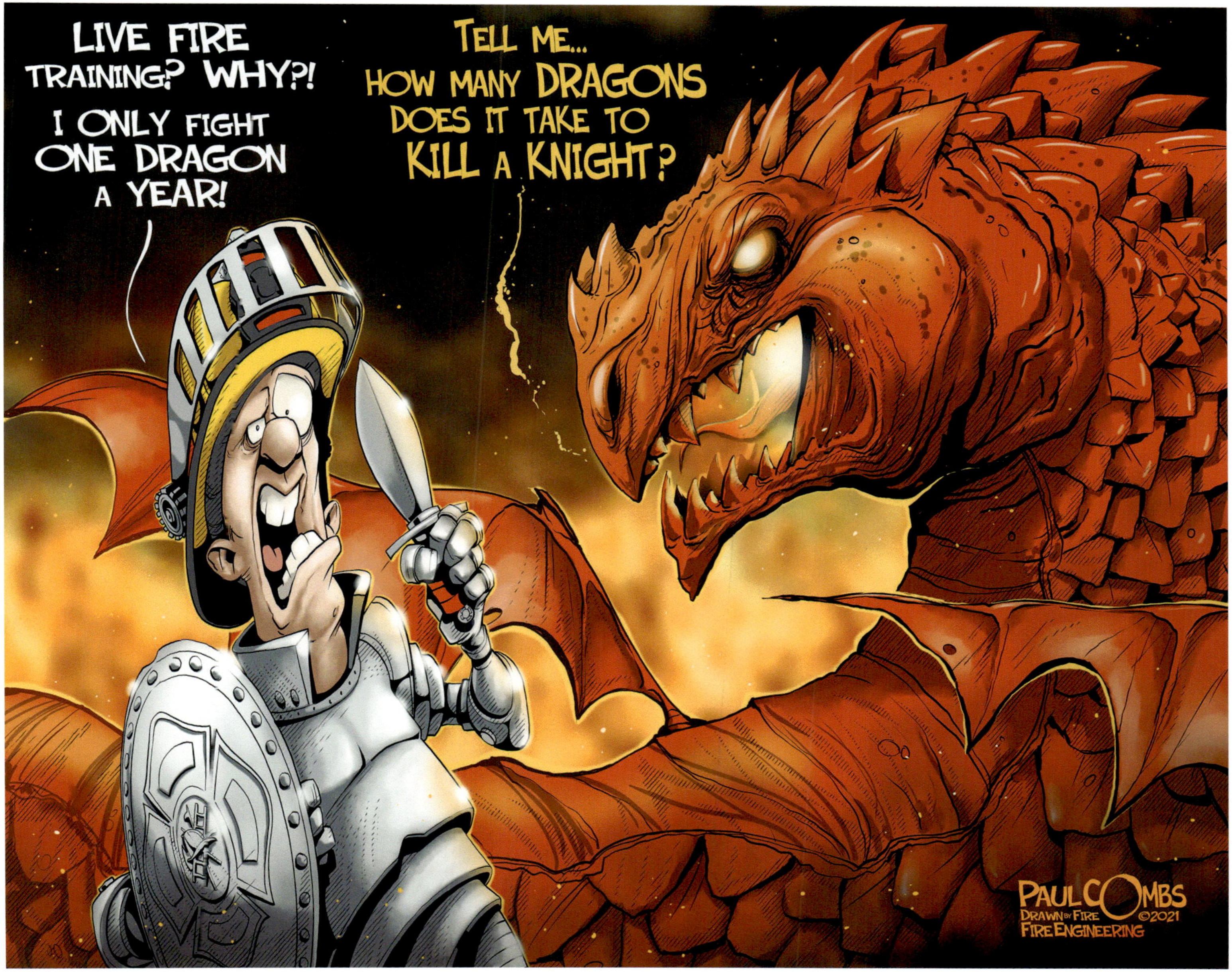
LIVE FIRE TRAINING? WHY?! I ONLY FIGHT ONE DRAGON A YEAR!
TELL ME... HOW MANY DRAGONS DOES IT TAKE TO KILL A KNIGHT?
PAUL COMBS
DRAWN BY FIRE ©2021
FIRE ENGINEERING

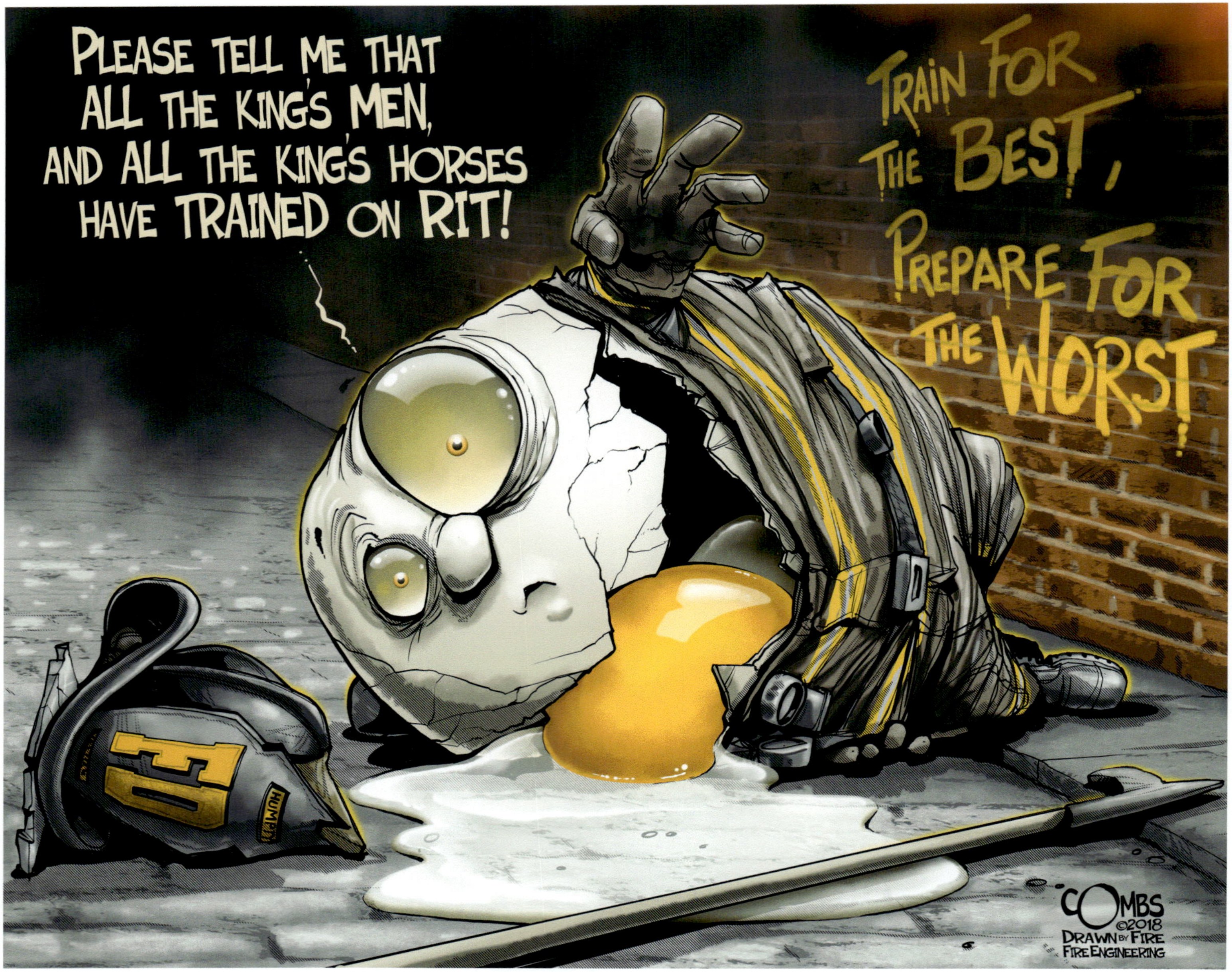
PLEASE TELL ME THAT ALL THE KING'S MEN, AND ALL THE KING'S HORSES HAVE TRAINED ON RIT!
TRAIN FOR THE BEST, PREPARE FOR THE WORST
FD
COMBS ©2018 DRAWN BY FIRE FIRE ENGINEERING

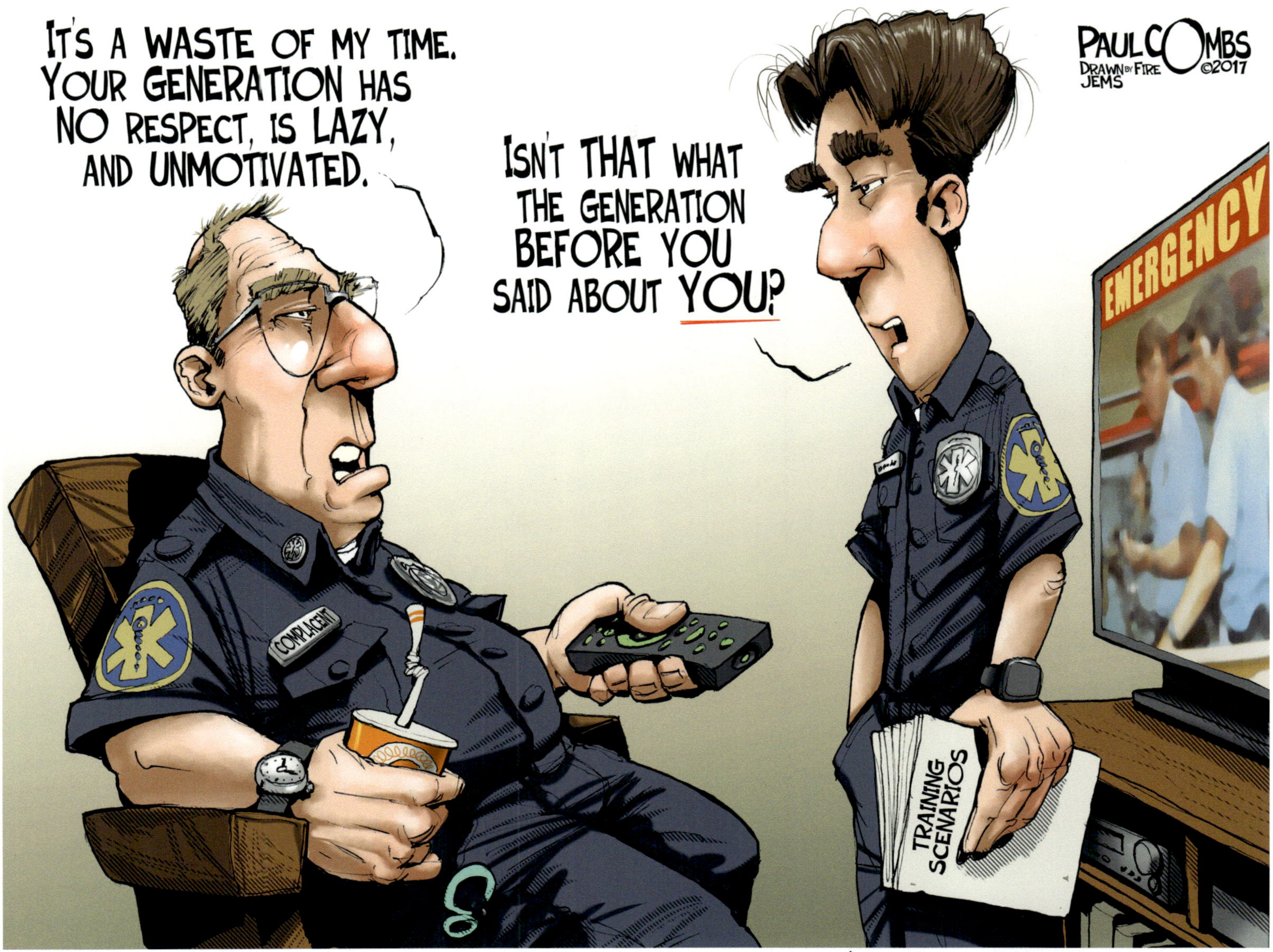

DON'T COMPLAIN ABOUT THE YOUNGER GENERATION IF YOU'RE UNWILLING TO LEAD THEM.

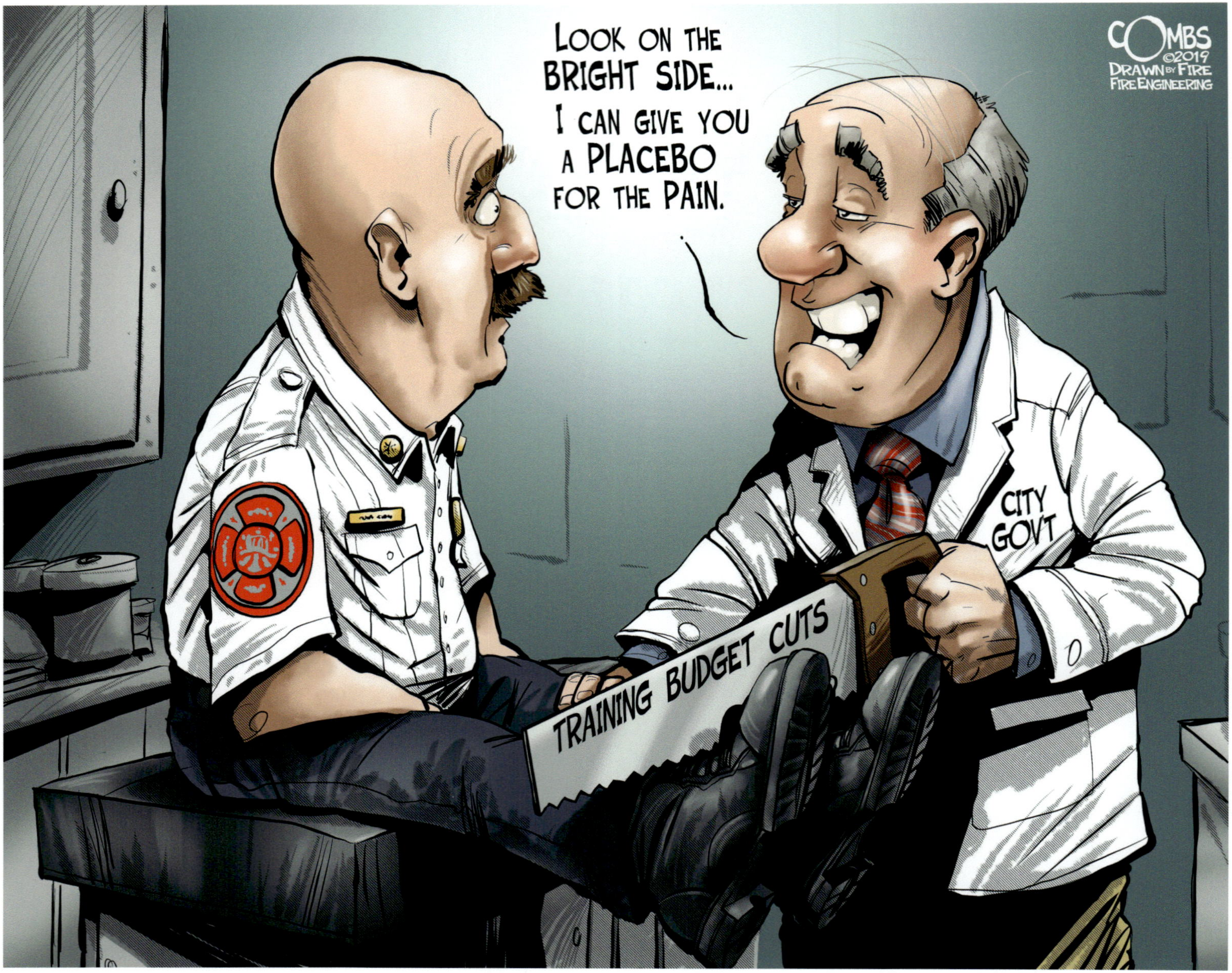
LOOK ON THE BRIGHT SIDE... I CAN GIVE YOU A PLACEBO FOR THE PAIN.
COMBS ©2019 DRAWN BY FIRE FIRE ENGINEERING
CITY GOV'T
TRAINING BUDGET CUTS

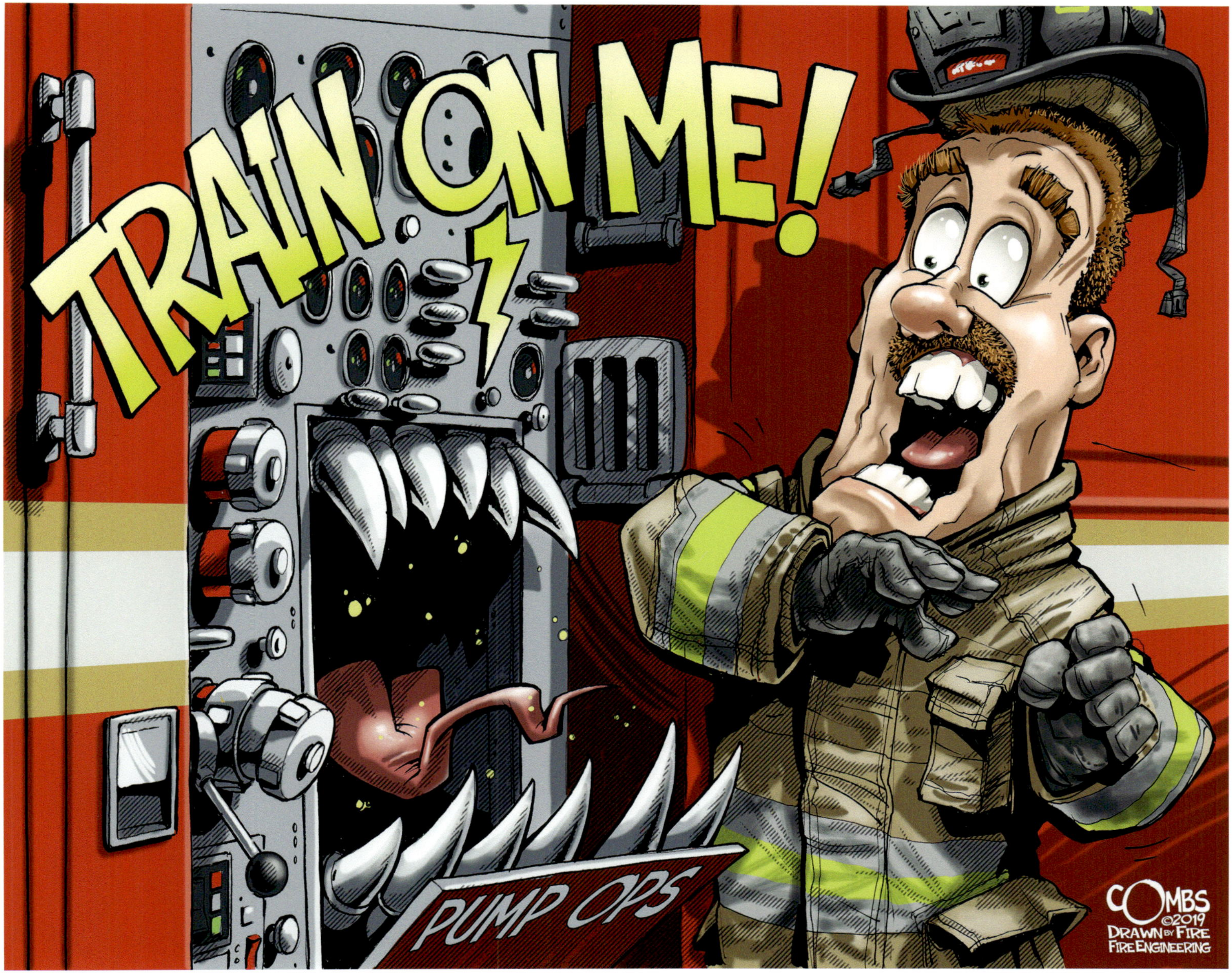
TRAIN ON ME!
PUMP OPS
COMBS
©2019
DRAWN BY FIRE
FIRE ENGINEERING

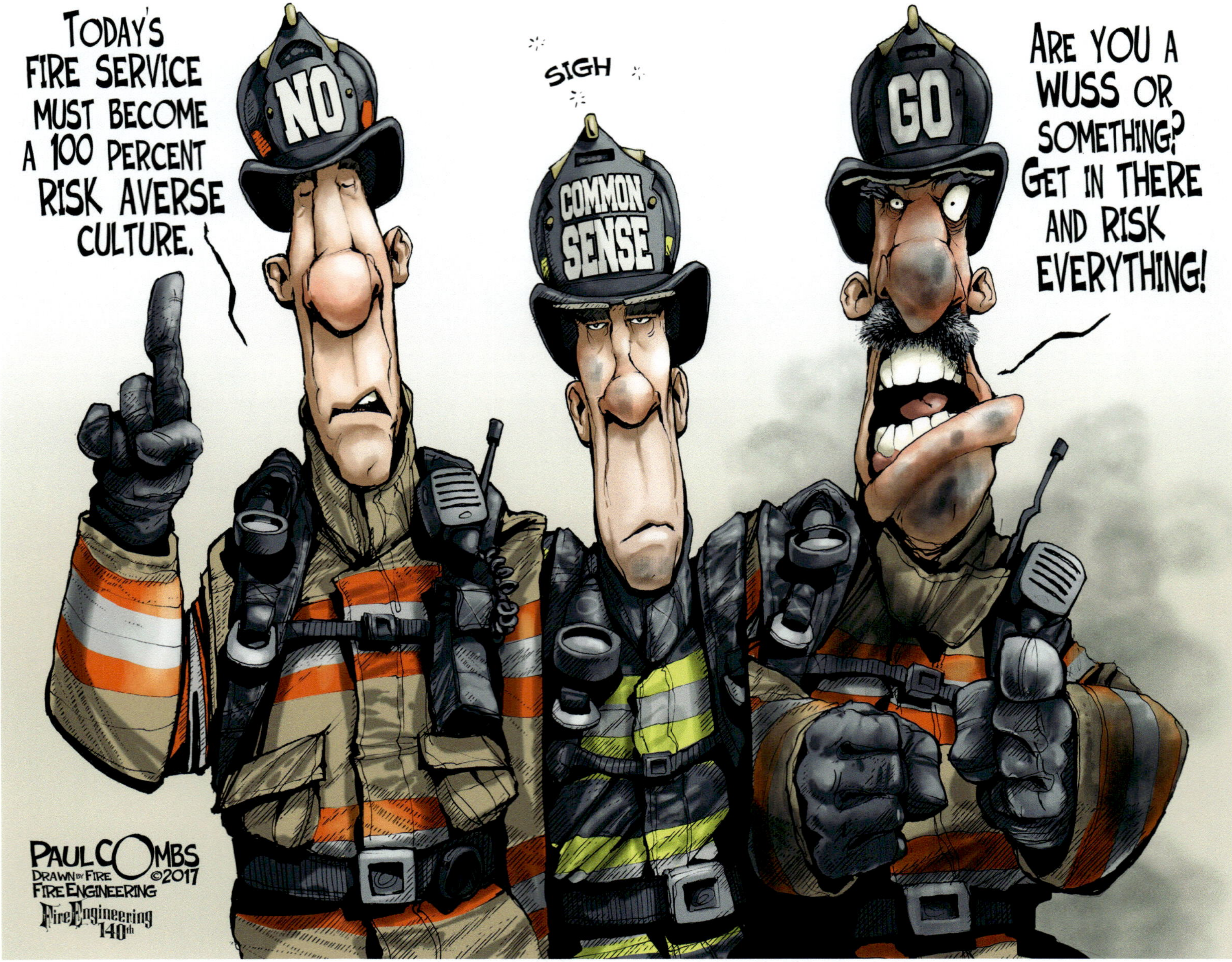
TODAY'S FIRE SERVICE MUST BECOME A 100 PERCENT RISK AVERSE CULTURE.
NO
SIGH
COMMON SENSE
GO
ARE YOU A WUSS OR SOMETHING? GET IN THERE AND RISK EVERYTHING!
PAUL COMBS
DRAWN BY FIRE ©2017
FIRE ENGINEERING
Fire Engineering 140th

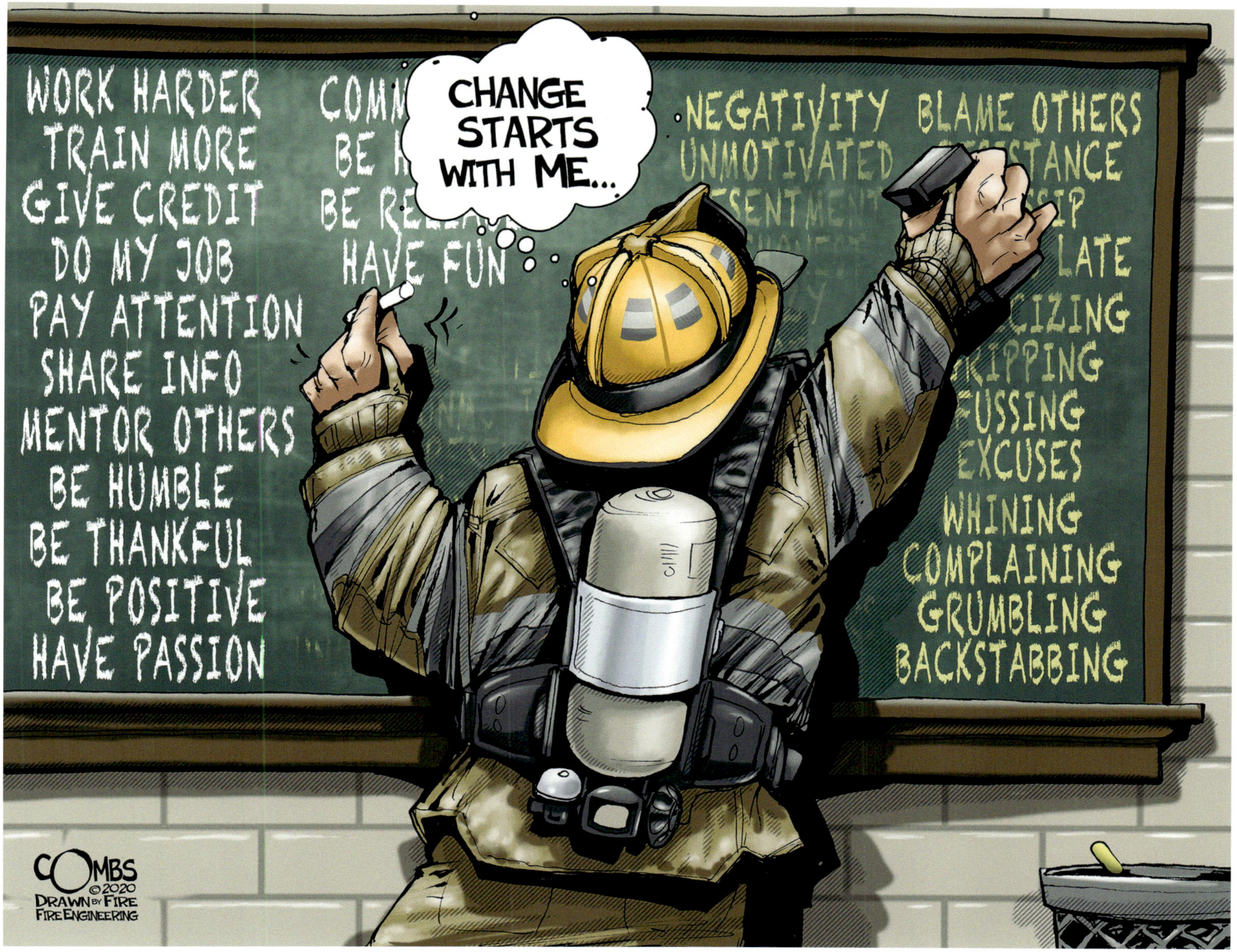
CHANGE STARTS WITH ME...
WORK HARDER
TRAIN MORE
GIVE CREDIT
DO MY JOB
PAY ATTENTION
SHARE INFO
MENTOR OTHERS
BE HUMBLE
BE THANKFUL
BE POSITIVE
HAVE PASSION
HAVE FUN
NEGATIVITY
BLAME OTHERS
UNMOTIVATED
FUSSING
EXCUSES
WHINING
COMPLAINING
GRUMBLING
BACKSTABBING
COMBS ©2020
DRAWN BY FIRE
FIRE ENGINEERING

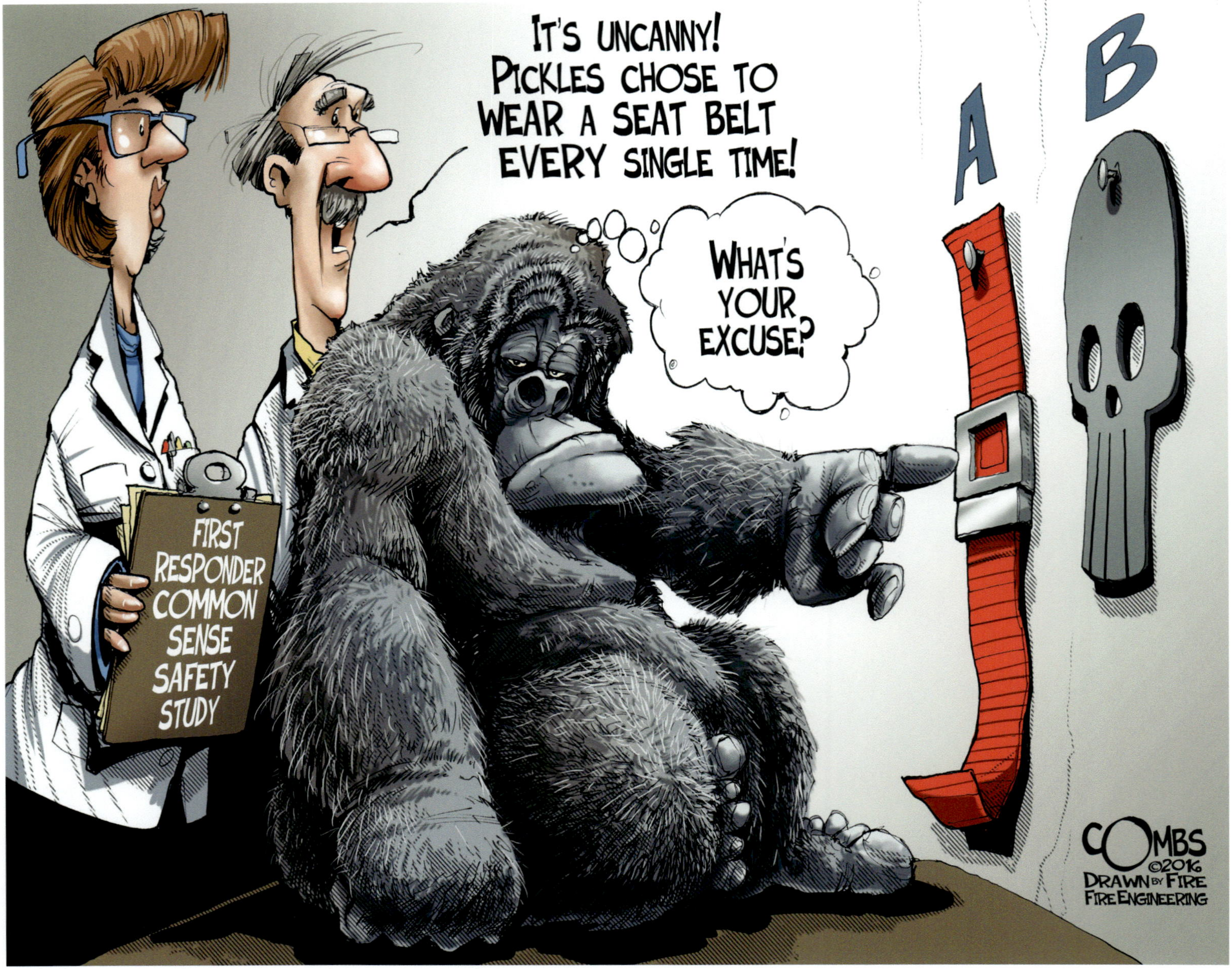
IT'S UNCANNY! PICKLES CHOSE TO WEAR A SEAT BELT EVERY SINGLE TIME!
WHAT'S YOUR EXCUSE?
FIRST RESPONDER COMMON SENSE SAFETY STUDY
A
B
COMBS ©2016 DRAWN BY FIRE FIRE ENGINEERING

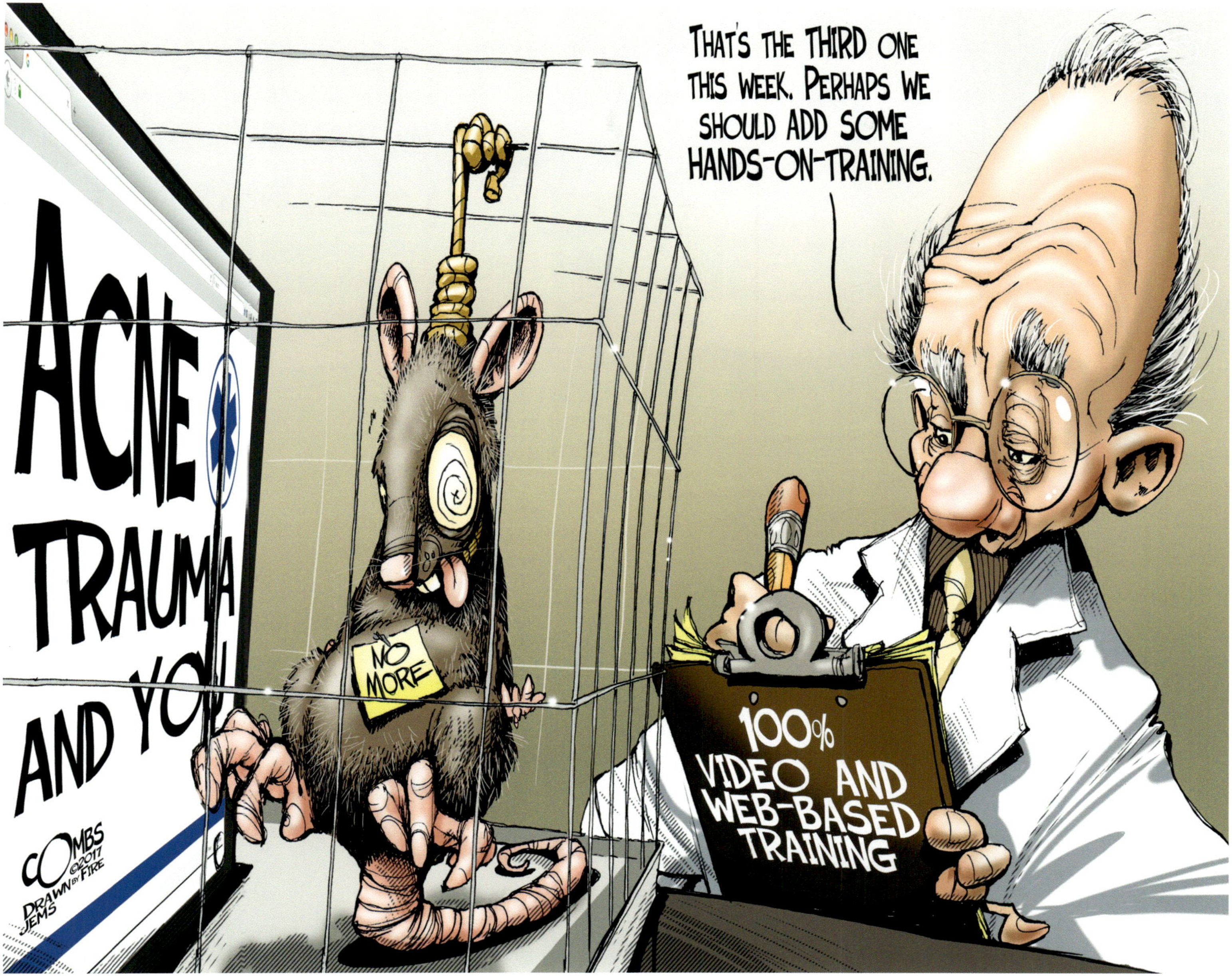
THAT'S THE THIRD ONE THIS WEEK. PERHAPS WE SHOULD ADD SOME HANDS-ON-TRAINING.
ACNE TRAUMA AND YOU
NO MORE
100% VIDEO AND WEB-BASED TRAINING
COMBS ©2017 DRAWN BY FIRE JEMS

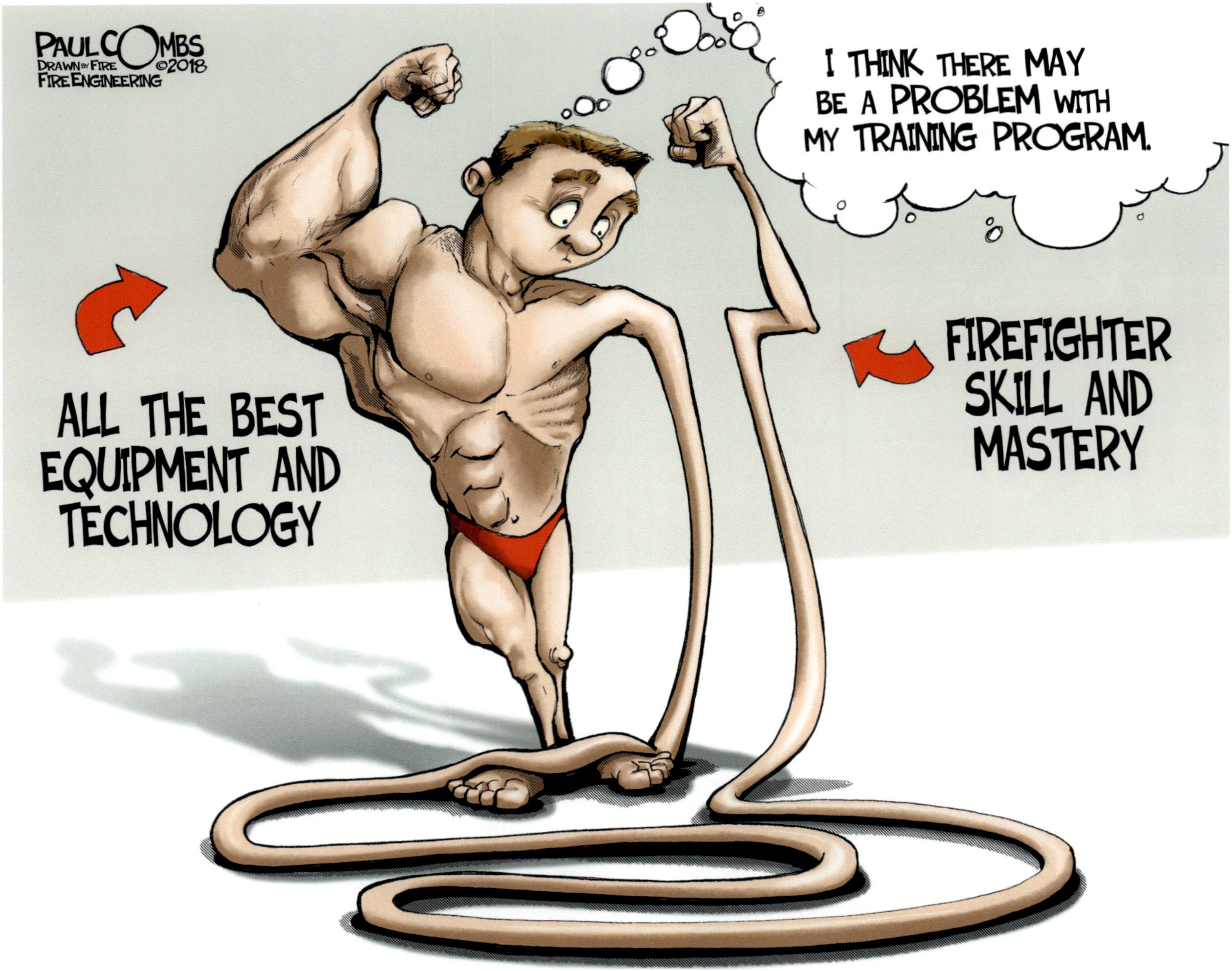
PAUL COMBS
DRAWN BY FIRE ©2018
FIRE ENGINEERING
I THINK THERE MAY BE A PROBLEM WITH MY TRAINING PROGRAM.
ALL THE BEST EQUIPMENT AND TECHNOLOGY
FIREFIGHTER SKILL AND MASTERY

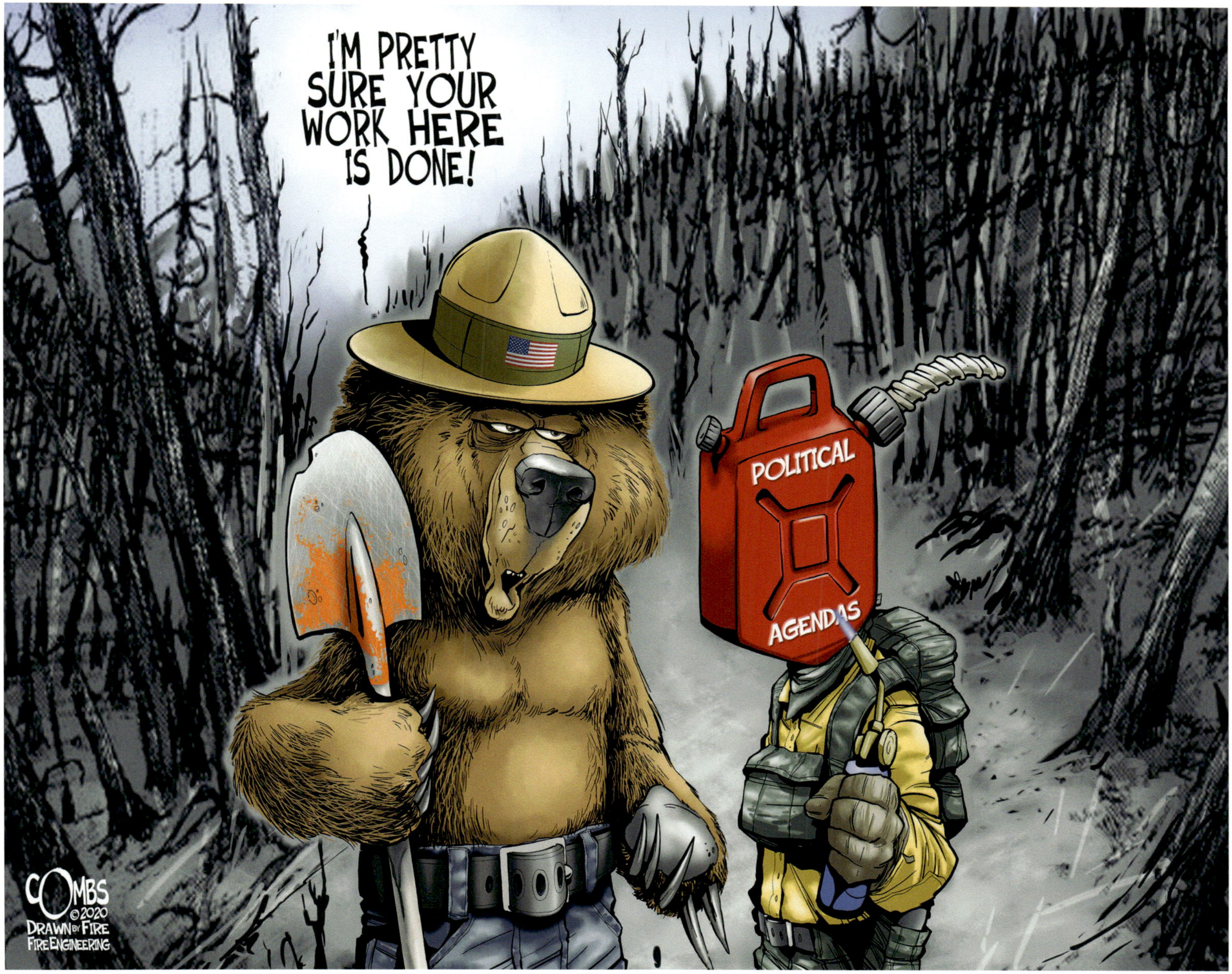
I'M PRETTY SURE YOUR WORK HERE IS DONE!
POLITICAL
AGENDAS
COMBS ©2020 DRAWN BY FIRE FIREENGINEERING

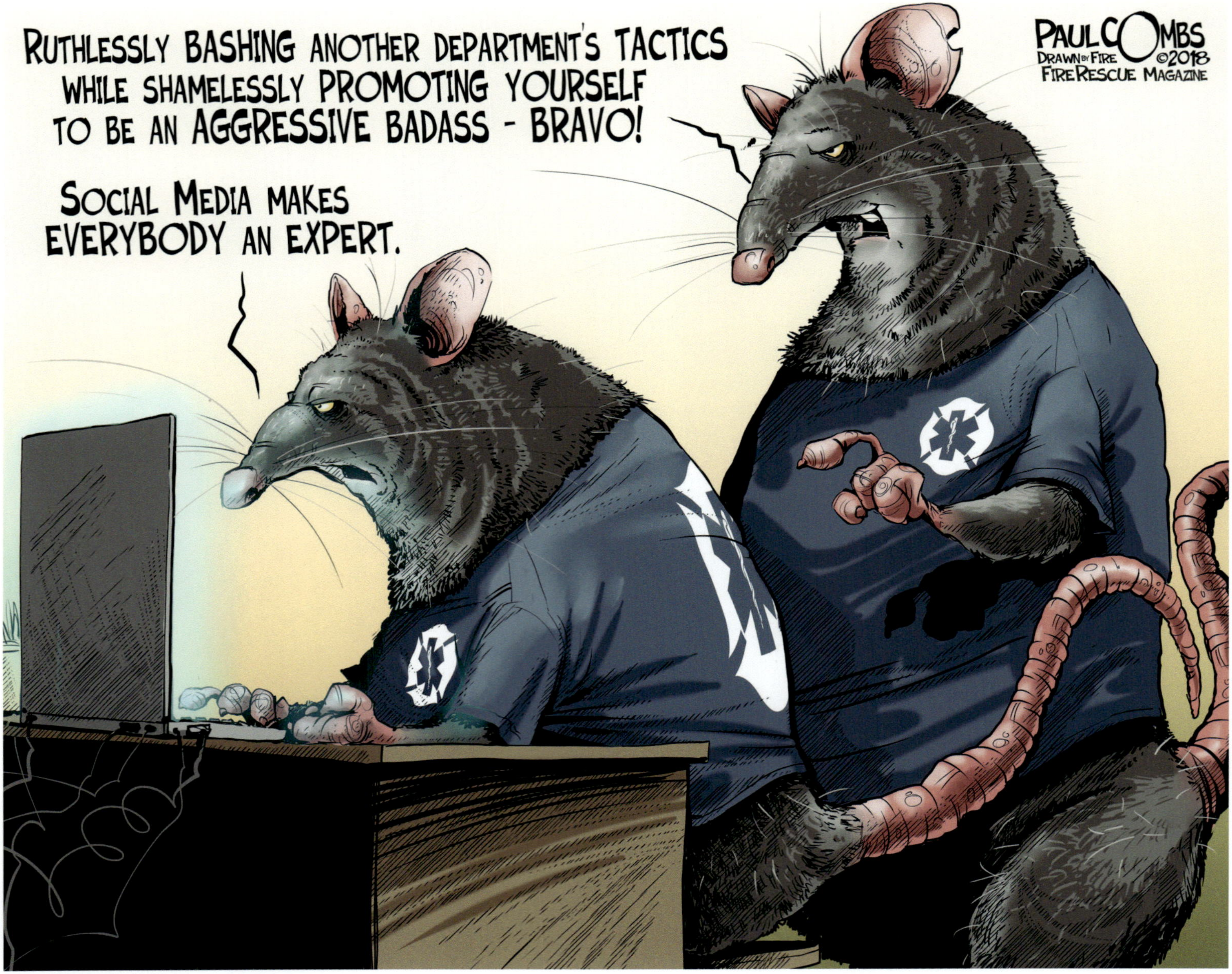
PAUL COMBS
DRAWN BY FIRE ©2018
FIRERESCUE MAGAZINE
RUTHLESSLY BASHING ANOTHER DEPARTMENT'S TACTICS WHILE SHAMELESSLY PROMOTING YOURSELF TO BE AN AGGRESSIVE BADASS - BRAVO!
SOCIAL MEDIA MAKES EVERYBODY AN EXPERT.

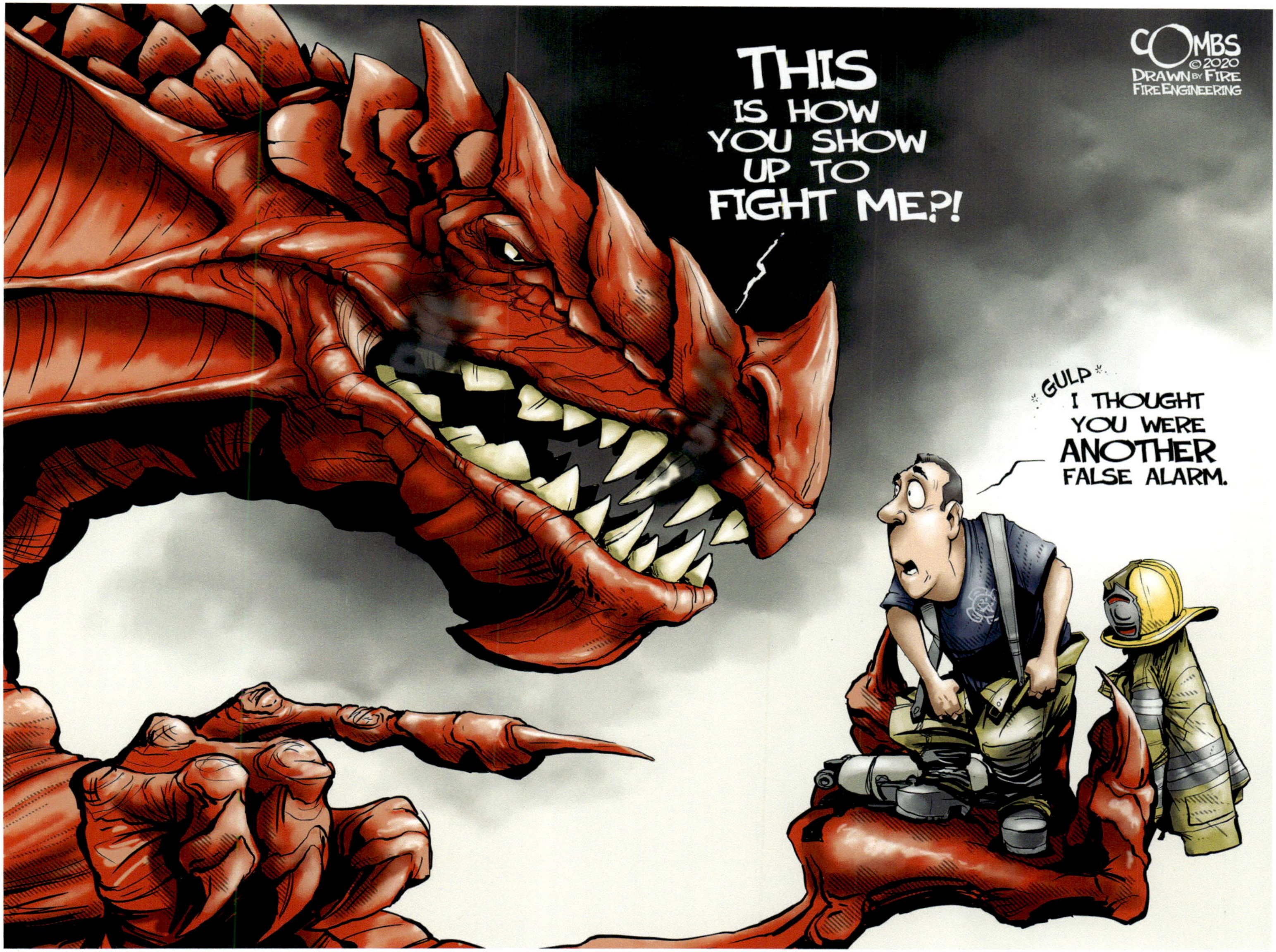

SECONDS COUNT - BE READY TO FIGHT!

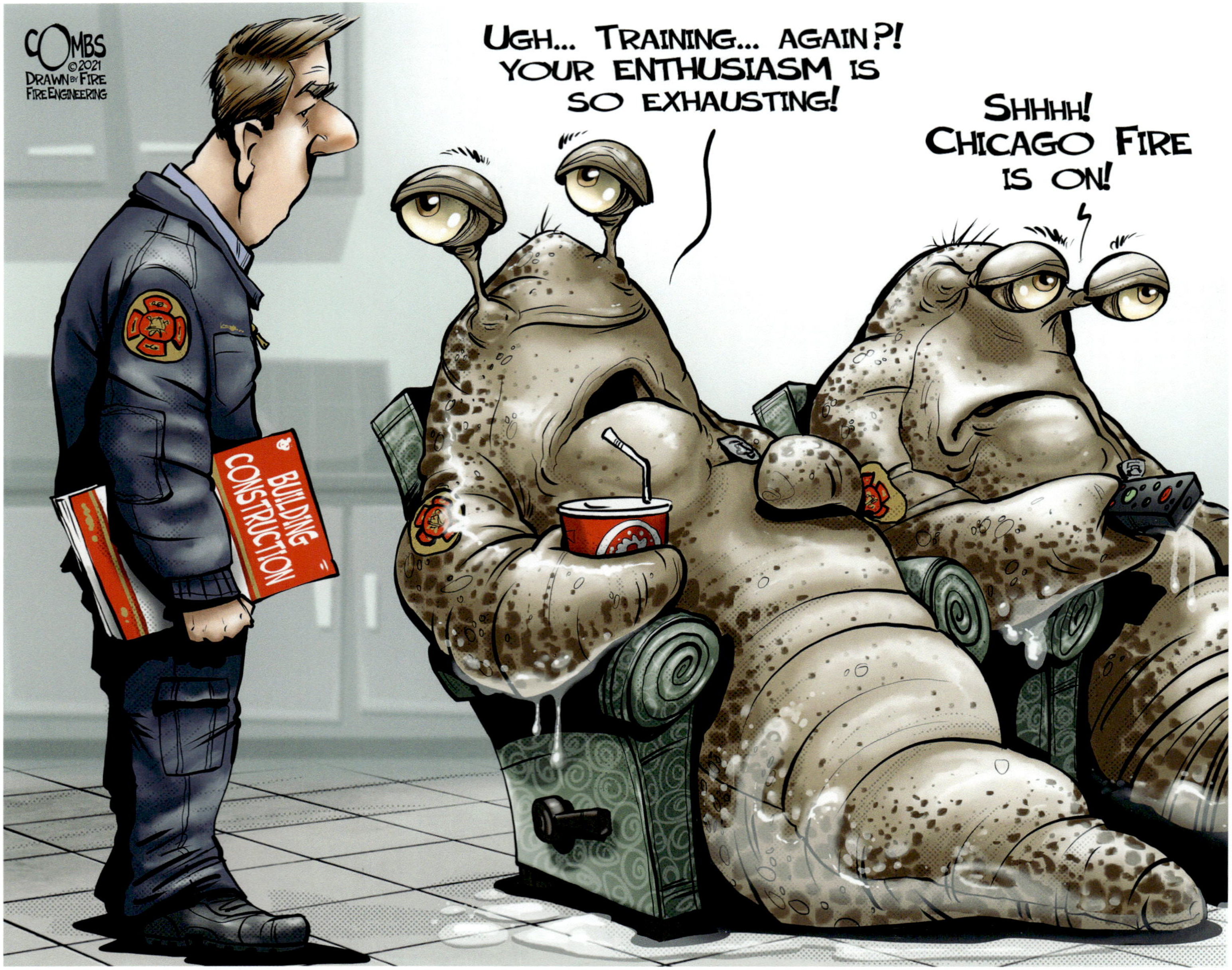
COMBS ©2021 DRAWN BY FIRE FIRE ENGINEERING
UGH... TRAINING... AGAIN?! YOUR ENTHUSIASM IS SO EXHAUSTING!
SHHHH! CHICAGO FIRE IS ON!
BUILDING CONSTRUCTION

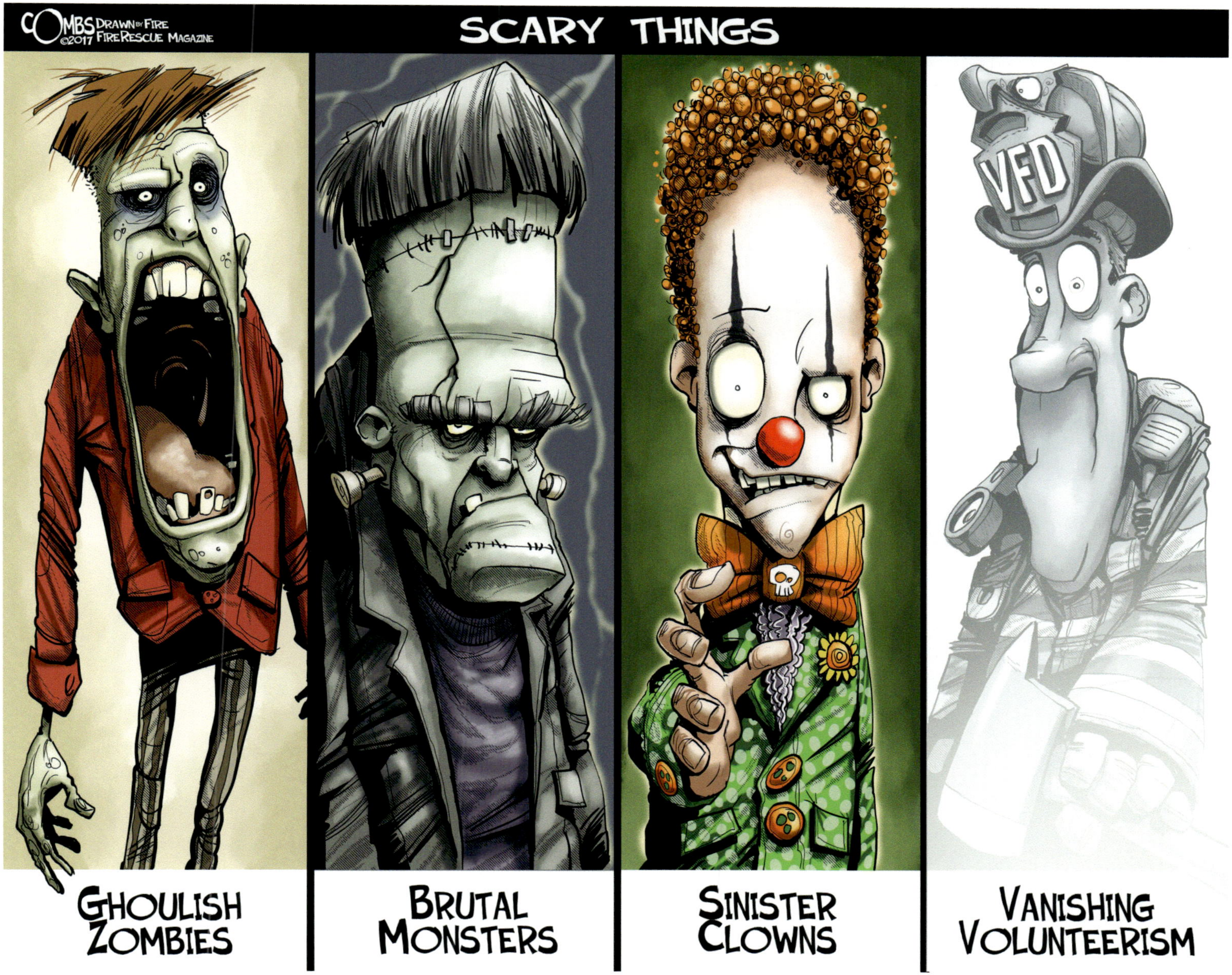
COMBS DRAWN BY FIRE
©2017 FIRERESCUE MAGAZINE
SCARY THINGS
VFD
GHOULISH ZOMBIES
BRUTAL MONSTERS
SINISTER CLOWNS
VANISHING VOLUNTEERISM

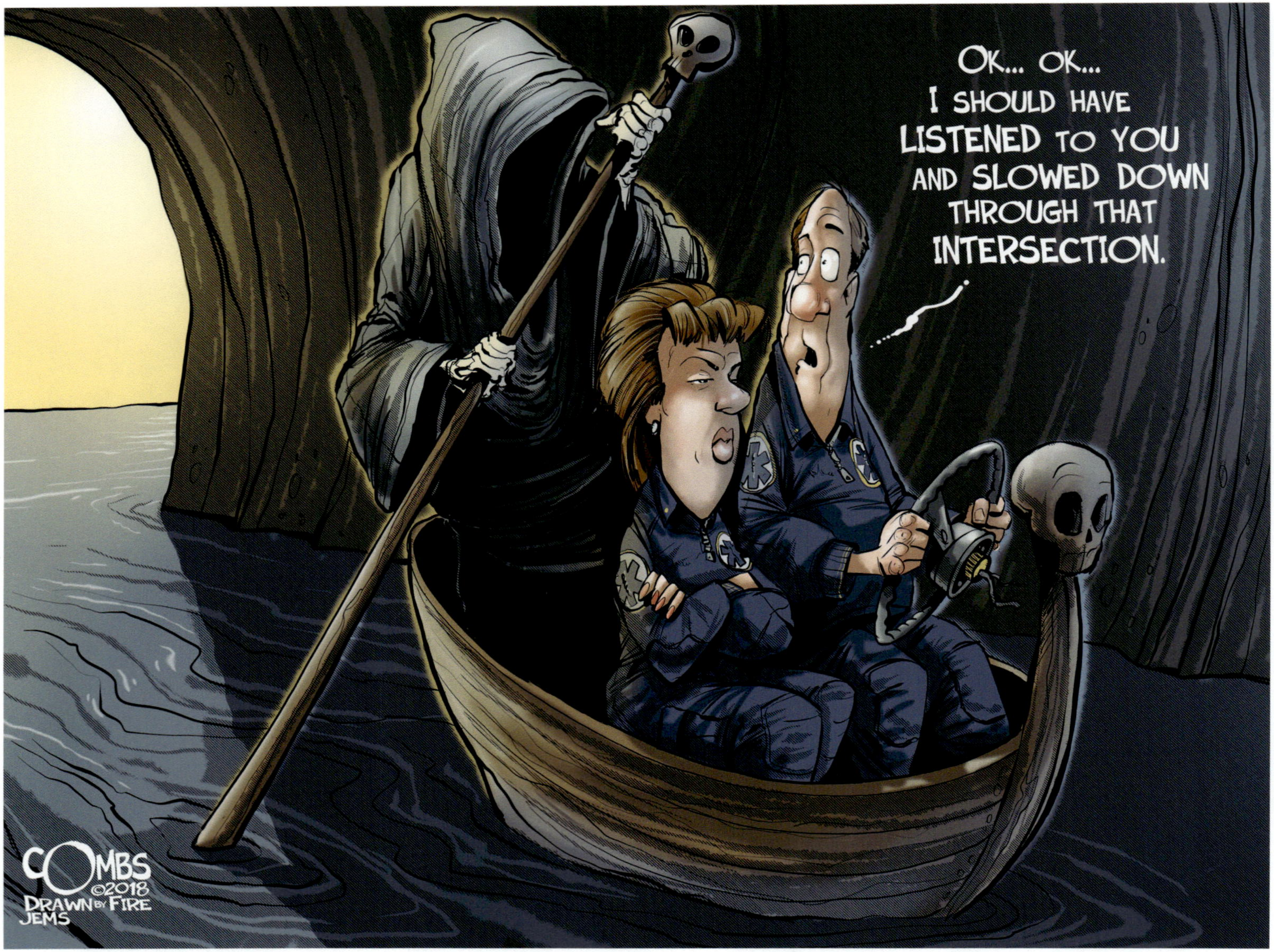

SLOW DOWN, AND DRIVE LIKE YOUR LIFE, AND THE LIVES OF OTHERS, DEPEND ON IT!

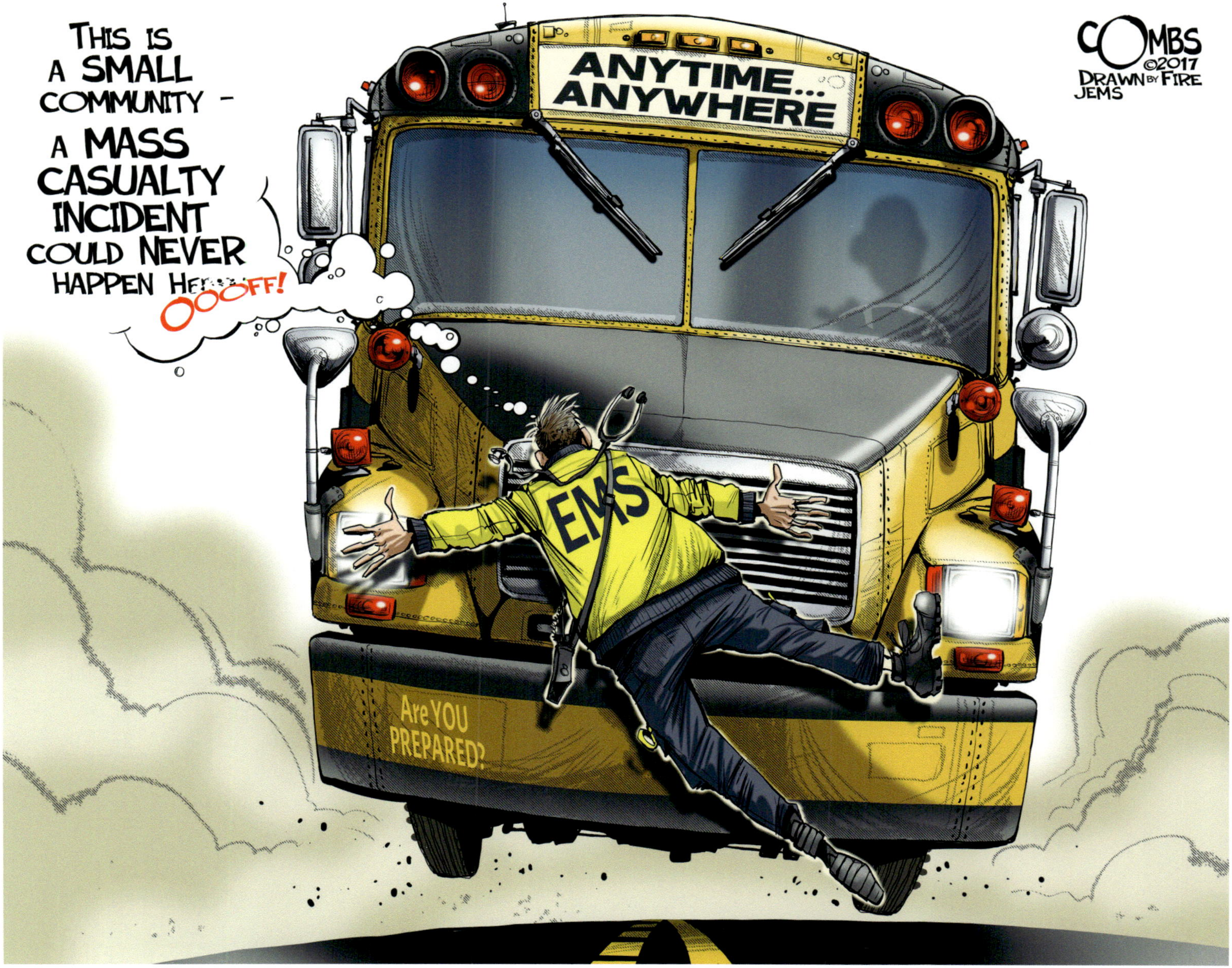
THIS IS A SMALL COMMUNITY - A MASS CASUALTY INCIDENT COULD NEVER HAPPEN HE
OOOFF!
ANYTIME... ANYWHERE
EMS
Are YOU PREPARED?
COMBS ©2017 DRAWN BY FIRE JEMS

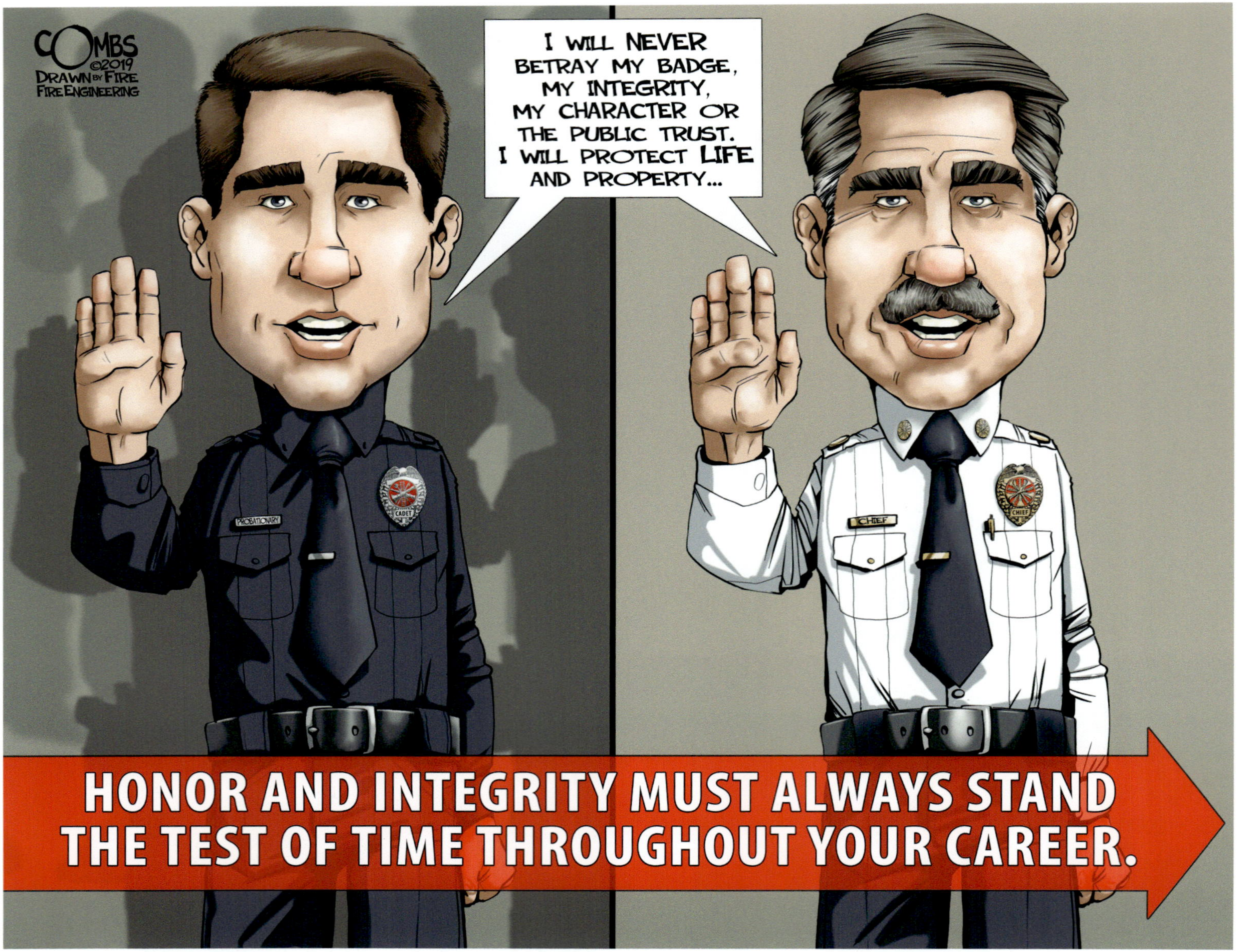
COMBS ©2019 DRAWN BY FIRE FIRE ENGINEERING
I WILL NEVER BETRAY MY BADGE, MY INTEGRITY, MY CHARACTER OR THE PUBLIC TRUST. I WILL PROTECT LIFE AND PROPERTY...
PROBATIONARY
CADET
CHIEF
CHIEF
HONOR AND INTEGRITY MUST ALWAYS STAND THE TEST OF TIME THROUGHOUT YOUR CAREER.

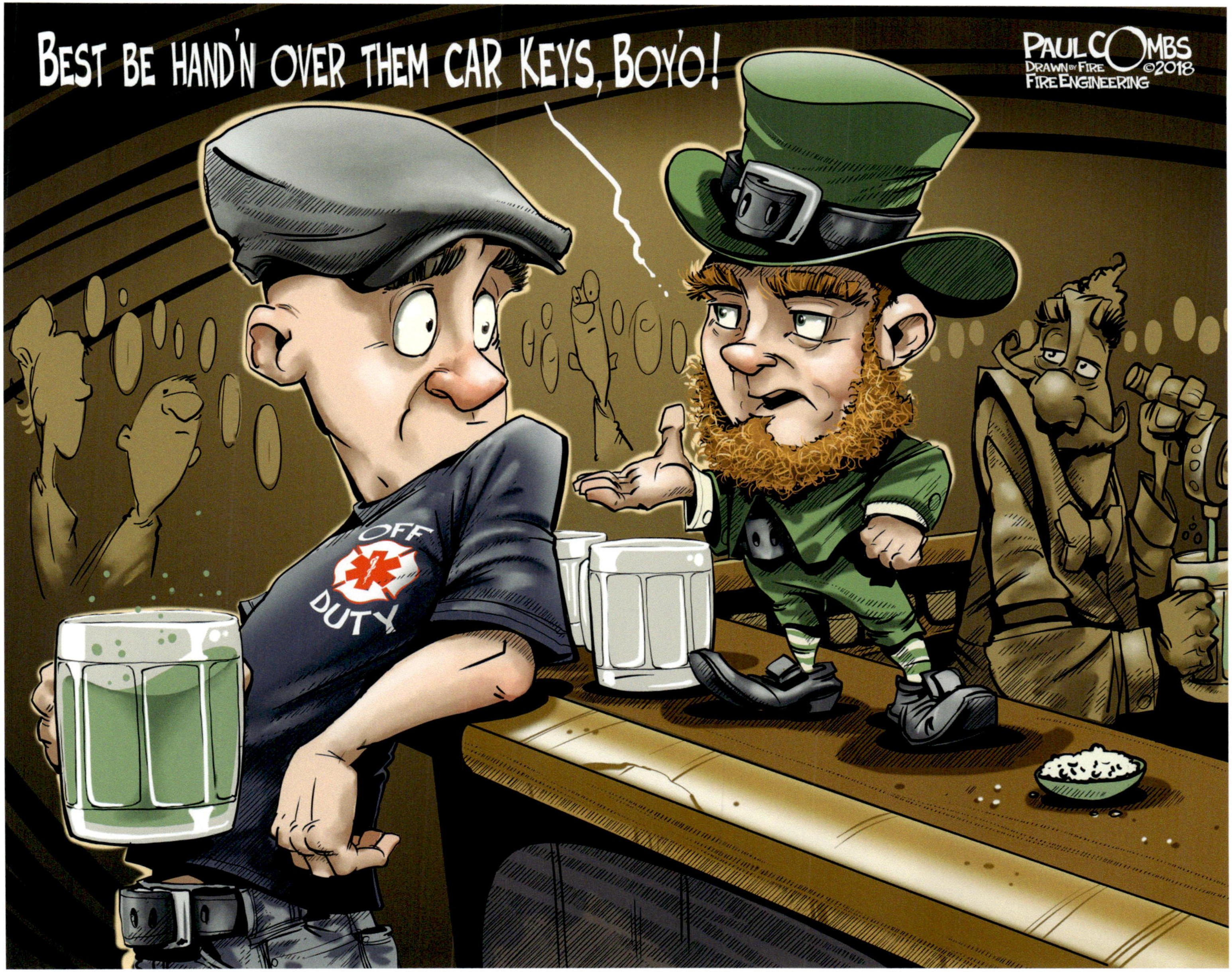
BEST BE HAND'N OVER THEM CAR KEYS, BOY'O!
PAUL COMBS
DRAWN BY FIRE ©2018
FIRE ENGINEERING
OFF DUTY

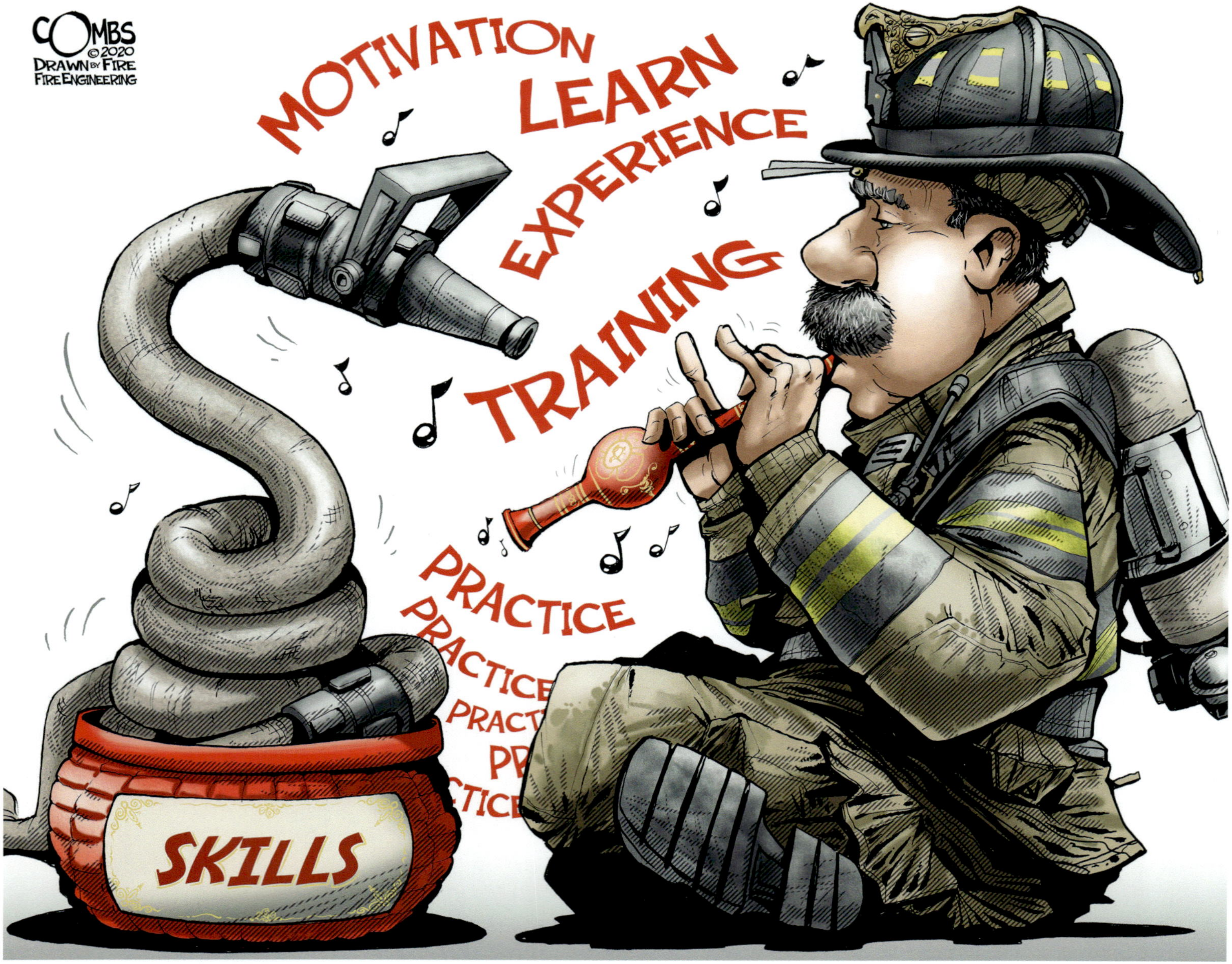
COMBS
©2020
DRAWN BY FIRE
FIRE ENGINEERING
MOTIVATION
LEARN
EXPERIENCE
TRAINING
PRACTICE
SKILLS

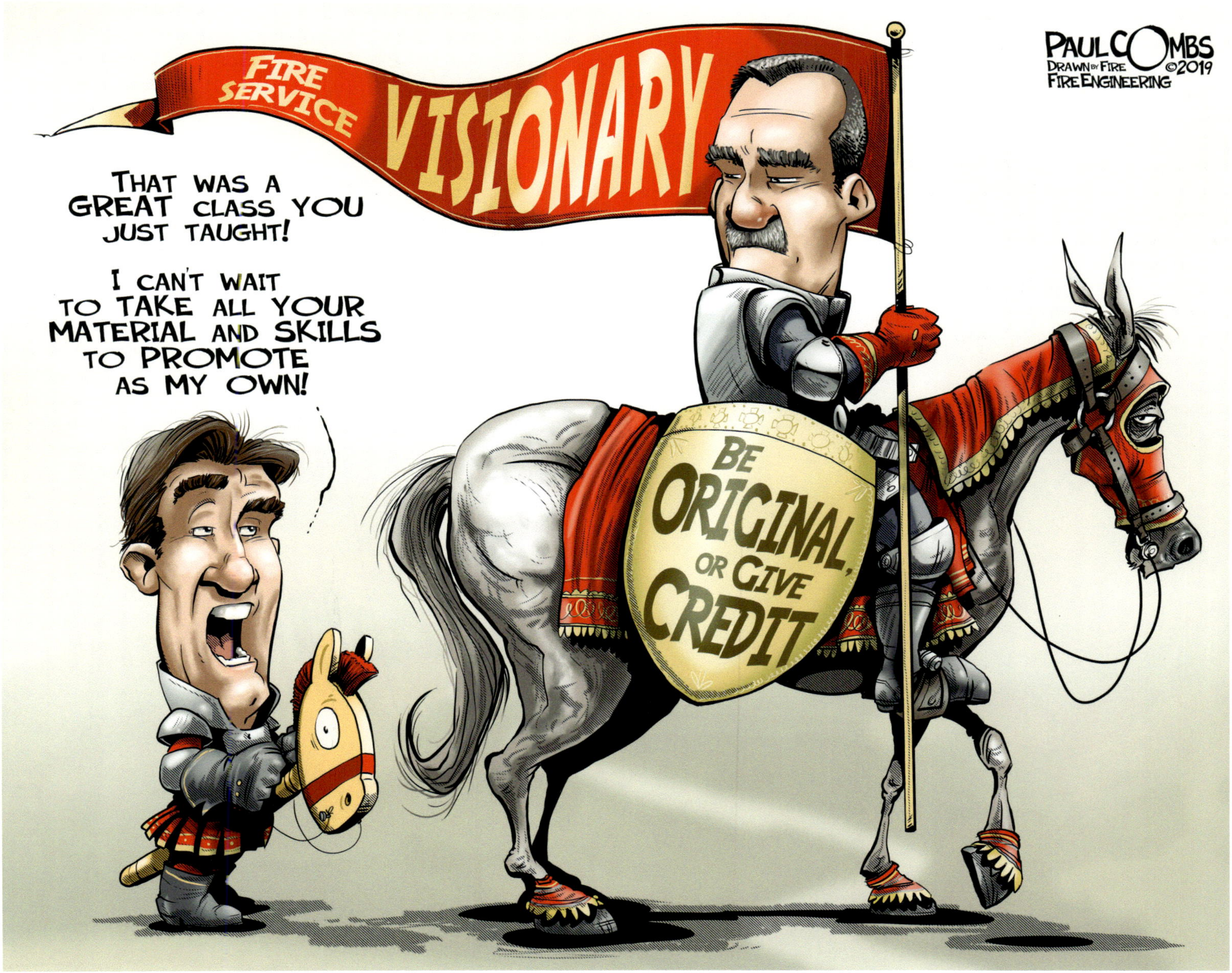
PAUL COMBS
DRAWN BY FIRE ©2019
FIRE ENGINEERING
FIRE SERVICE VISIONARY
THAT WAS A GREAT CLASS YOU JUST TAUGHT!
I CAN'T WAIT TO TAKE ALL YOUR MATERIAL AND SKILLS TO PROMOTE AS MY OWN!
BE ORIGINAL, OR GIVE CREDIT

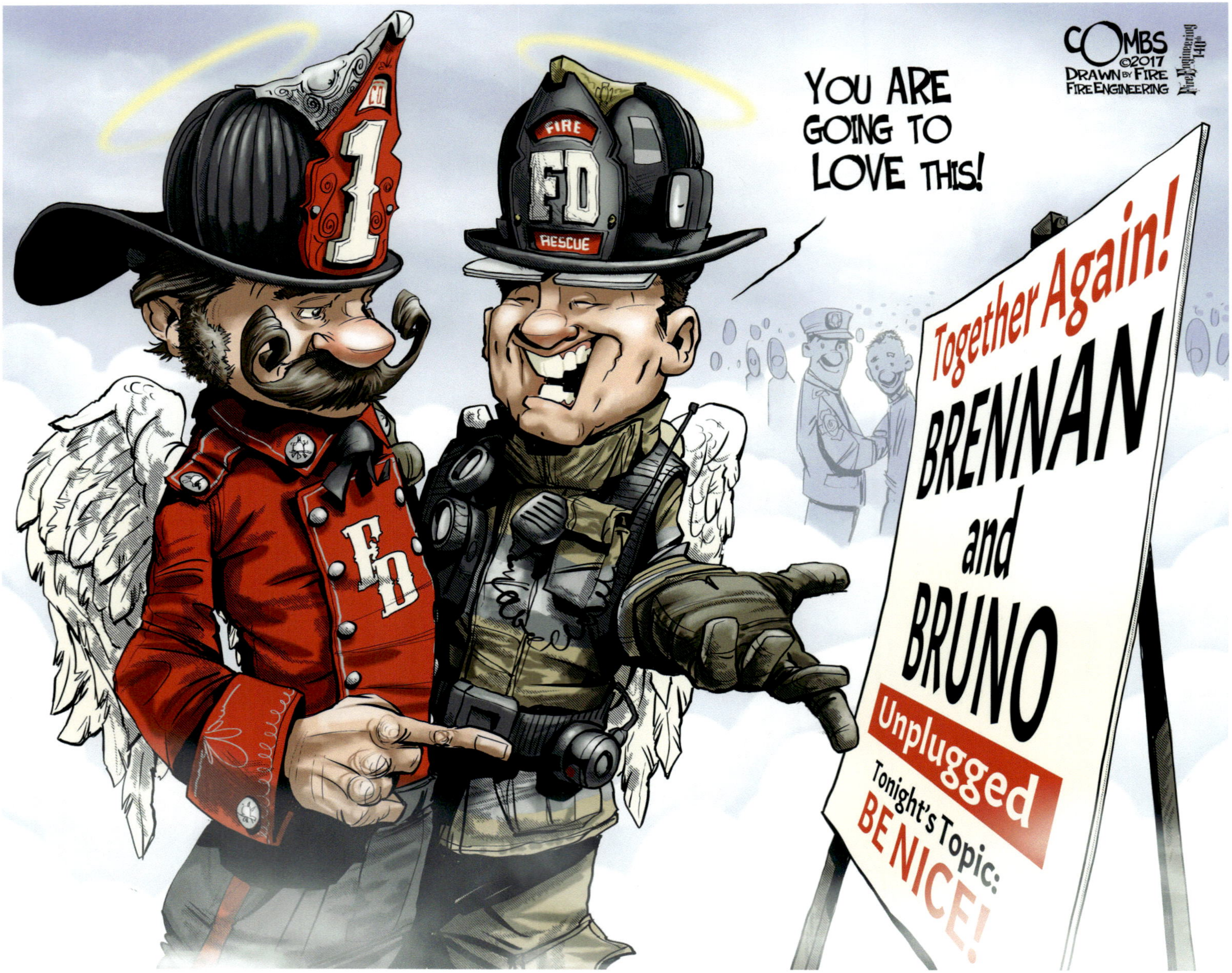
COMBS ©2017 DRAWN BY FIRE FIRE ENGINEERING
YOU ARE GOING TO LOVE THIS!
FIRE
FD
RESCUE
FD
Together Again!
BRENNAN
and
BRUNO
Unplugged
Tonight's Topic:
BE NICE!

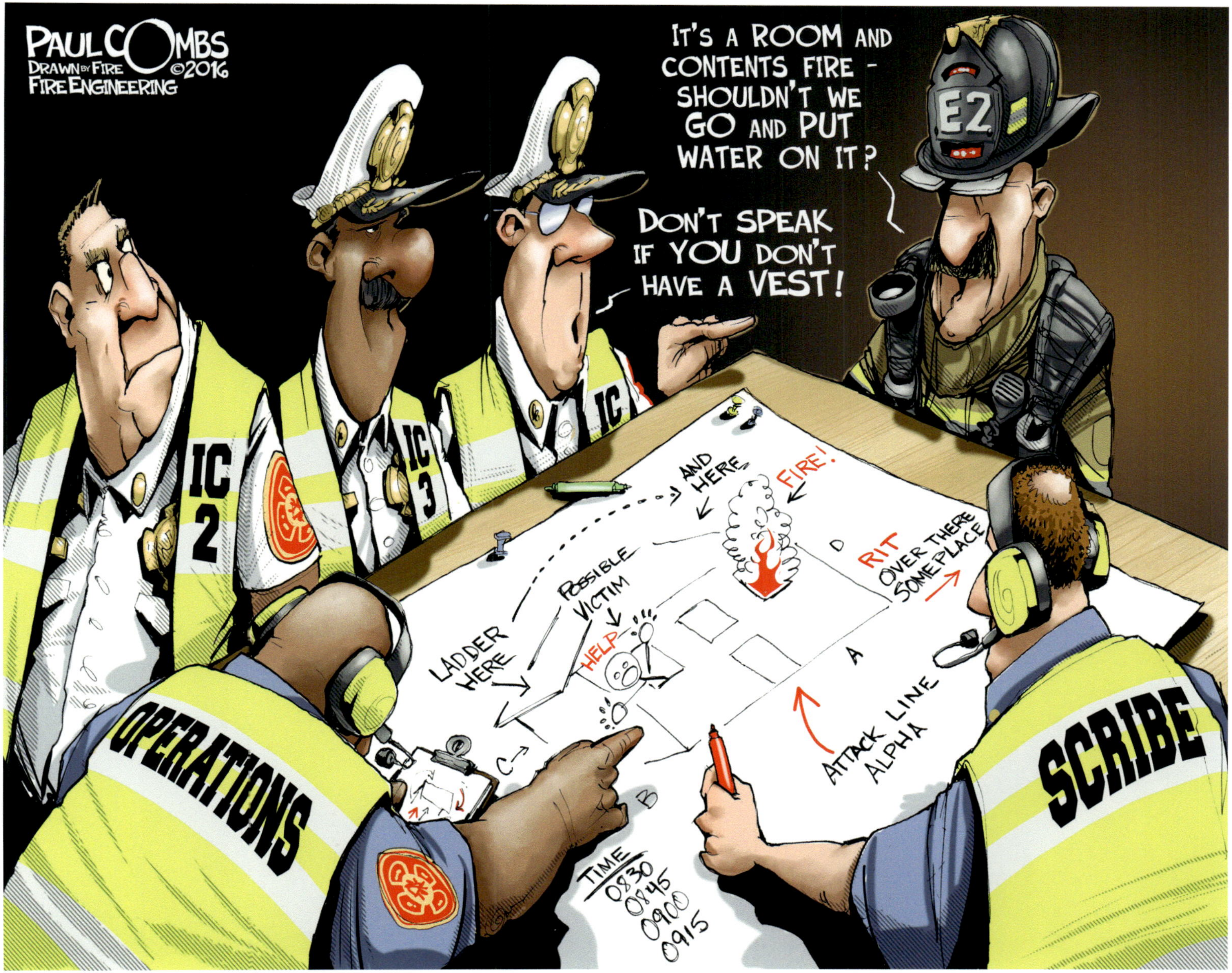
PAUL COMBS
DRAWN BY FIRE ©2016
FIRE ENGINEERING
IT'S A ROOM AND CONTENTS FIRE - SHOULDN'T WE GO AND PUT WATER ON IT?
DON'T SPEAK IF YOU DON'T HAVE A VEST!
E2
IC 2
IC 3
IC
AND HERE
FIRE!
POSSIBLE VICTIM
HELP
LADDER HERE
RIT OVER THERE SOMEPLACE
ATTACK LINE ALPHA
A
B
C
D
TIME
0830
0845
0900
0915
OPERATIONS
SCRIBE

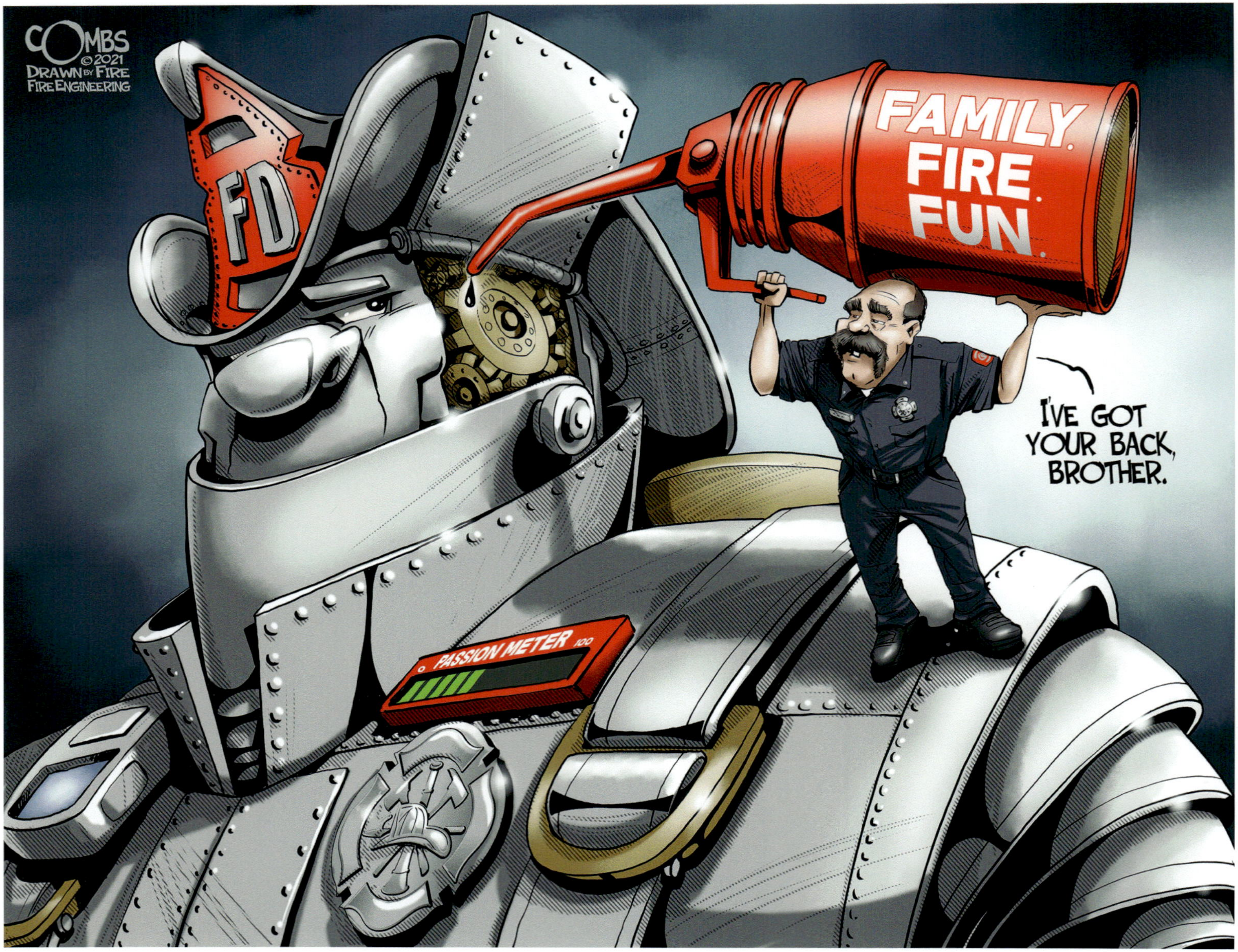
COMBS ©2021 DRAWN BY FIRE FIRE ENGINEERING
FD
FAMILY.
FIRE.
FUN.
I'VE GOT YOUR BACK, BROTHER.
PASSION METER
100

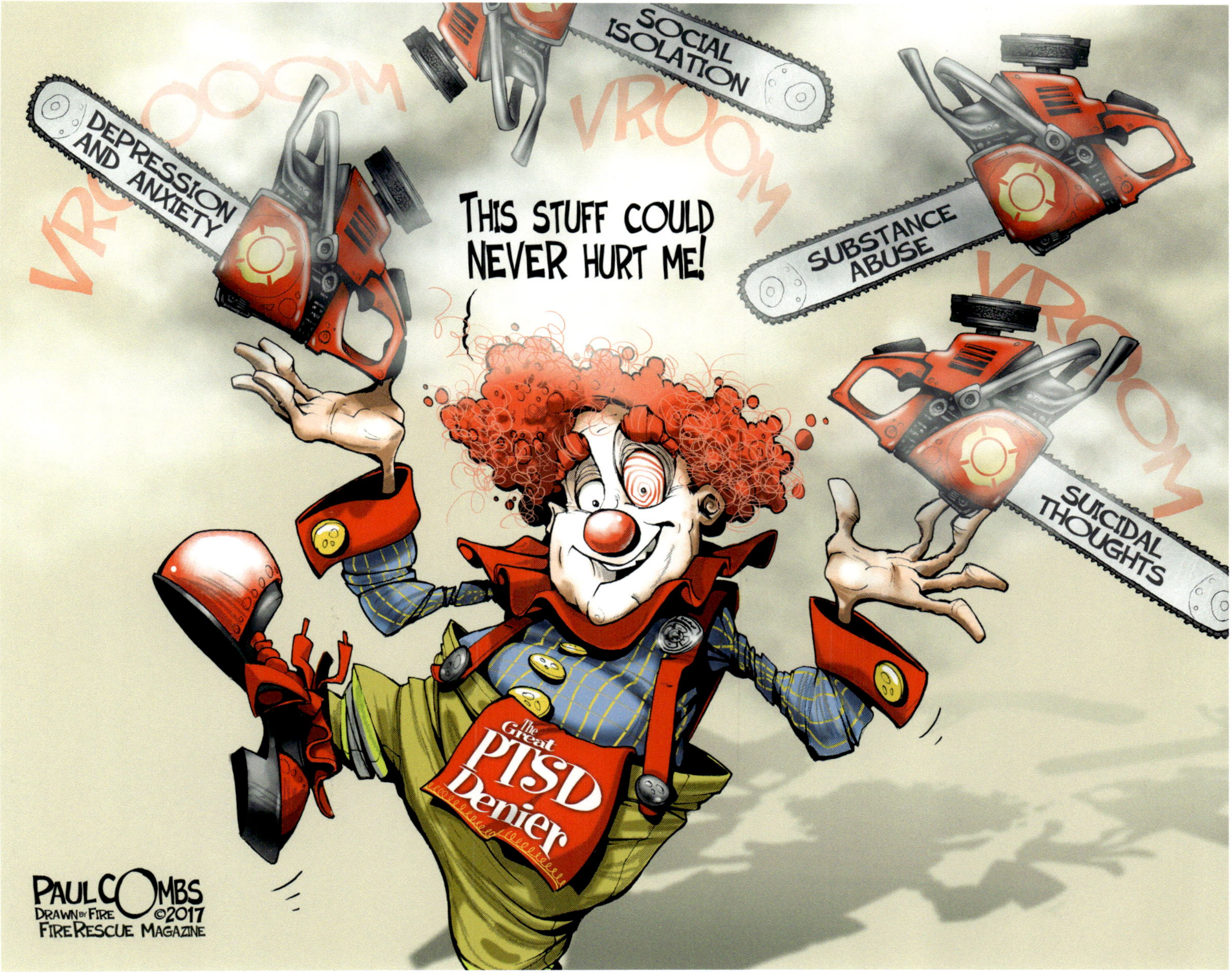
VROOOOM
VROOM
VROOM
DEPRESSION AND ANXIETY
SOCIAL ISOLATION
SUBSTANCE ABUSE
SUICIDAL THOUGHTS
THIS STUFF COULD NEVER HURT ME!
The Great PTSD Denier
PAUL COMBS
DRAWN BY FIRE ©2017
FIRERESCUE MAGAZINE

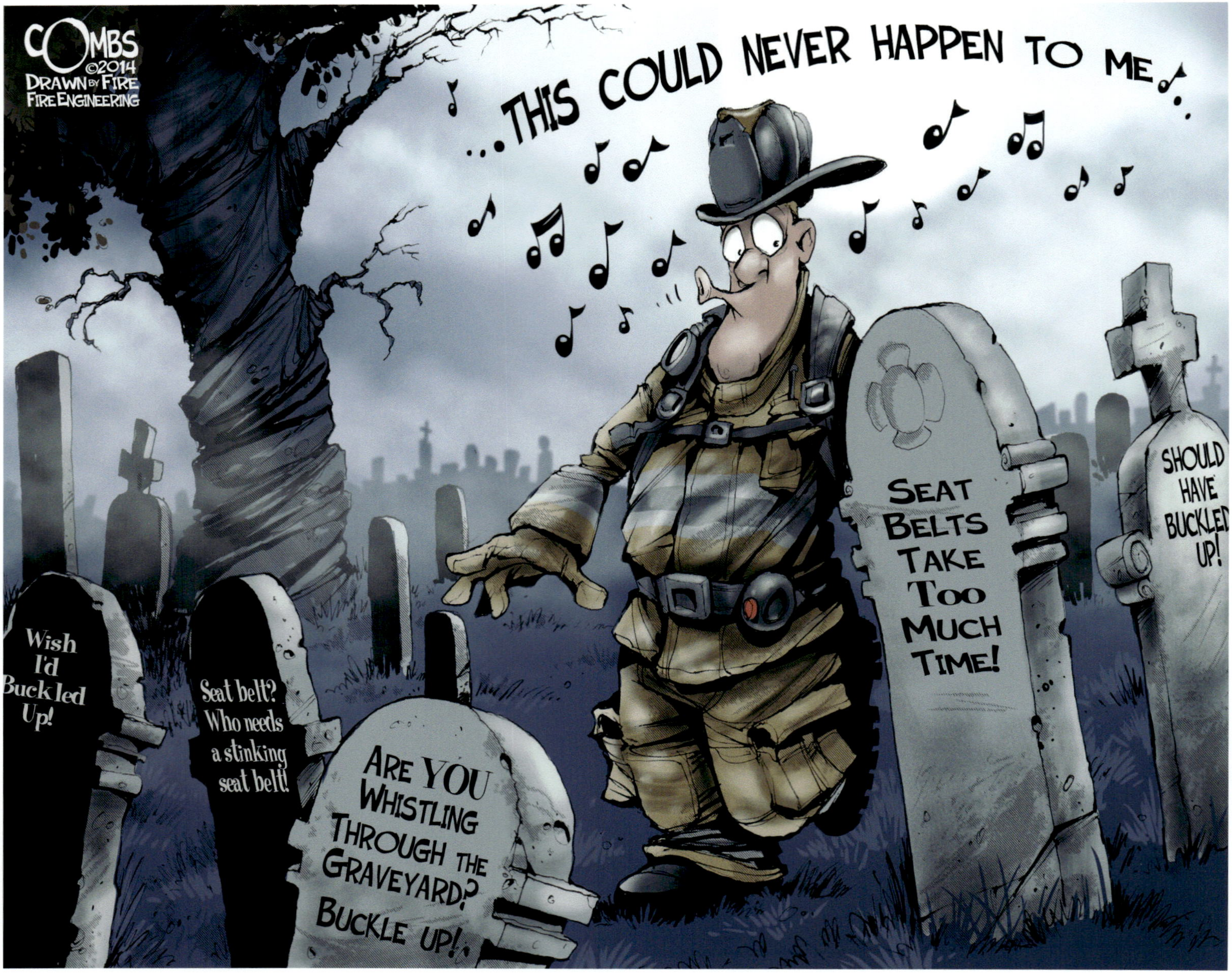
COMBS ©2014 DRAWN BY FIRE FIRE ENGINEERING
...THIS COULD NEVER HAPPEN TO ME...
Wish I'd Buckled Up!
Seat belt? Who needs a stinking seat belt!
ARE YOU WHISTLING THROUGH THE GRAVEYARD? BUCKLE UP!
SEAT BELTS TAKE TOO MUCH TIME!
SHOULD HAVE BUCKLED UP!

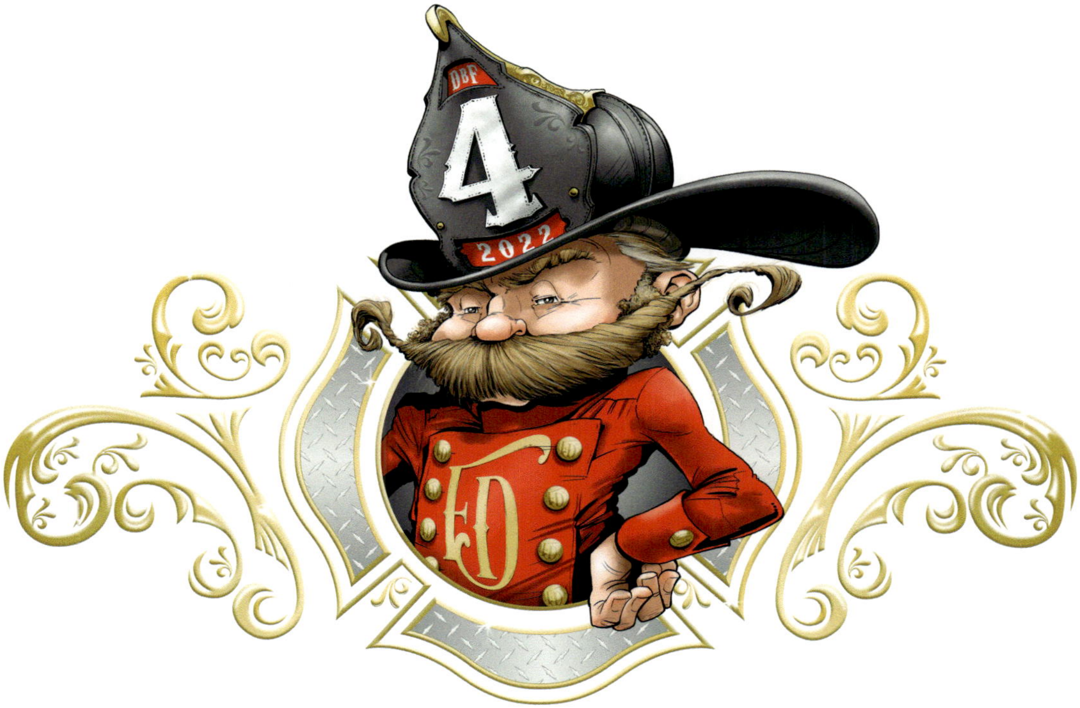

Humor Is Good Medicine

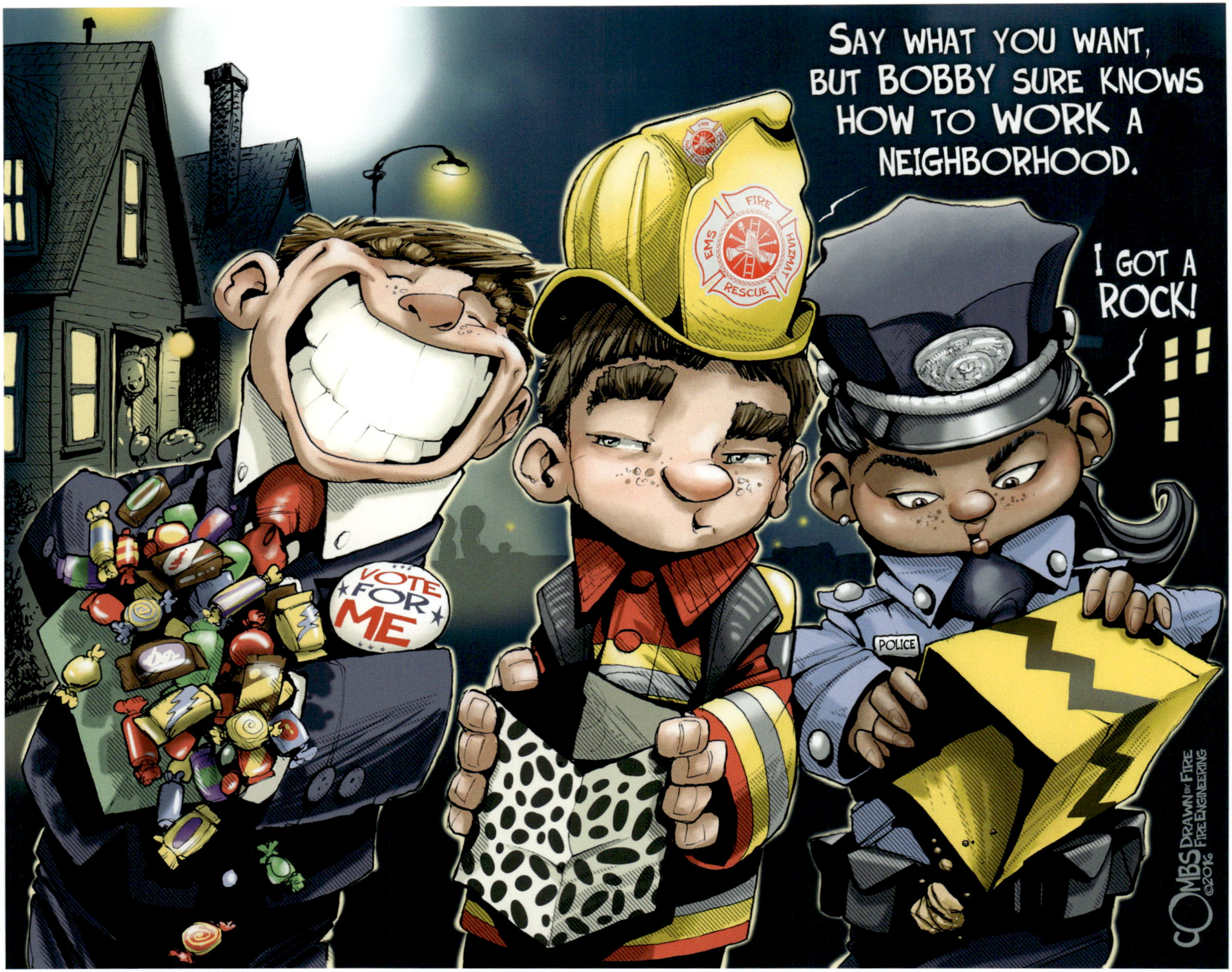
SAY WHAT YOU WANT, BUT BOBBY SURE KNOWS HOW TO WORK A NEIGHBORHOOD.
I GOT A ROCK!
VOTE FOR ME
POLICE
FIRE EMS HAZMAT RESCUE
COMBS DRAWN BY FIRE FIREENGINEERING ©2016

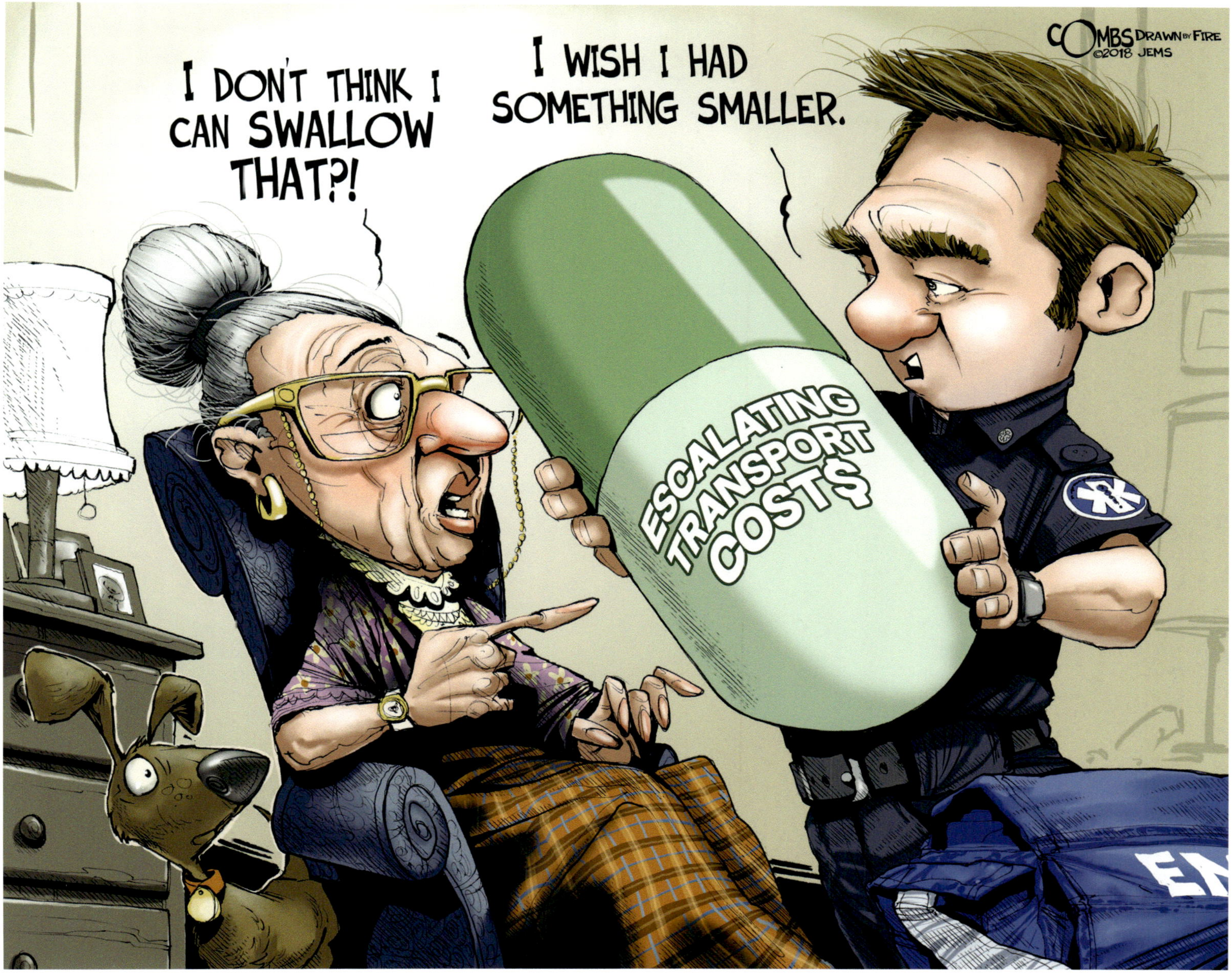
COMBS DRAWN BY FIRE ©2018 JEMS
I DON'T THINK I CAN SWALLOW THAT?!
I WISH I HAD SOMETHING SMALLER.
ESCALATING TRANSPORT COST$

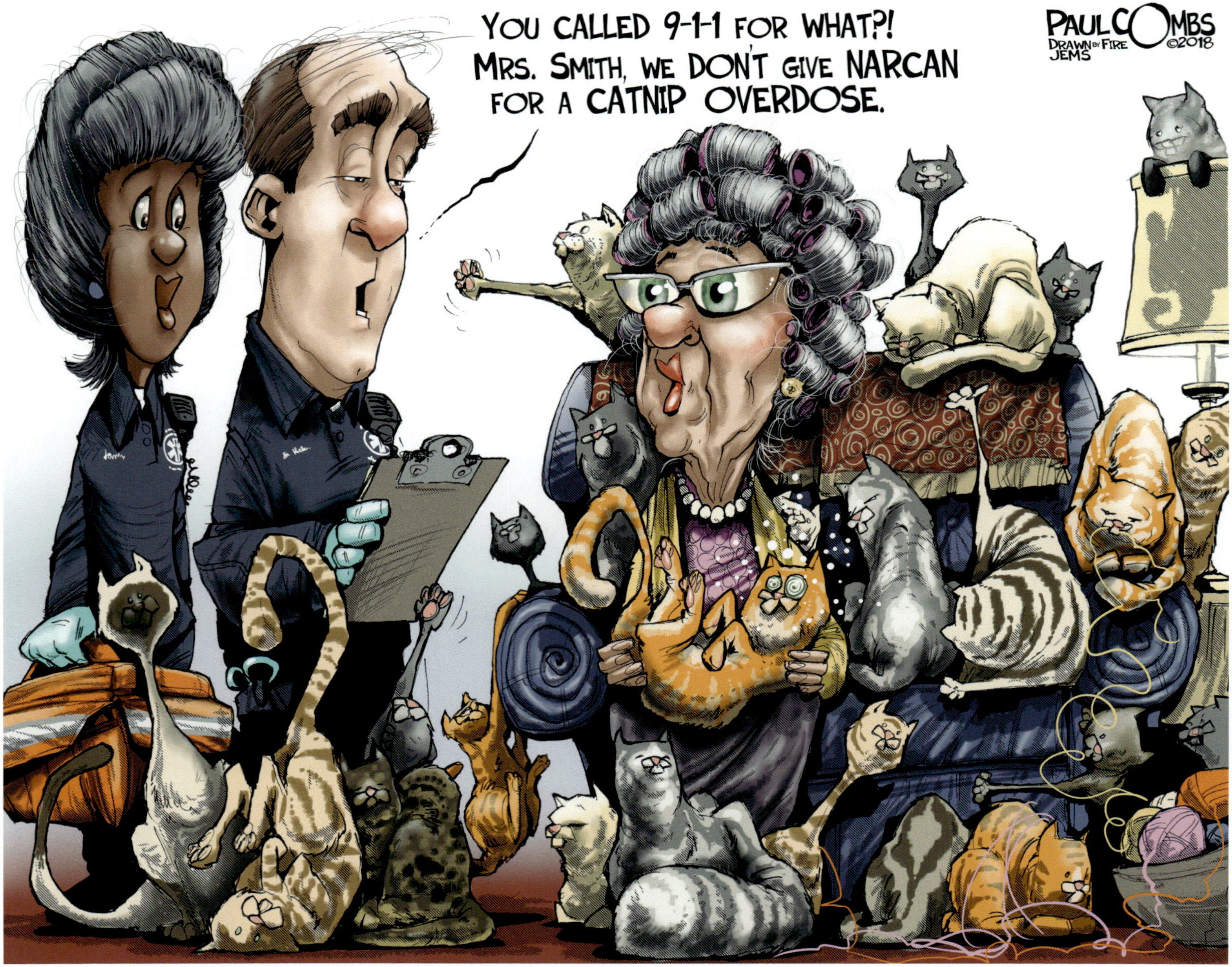
YOU CALLED 9-1-1 FOR WHAT?!
MRS. SMITH, WE DON'T GIVE NARCAN
FOR A CATNIP OVERDOSE.
PAUL COMBS
DRAWN BY FIRE
JEMS ©2018

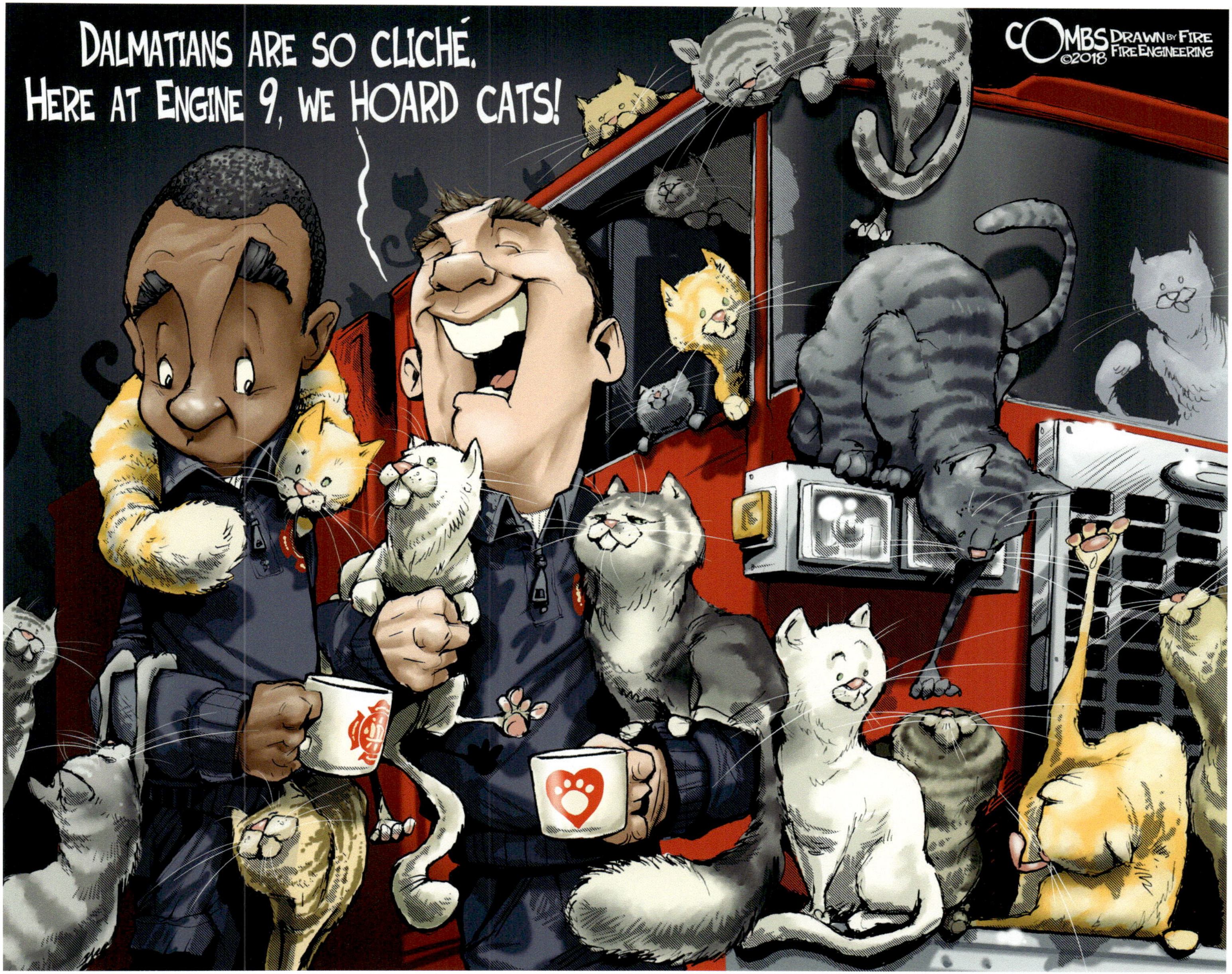
DALMATIANS ARE SO CLICHÉ.
HERE AT ENGINE 9, WE HOARD CATS!
COMBS DRAWN BY FIRE
©2018 FIRE ENGINEERING

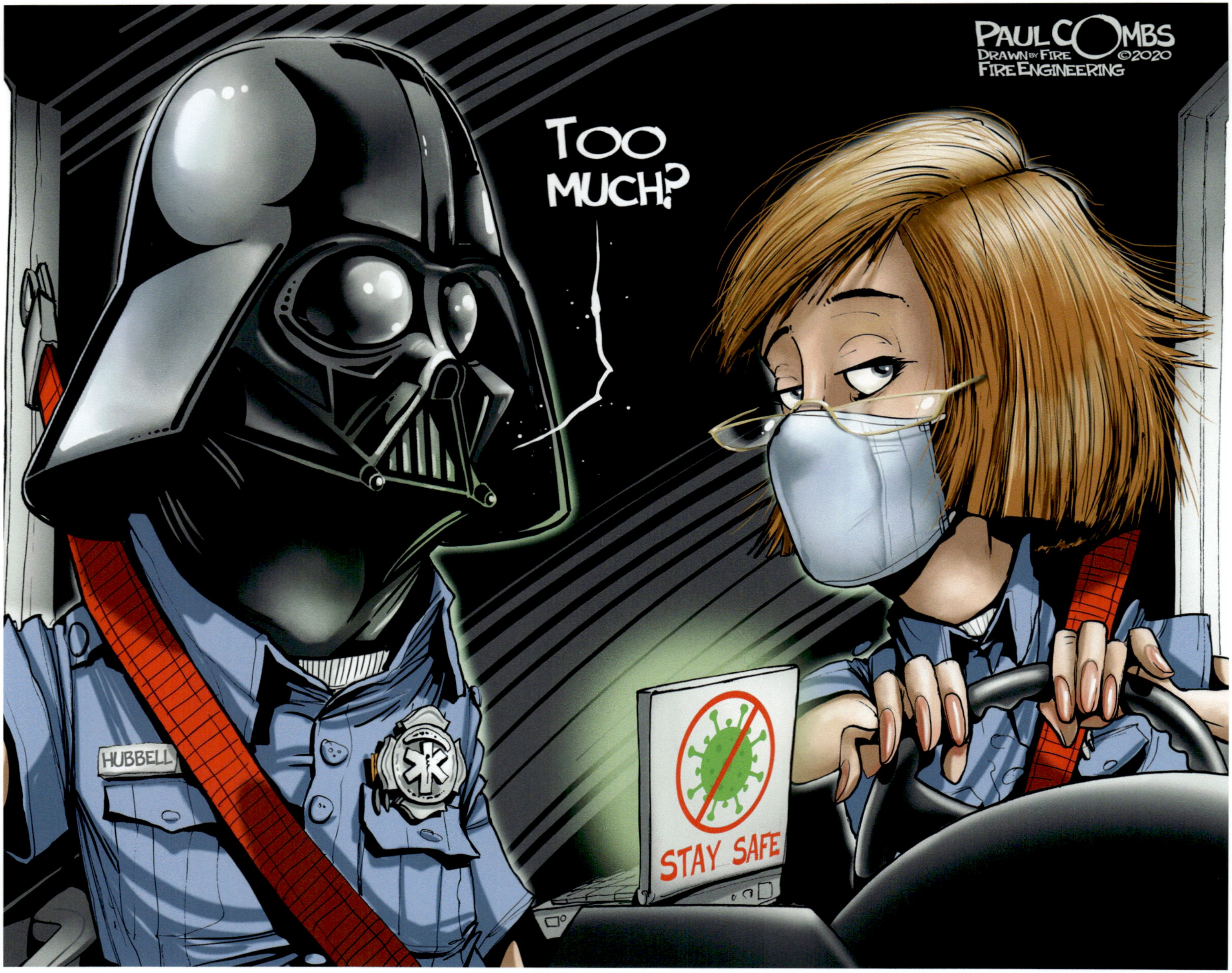
PAUL COMBS
DRAWN BY FIRE ©2020
FIRE ENGINEERING
TOO MUCH?
HUBBELL
STAY SAFE

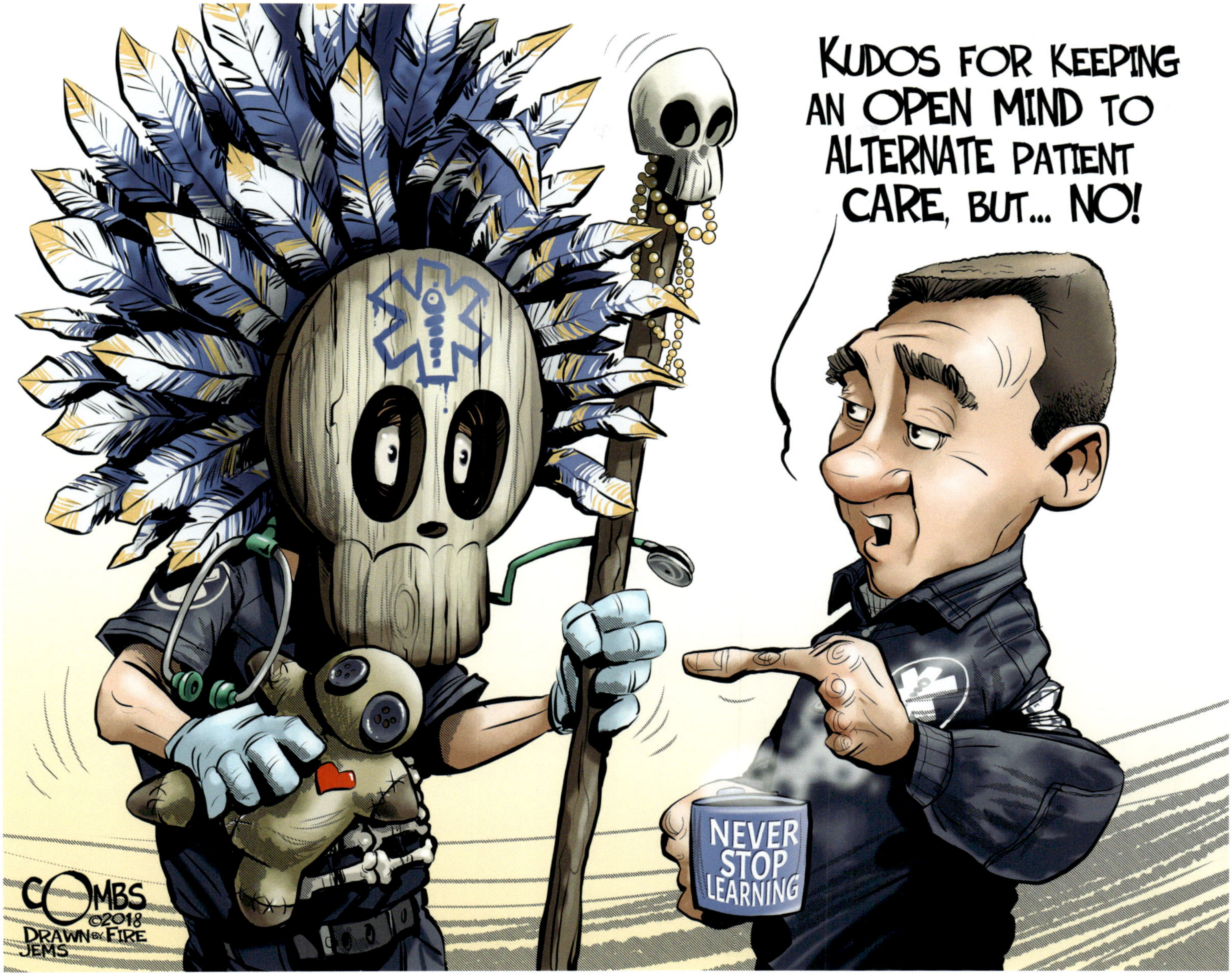
KUDOS FOR KEEPING AN OPEN MIND TO ALTERNATE PATIENT CARE, BUT... NO!
NEVER STOP LEARNING
COMBS ©2018 DRAWN BY FIRE JEMS

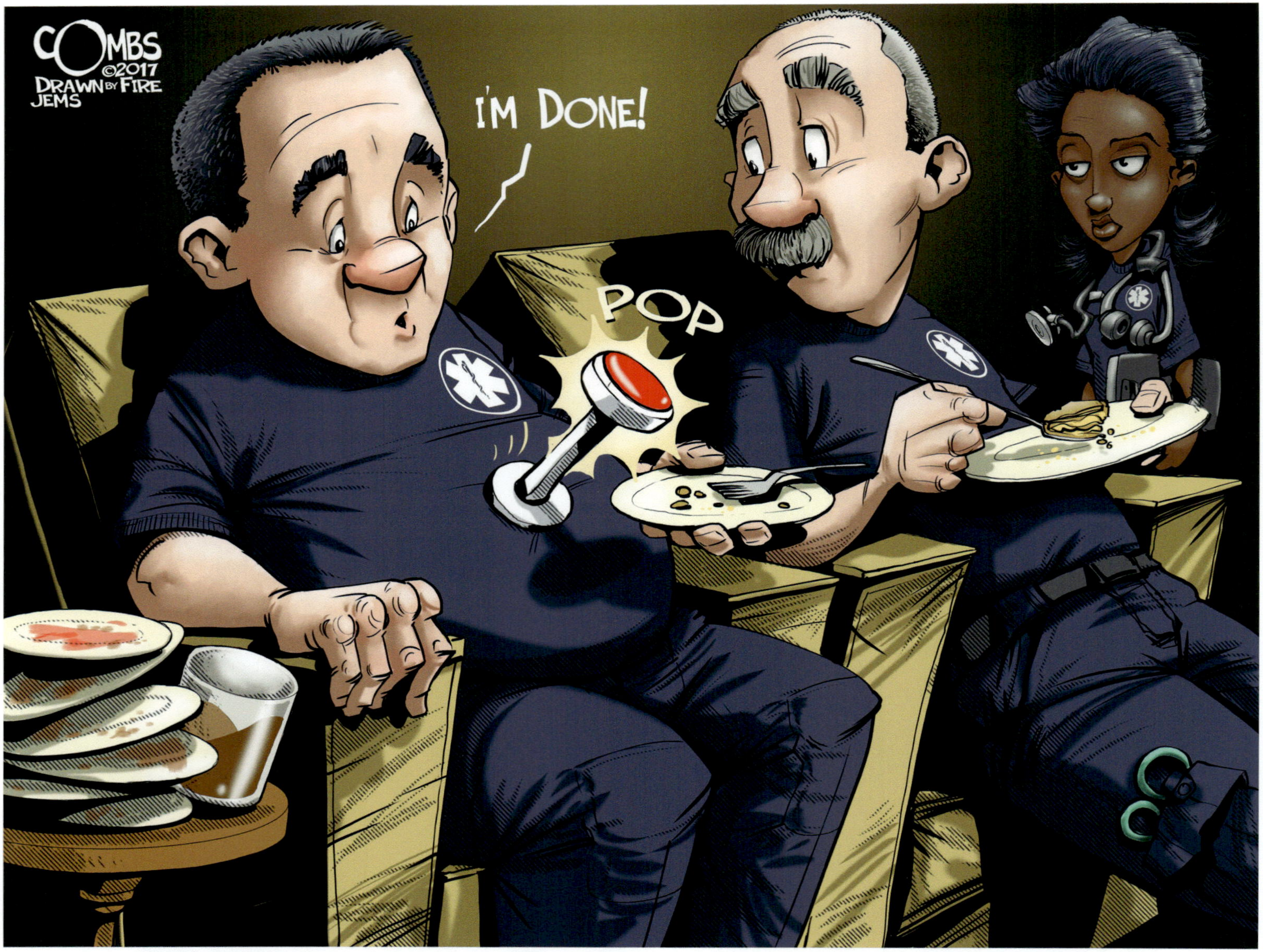

MAY WE ALL HAVE A SAFE AND BLESSED HOLIDAY SEASON!

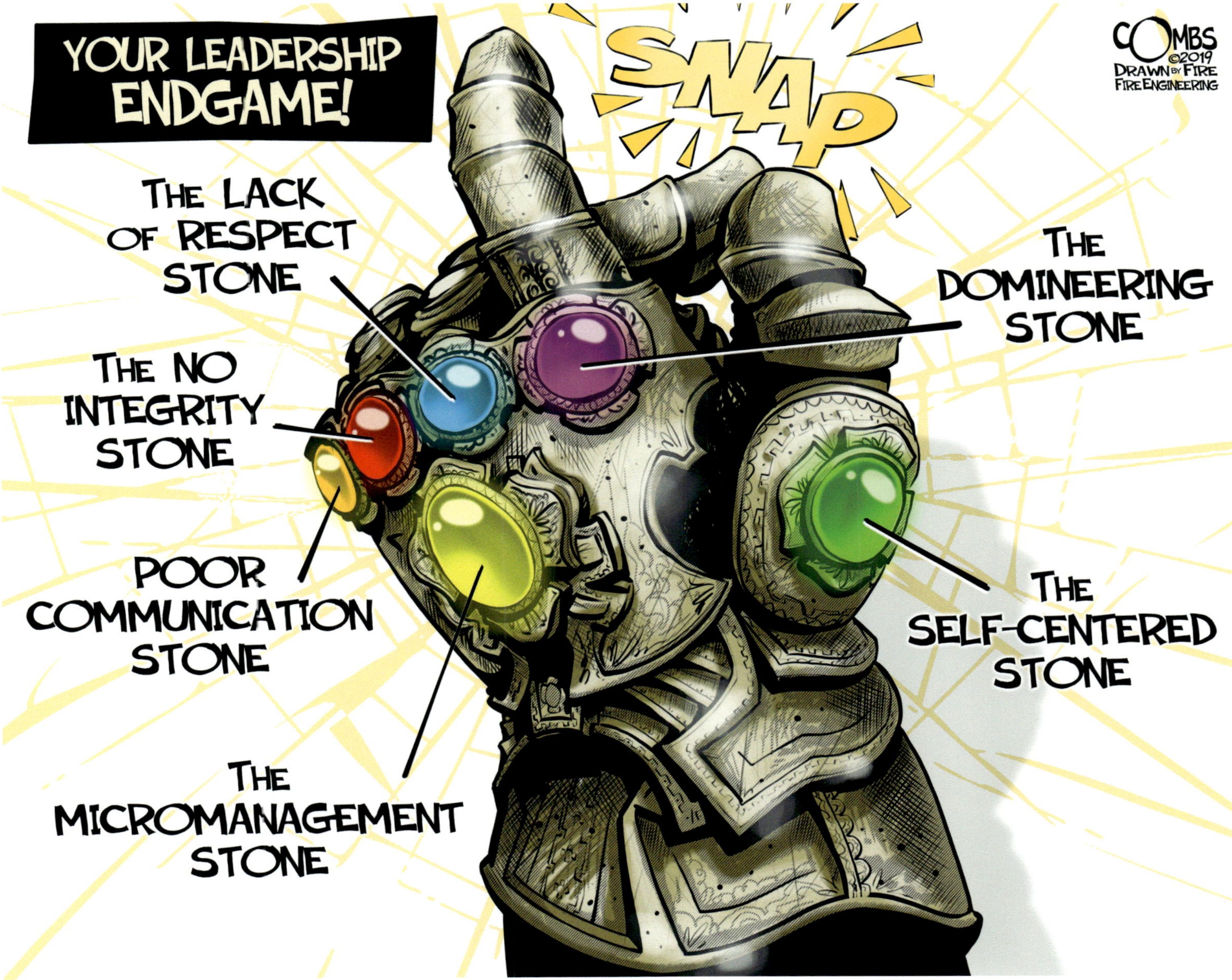
YOUR LEADERSHIP ENDGAME!
SNAP
COMBS ©2019 DRAWN BY FIRE FIRE ENGINEERING
THE LACK OF RESPECT STONE
THE DOMINEERING STONE
THE NO INTEGRITY STONE
POOR COMMUNICATION STONE
THE SELF-CENTERED STONE
THE MICROMANAGEMENT STONE

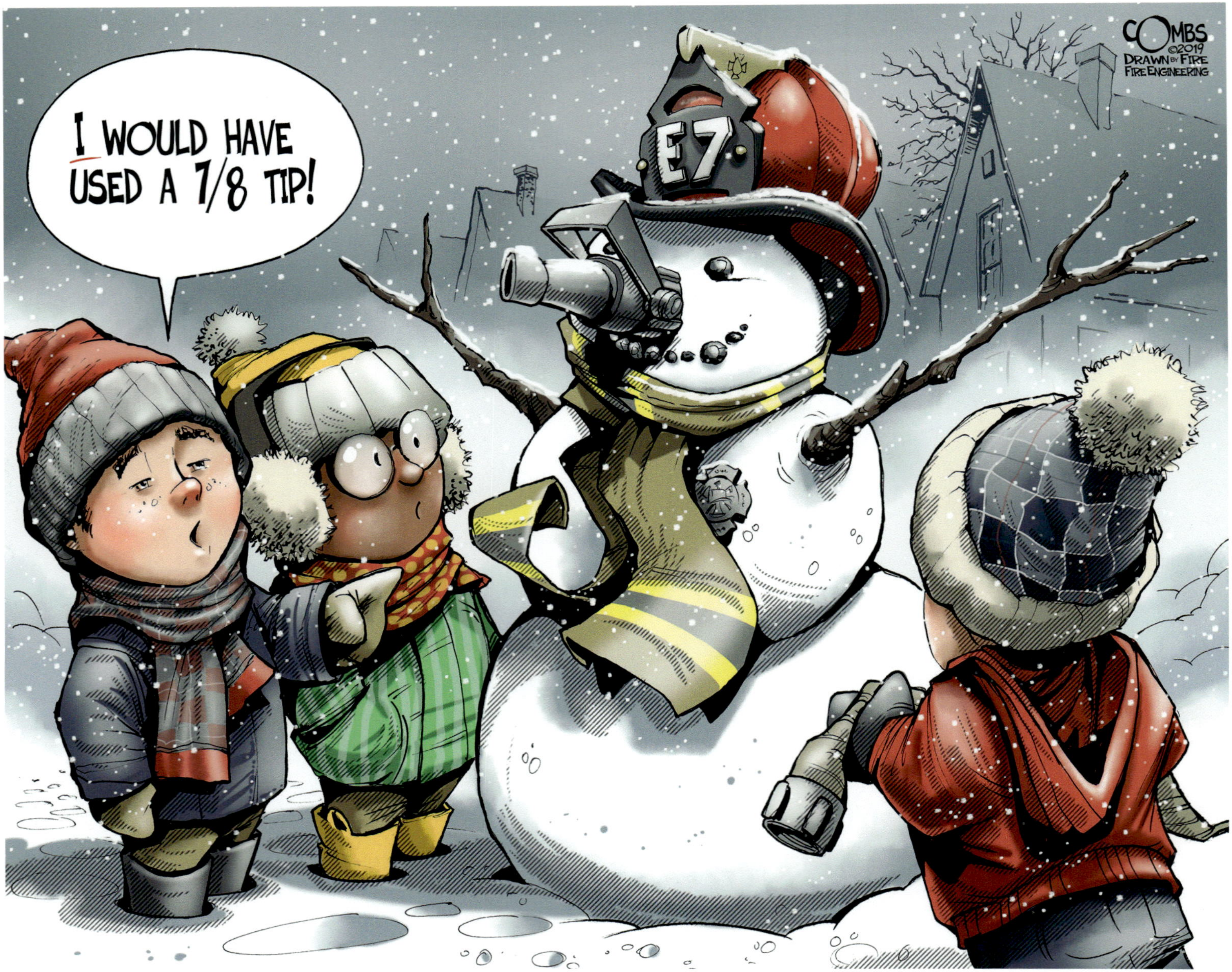
I WOULD HAVE USED A 7/8 TIP!
E7
COMBS ©2019 DRAWN BY FIRE FIRE ENGINEERING

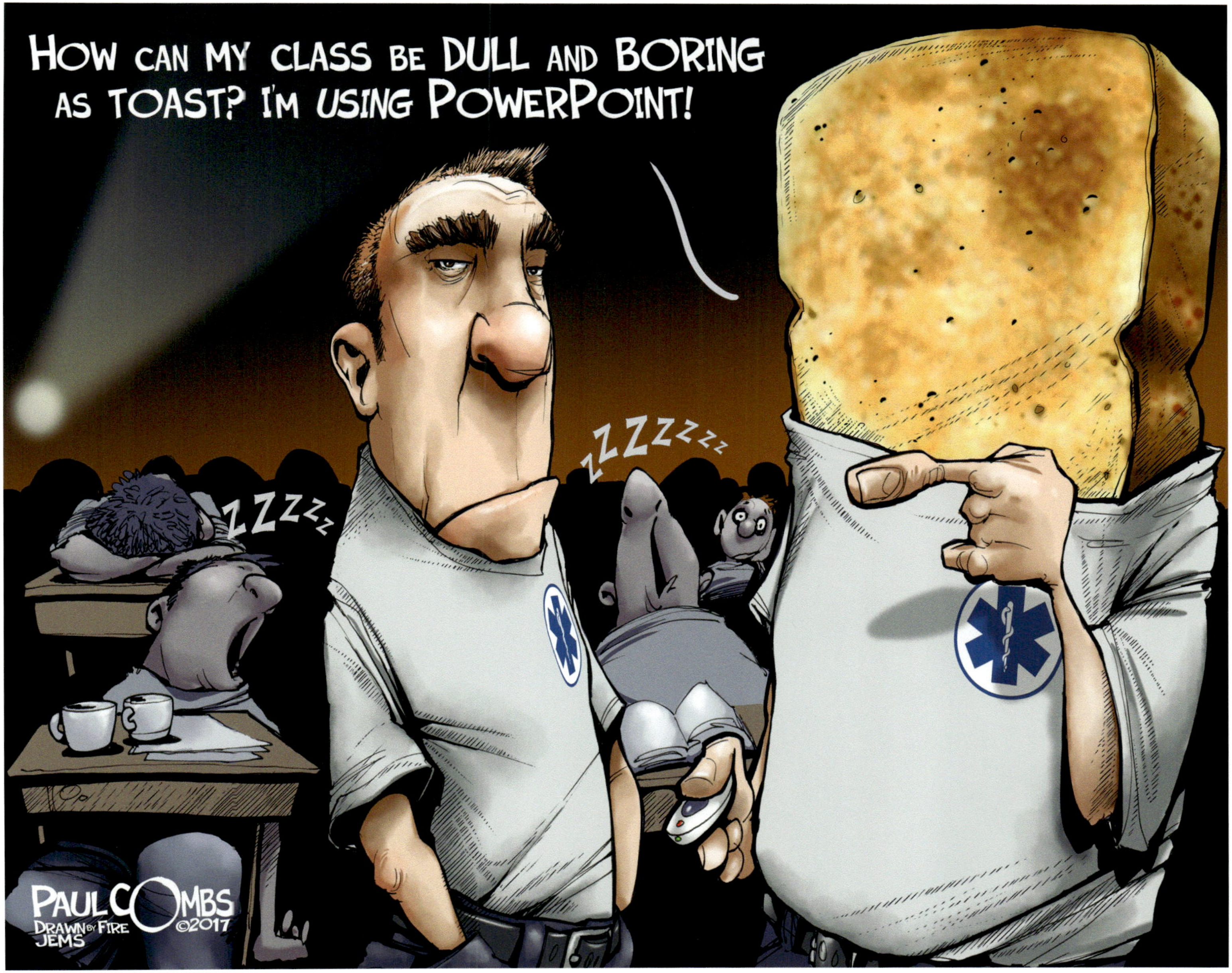
HOW CAN MY CLASS BE DULL AND BORING AS TOAST? I'M USING POWERPOINT!
ZZZZzzz
ZZZzz
PAUL COMBS
DRAWN BY FIRE JEMS ©2017

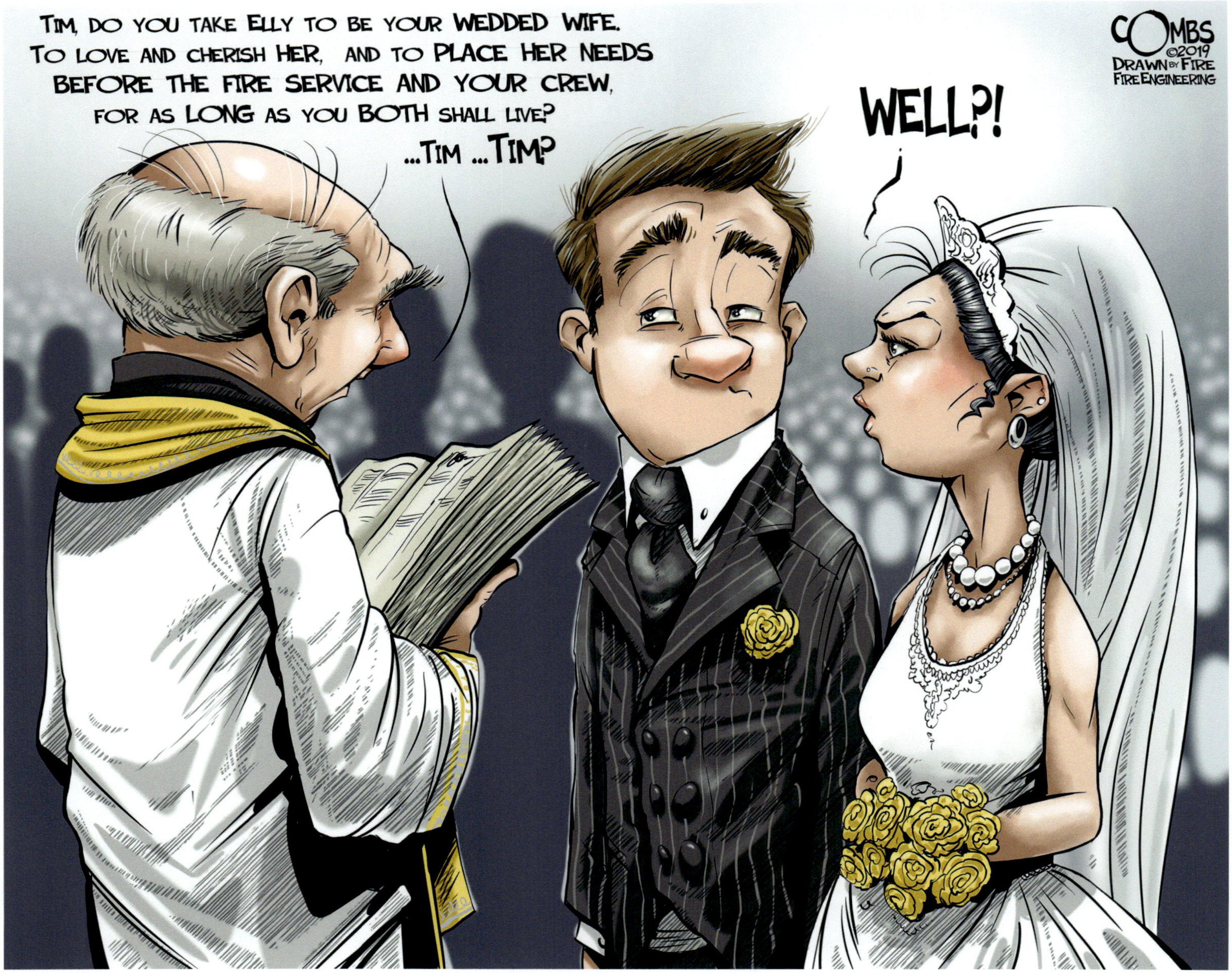
TIM, DO YOU TAKE ELLY TO BE YOUR WEDDED WIFE.
TO LOVE AND CHERISH HER, AND TO PLACE HER NEEDS
BEFORE THE FIRE SERVICE AND YOUR CREW,
FOR AS LONG AS YOU BOTH SHALL LIVE?
...TIM ...TIM?
WELL?!
COMBS ©2019 DRAWN BY FIRE FIRE ENGINEERING

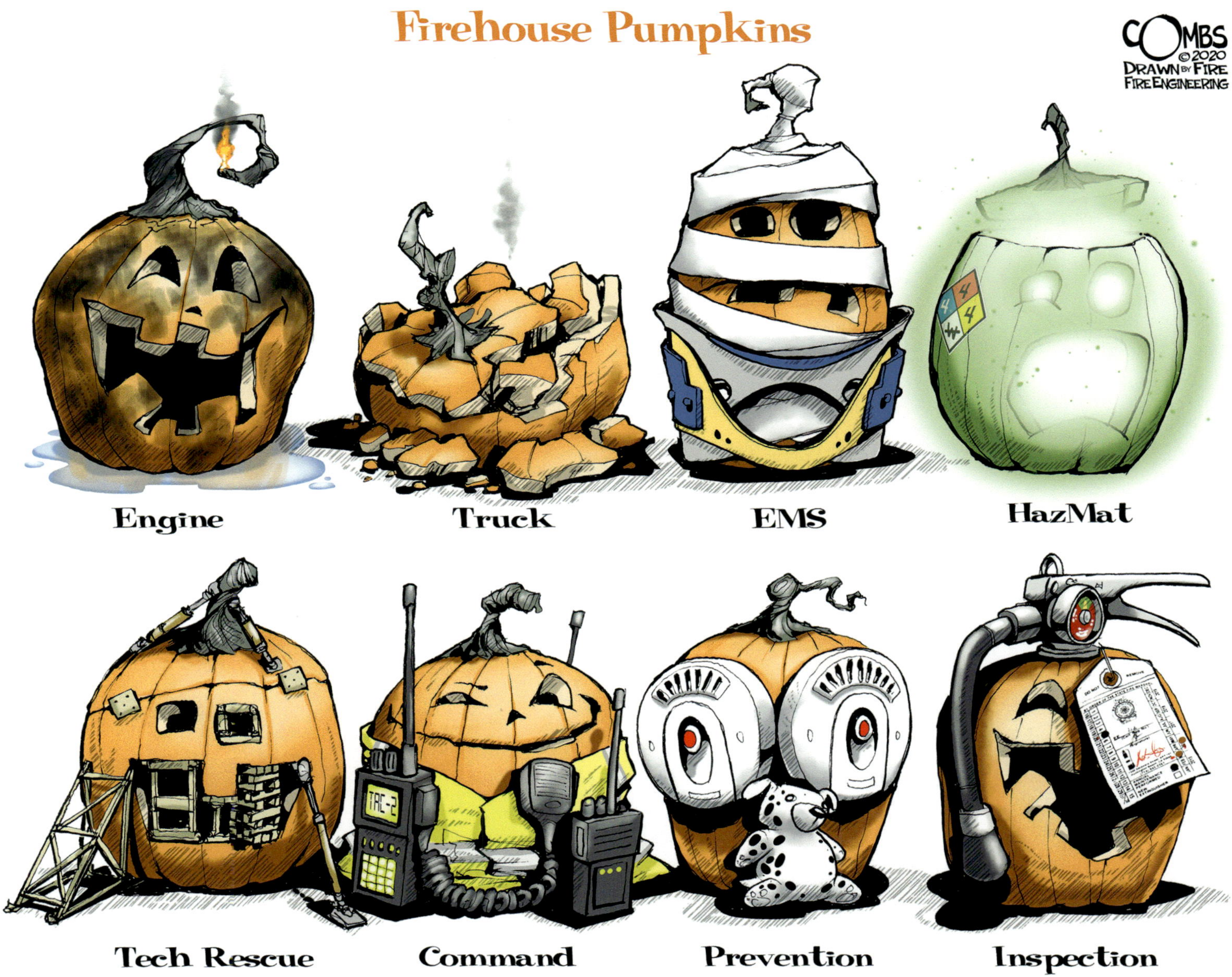
Firehouse Pumpkins
COMBS
©2020
DRAWN BY FIRE
FIRE ENGINEERING
Engine
Truck
EMS
HazMat
Tech Rescue
Command
Prevention
Inspection
TAC-2

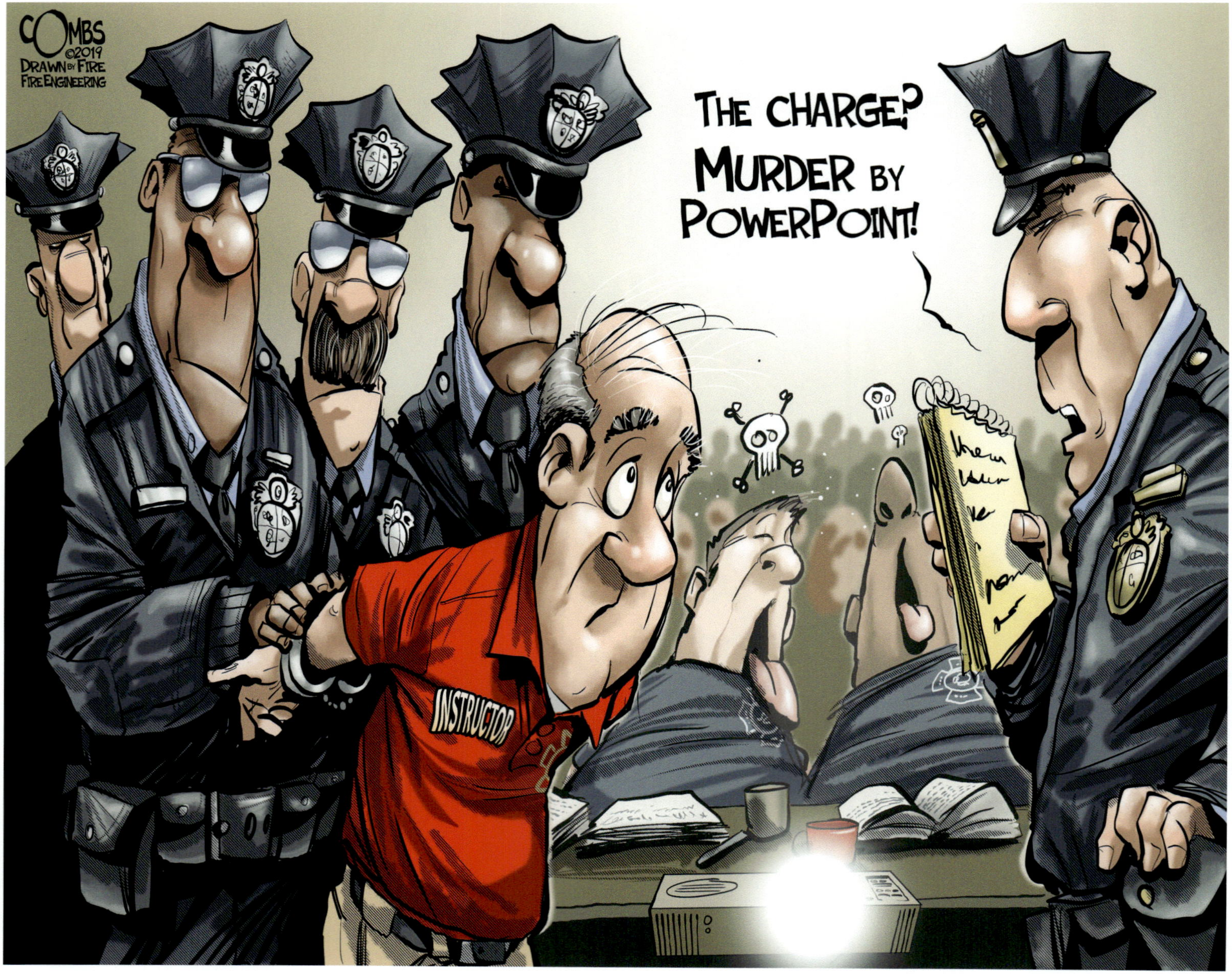
COMBS ©2019 DRAWN BY FIRE FIRE ENGINEERING
THE CHARGE? MURDER BY POWERPOINT!
INSTRUCTOR

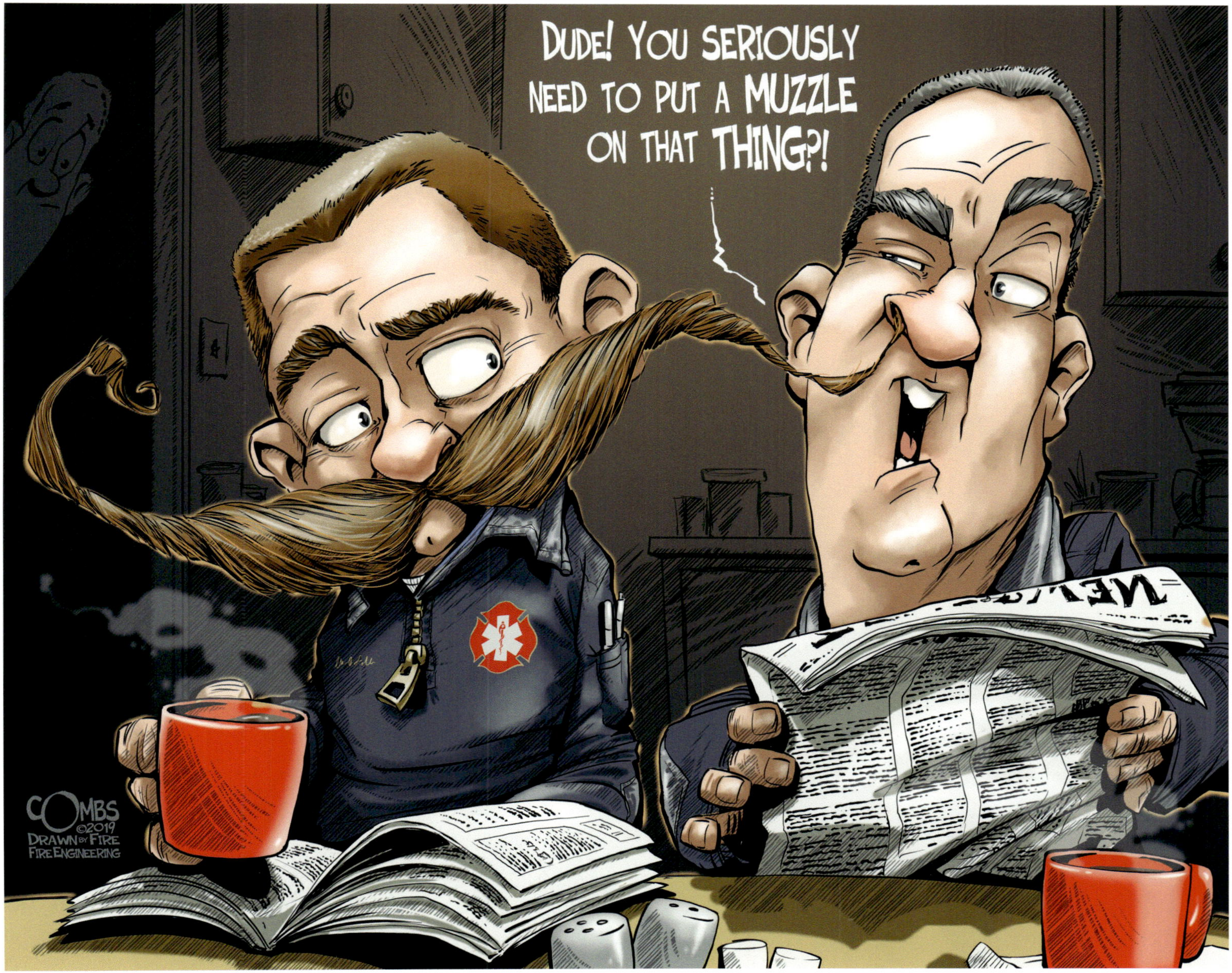
DUDE! YOU SERIOUSLY NEED TO PUT A MUZZLE ON THAT THING?!
COMBS ©2019 DRAWN BY FIRE FIRE ENGINEERING
NEWS

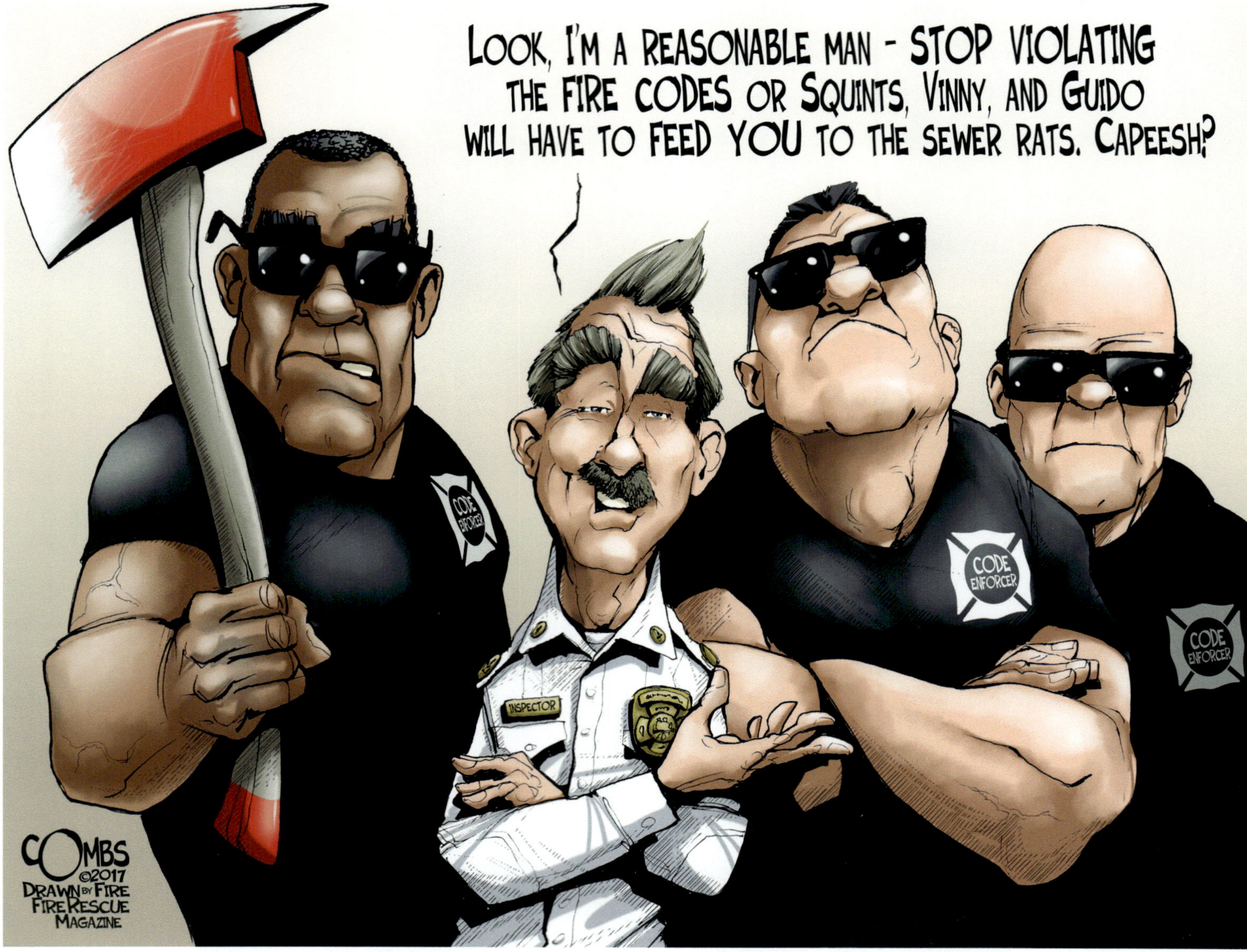

WHEN FIRE INSPECTORS FANTASIZE

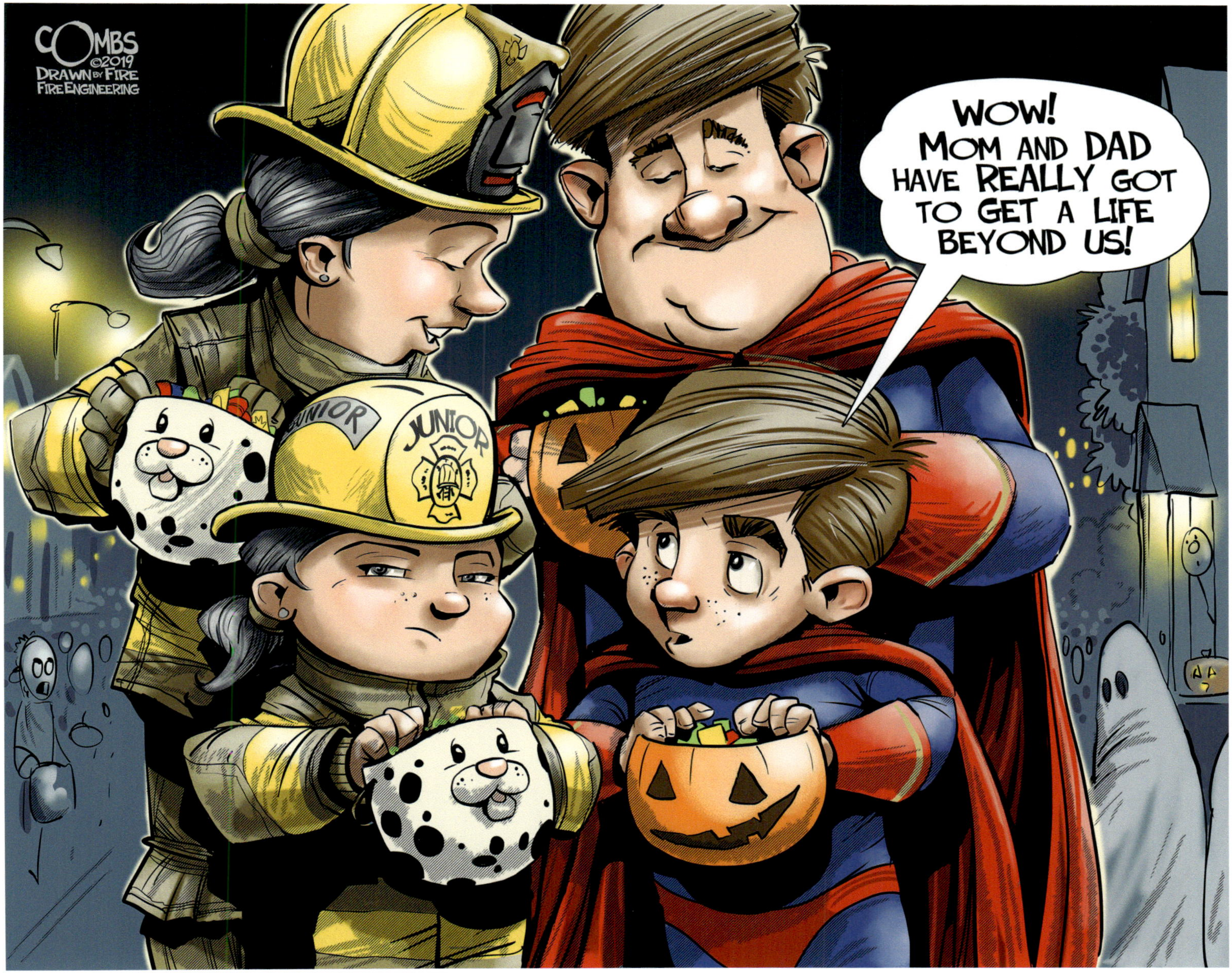
COMBS ©2019 DRAWN BY FIRE FIRE ENGINEERING
WOW! MOM AND DAD HAVE REALLY GOT TO GET A LIFE BEYOND US!
JUNIOR

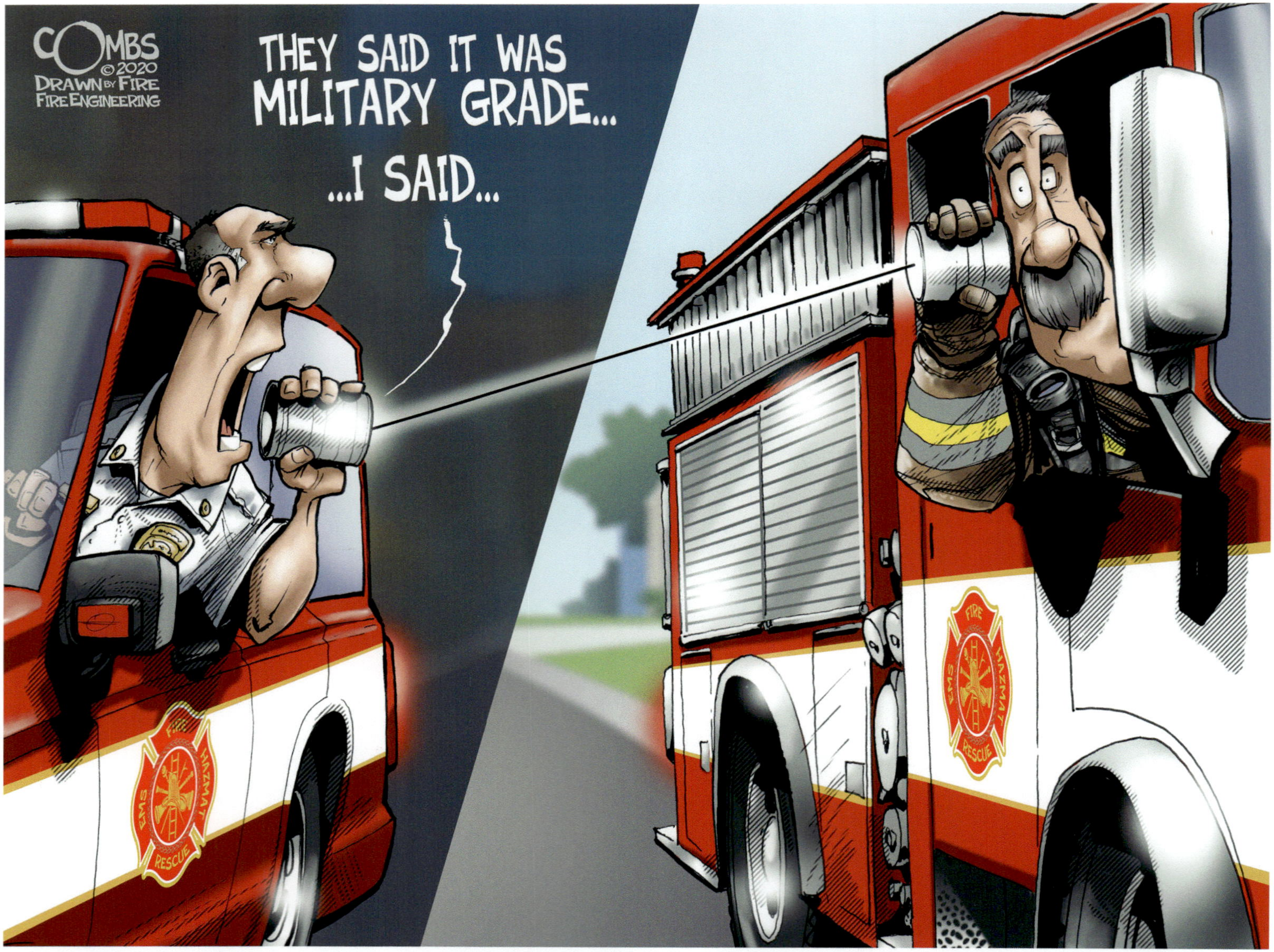

WHEN YOU BUY YOUR COMMUNICATIONS SYSTEM FROM THE LOWEST BIDDER...

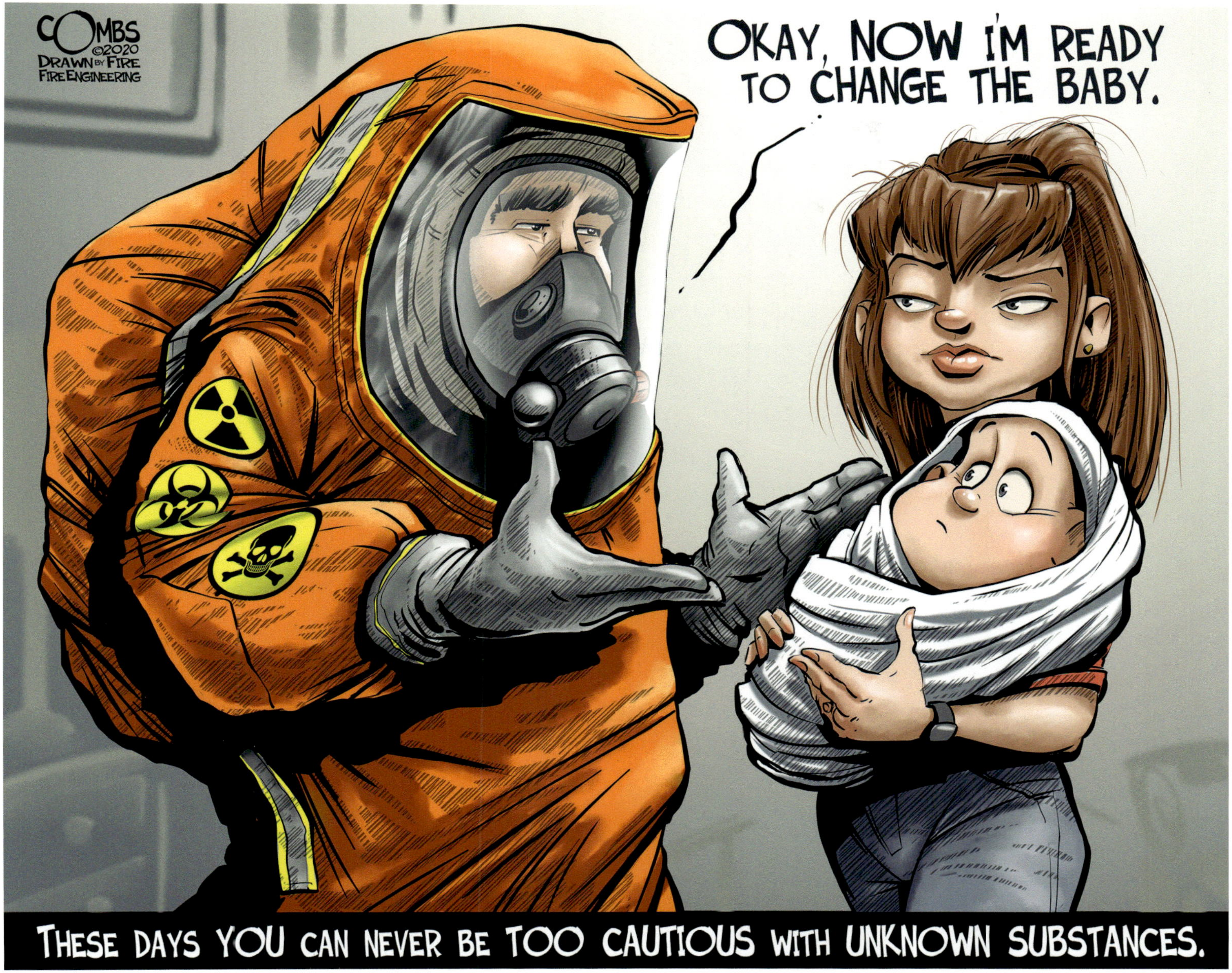
COMBS ©2020 DRAWN BY FIRE FIRE ENGINEERING
OKAY, NOW I'M READY TO CHANGE THE BABY.
THESE DAYS YOU CAN NEVER BE TOO CAUTIOUS WITH UNKNOWN SUBSTANCES.

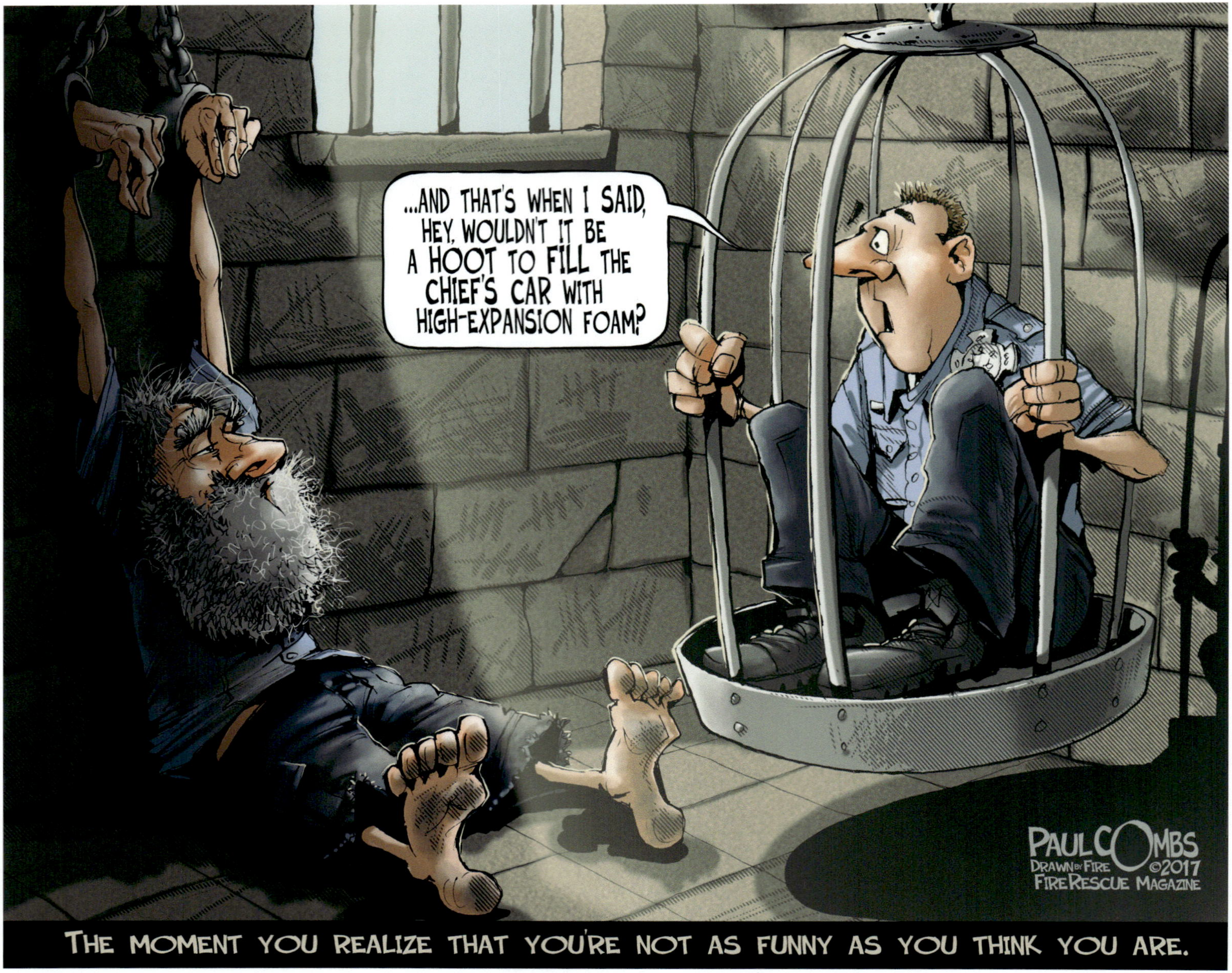

THE MOMENT YOU REALIZE THAT YOU'RE NOT AS FUNNY AS YOU THINK YOU ARE.

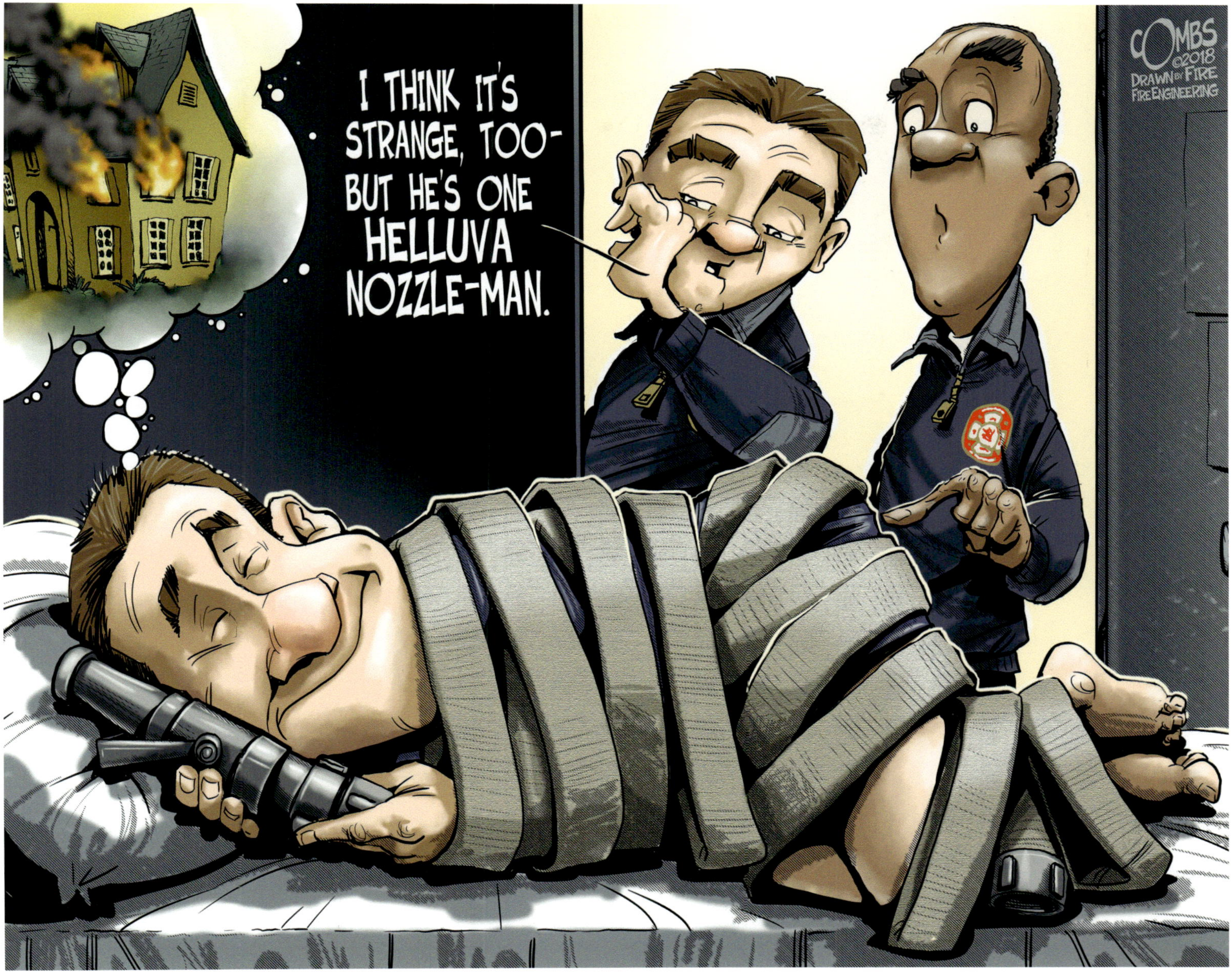
I THINK IT'S STRANGE, TOO- BUT HE'S ONE HELLUVA NOZZLE-MAN.
COMBS ©2018 DRAWN BY FIRE FIRE ENGINEERING

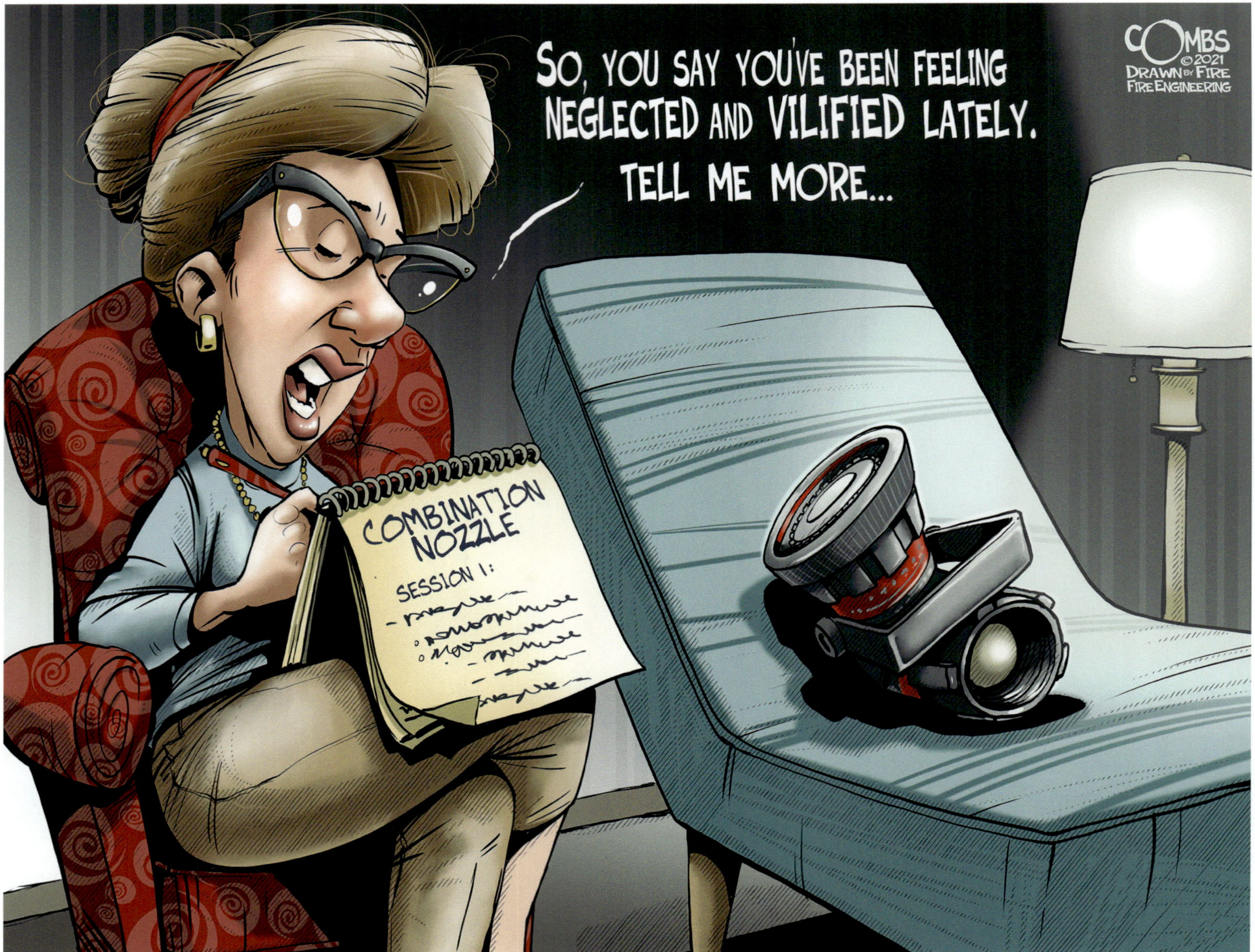
SO, YOU SAY YOU'VE BEEN FEELING NEGLECTED AND VILIFIED LATELY. TELL ME MORE...
COMBINATION NOZZLE
SESSION 1:
COMBS ©2021 DRAWN BY FIRE FIRE ENGINEERING

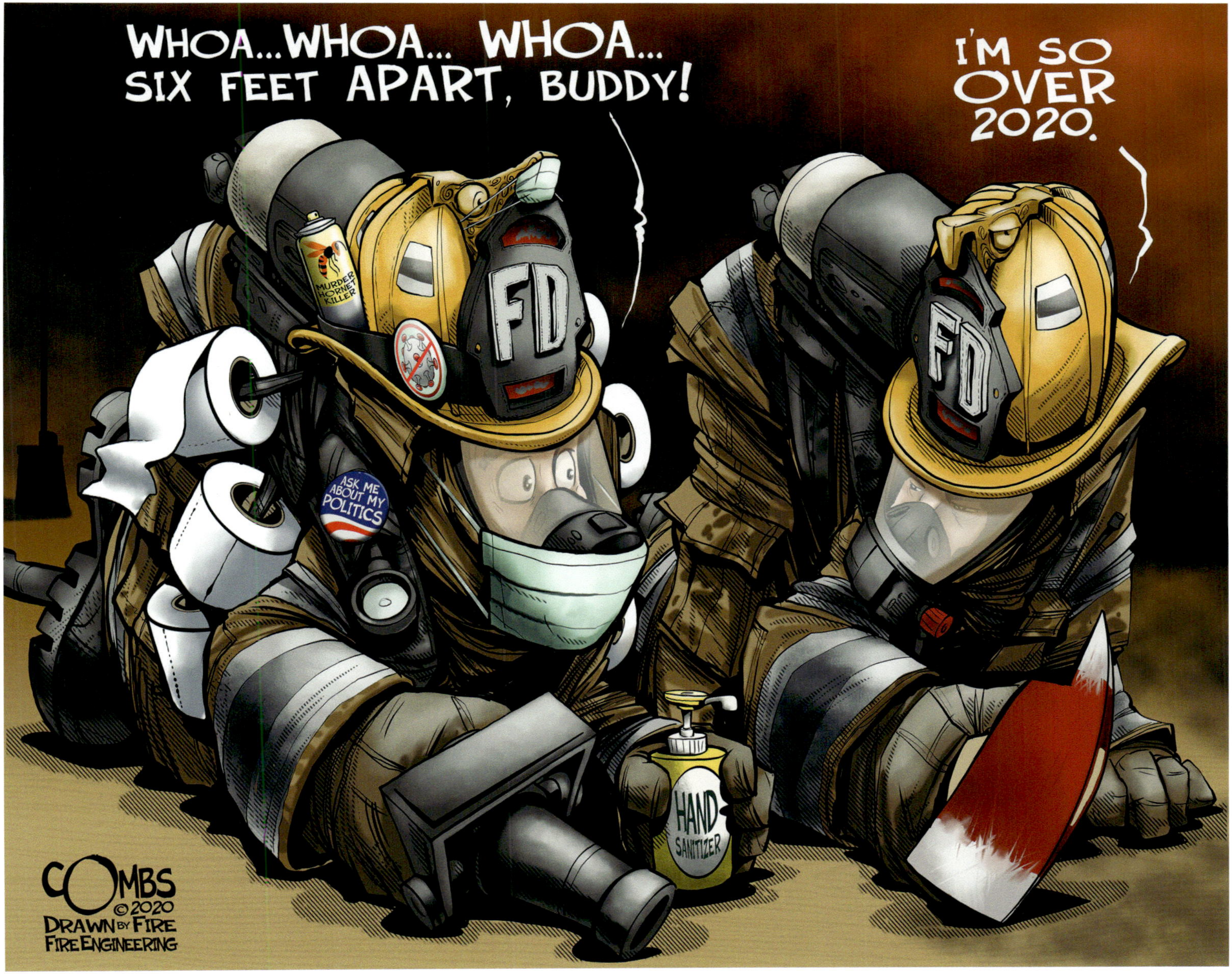
WHOA...WHOA... WHOA...
SIX FEET APART, BUDDY!
I'M SO
OVER
2020.
MURDER
HORNET
KILLER
FD
ASK ME
ABOUT MY
POLITICS
HAND
SANITIZER
COMBS
©2020
DRAWN BY FIRE
FIRE ENGINEERING

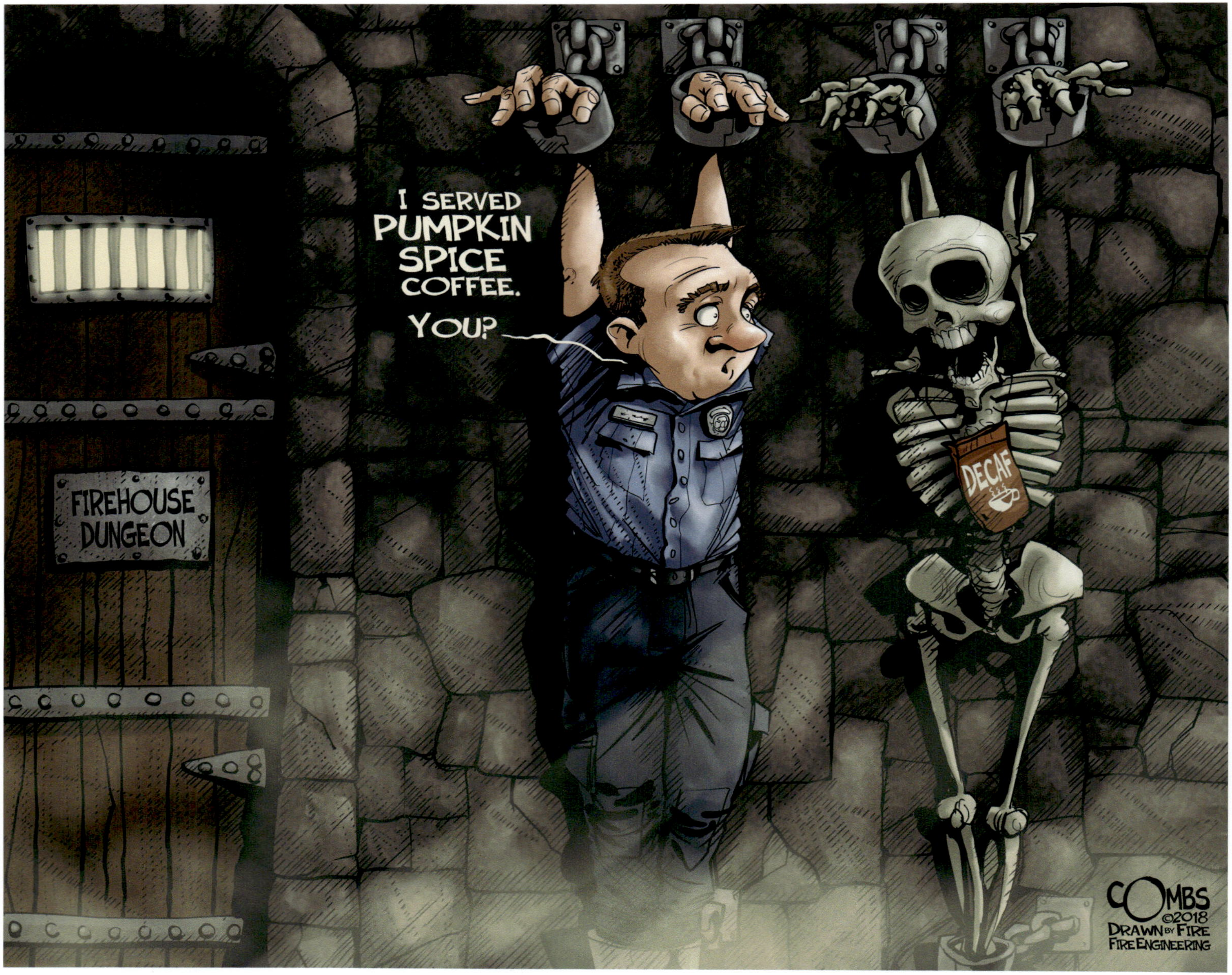
I SERVED PUMPKIN SPICE COFFEE.
YOU?
FIREHOUSE DUNGEON
DECAF
COMBS ©2018 DRAWN BY FIRE FIRE ENGINEERING

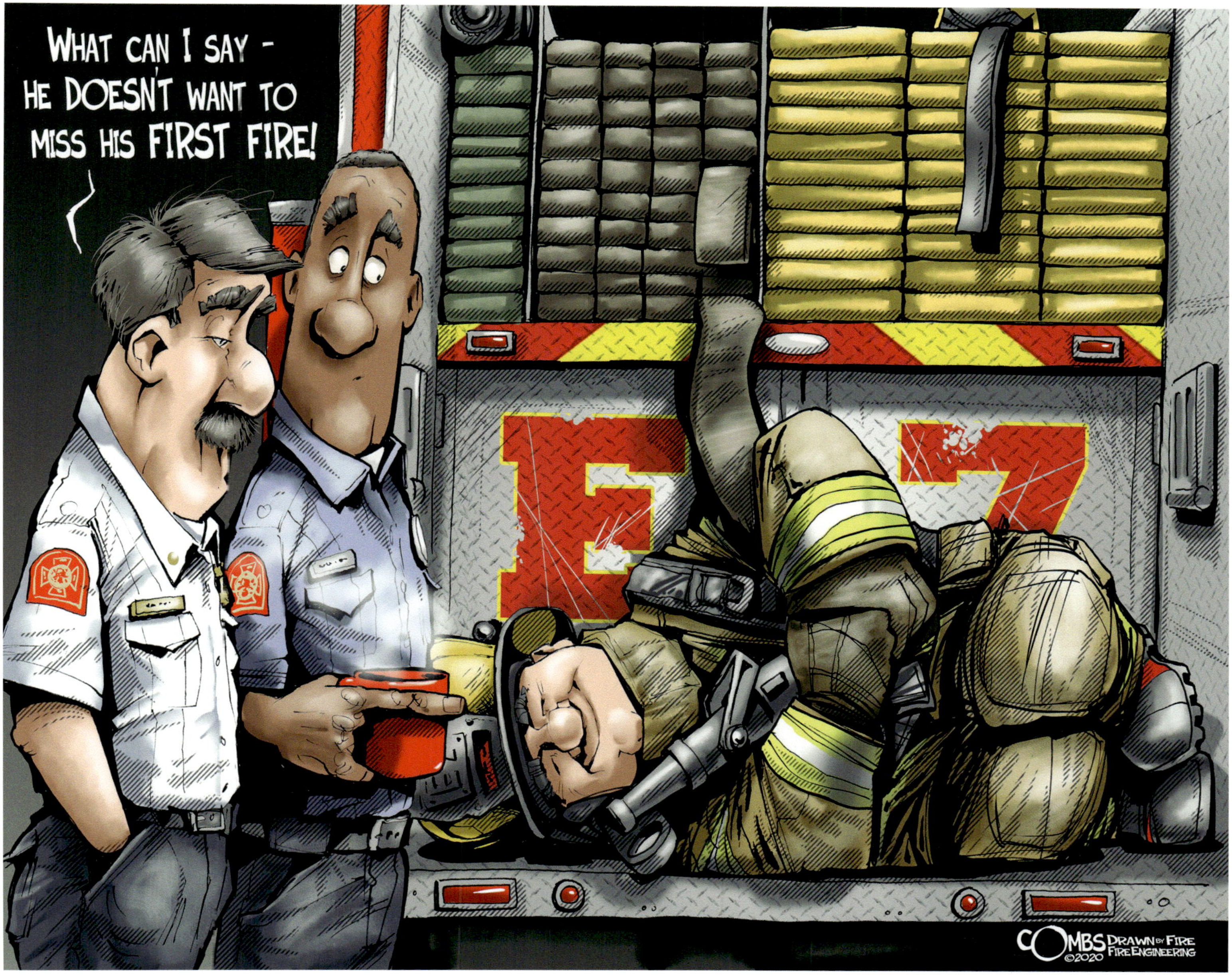
WHAT CAN I SAY - HE DOESN'T WANT TO MISS HIS FIRST FIRE!
COMBS DRAWN BY FIRE ©2020 FIRE ENGINEERING

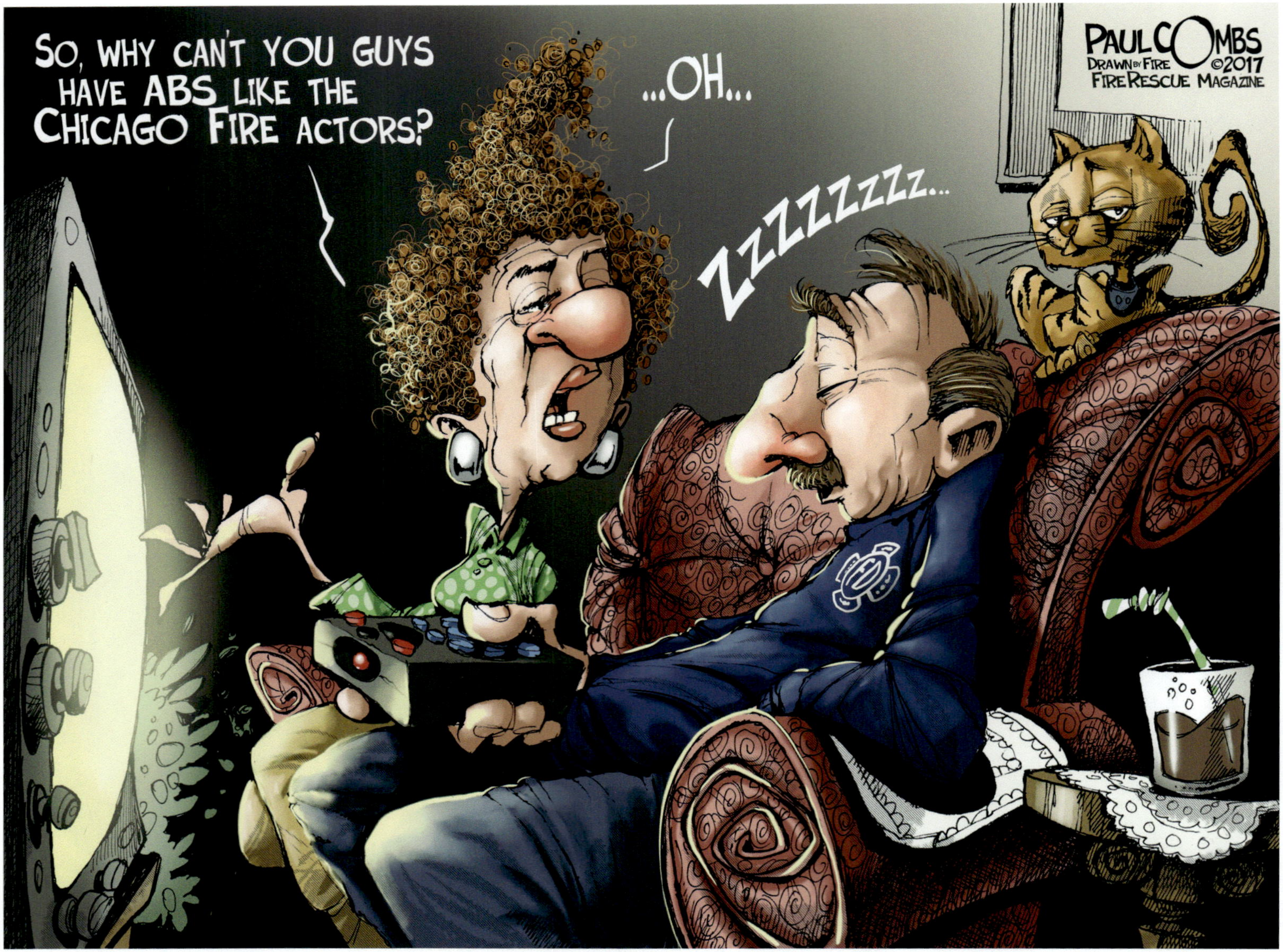

REALITY CHECK

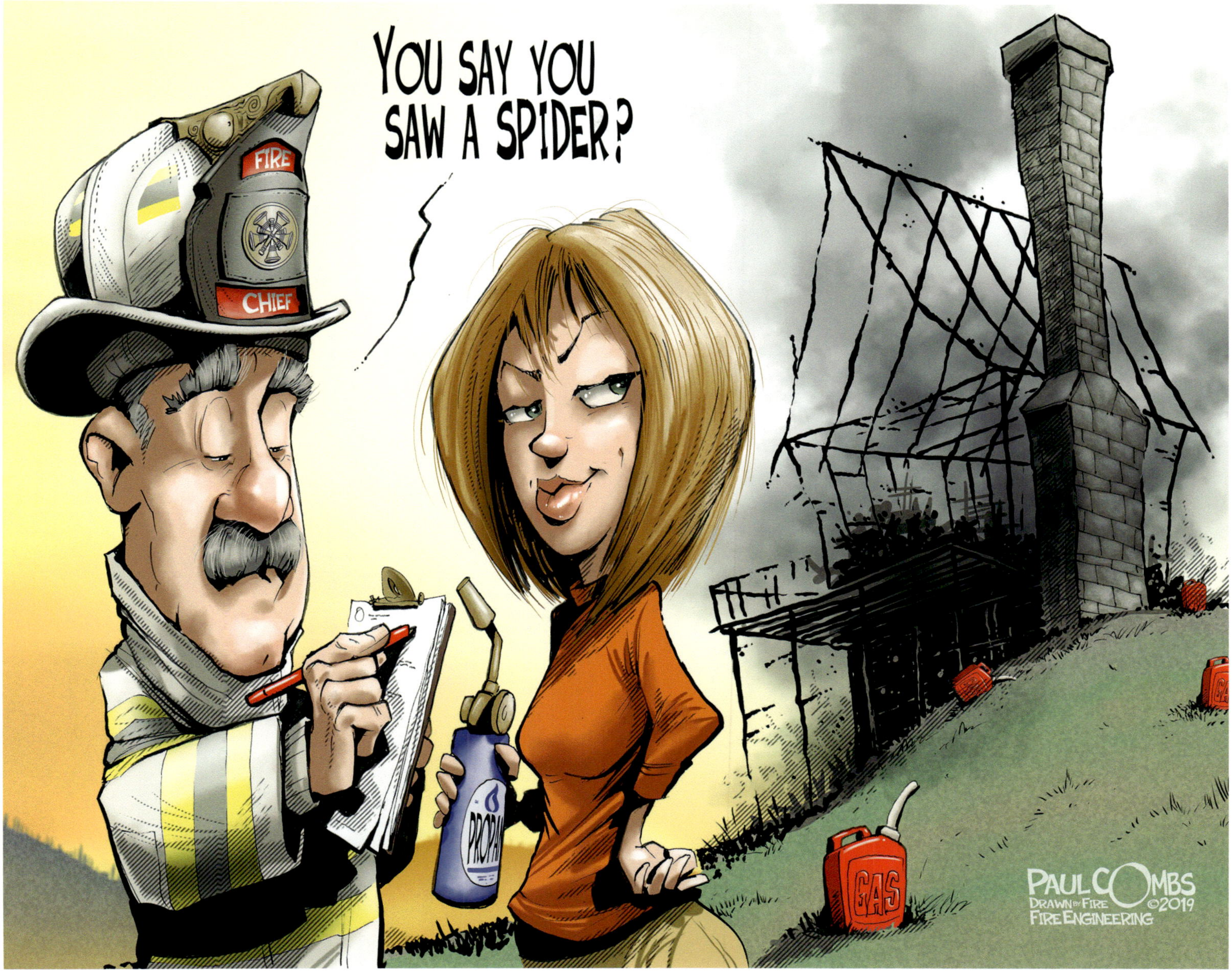

YOU SAY YOU SAW A SPIDER?
FIRE
CHIEF
PROPANE
GAS
PAUL COMBS
DRAWN BY FIRE ©2019
FIRE ENGINEERING

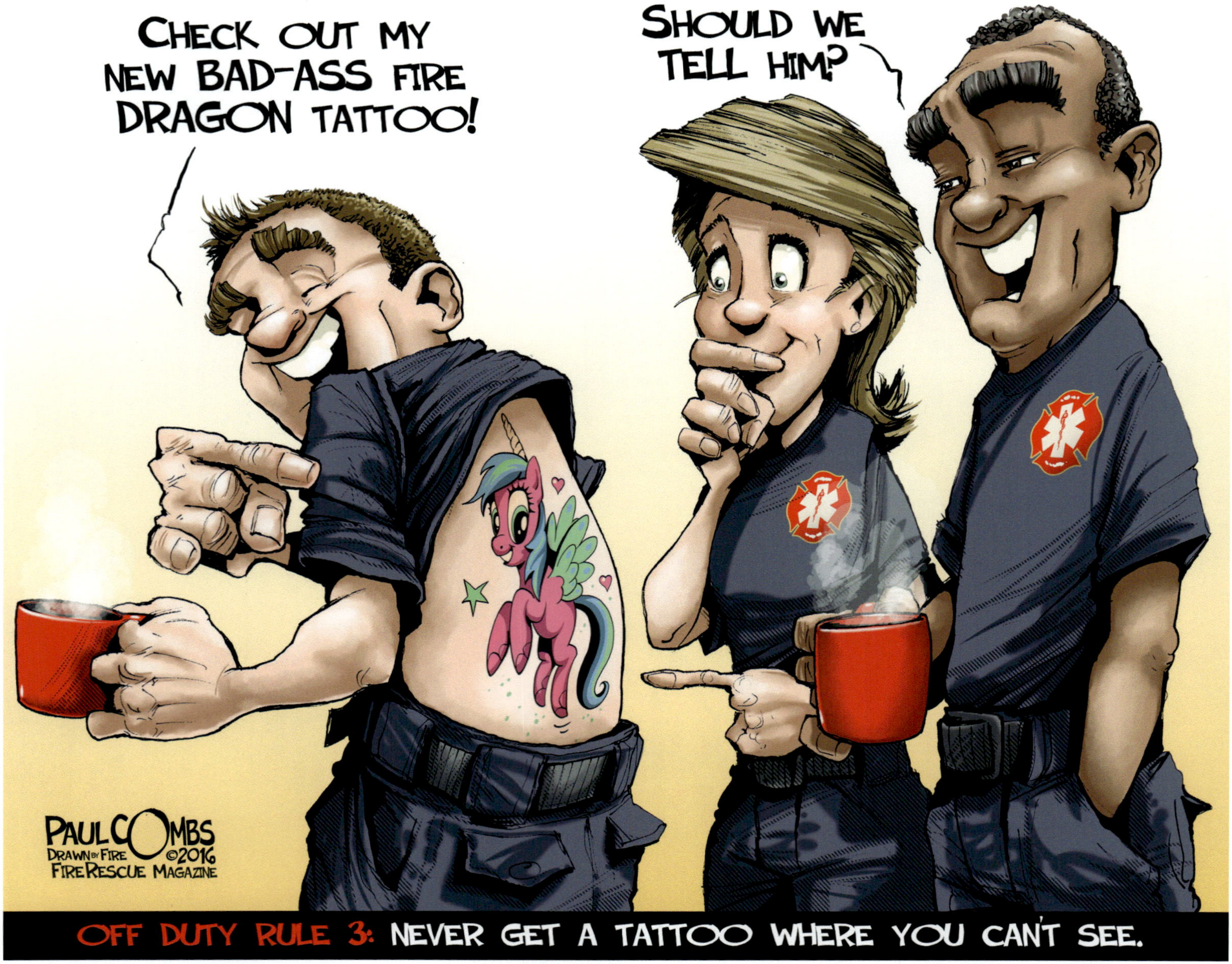
CHECK OUT MY NEW BAD-ASS FIRE DRAGON TATTOO!
SHOULD WE TELL HIM?
PAUL COMBS
DRAWN BY FIRE ©2016
FIRERESCUE MAGAZINE
OFF DUTY RULE 3: NEVER GET A TATTOO WHERE YOU CAN'T SEE.

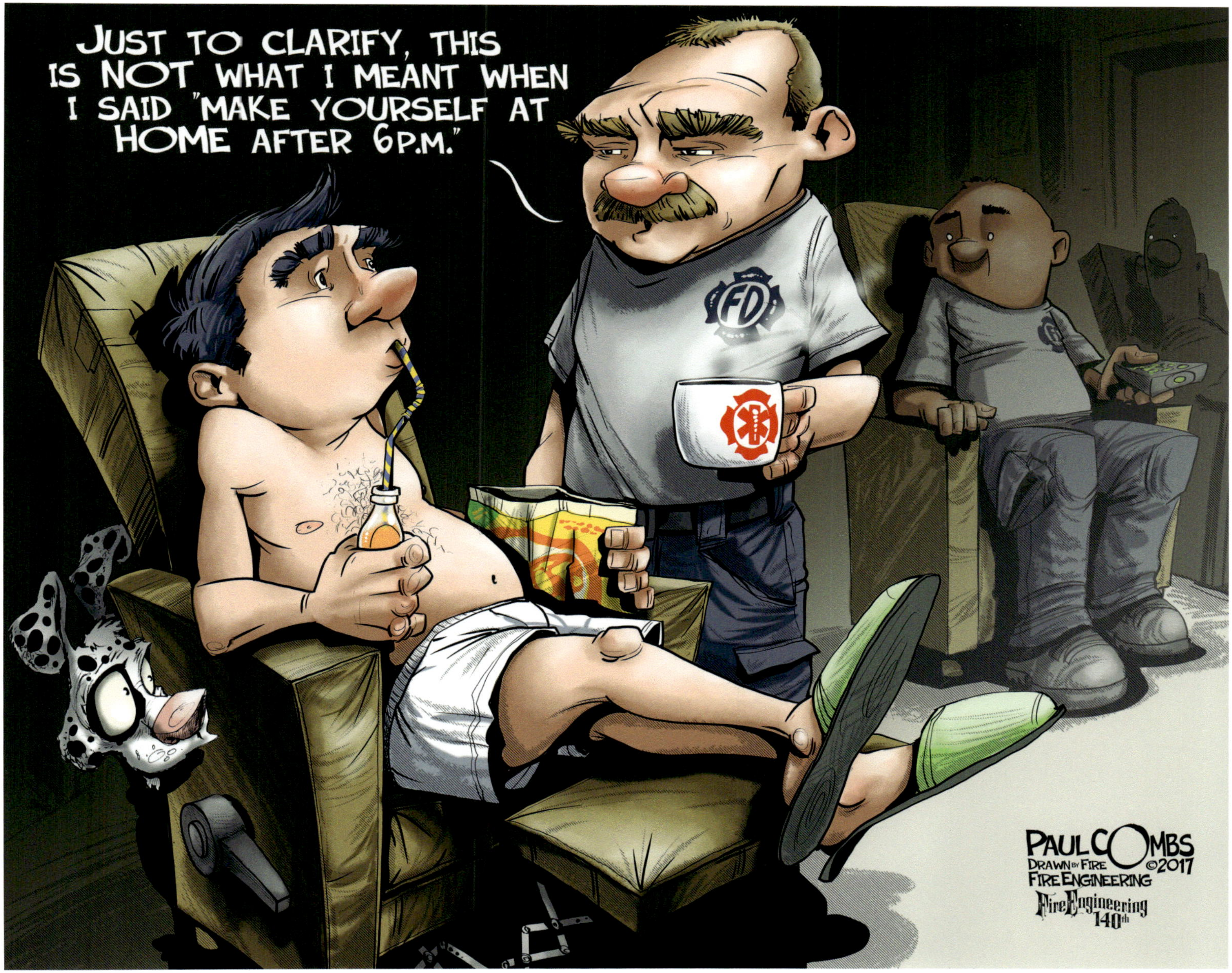
JUST TO CLARIFY, THIS IS NOT WHAT I MEANT WHEN I SAID "MAKE YOURSELF AT HOME AFTER 6P.M."
FD
PAUL COMBS
DRAWN BY FIRE ©2017
FIRE ENGINEERING
Fire Engineering 140th

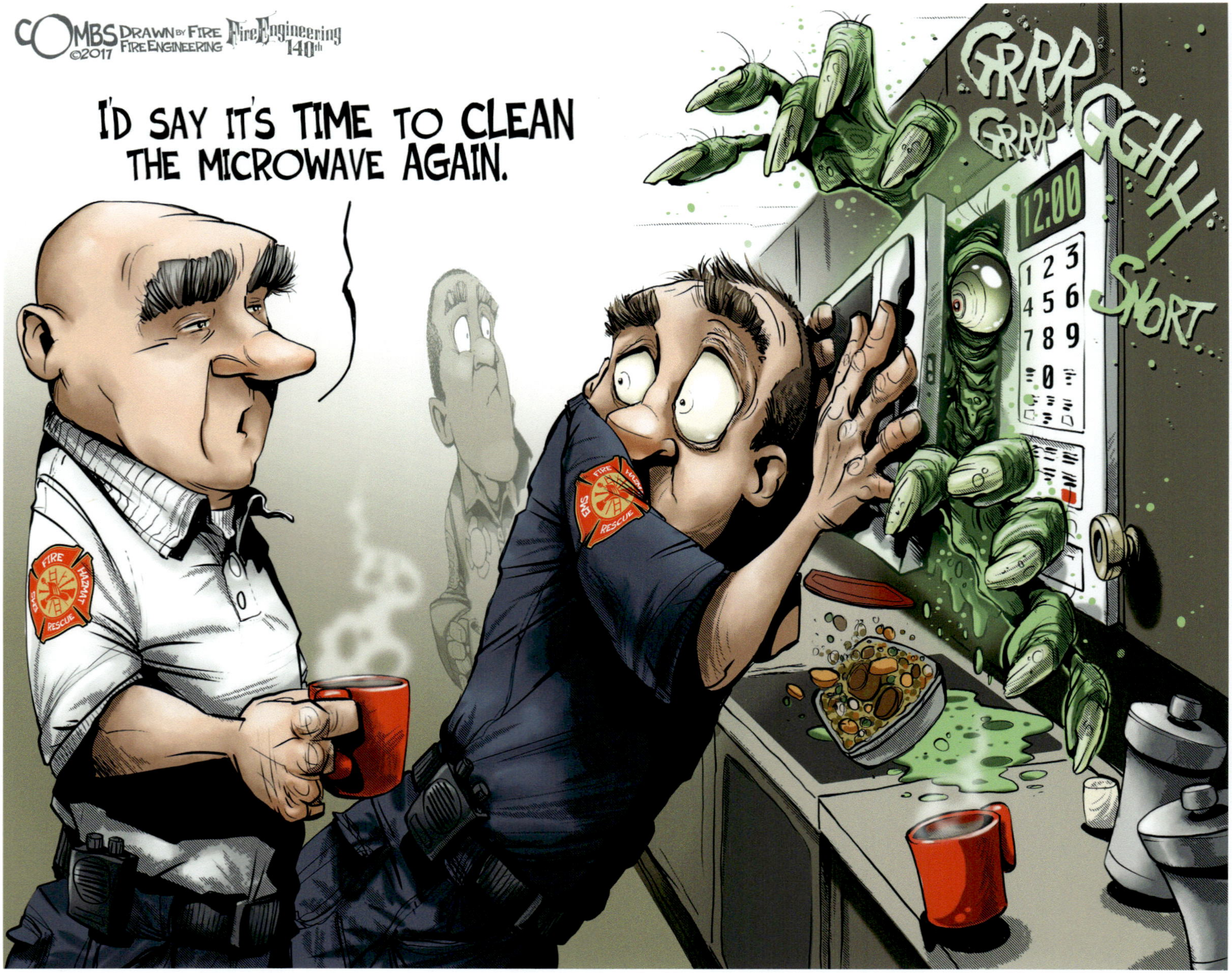
COMBS ©2017 DRAWN BY FIRE FIRE ENGINEERING Fire Engineering 140th
I'D SAY IT'S TIME TO CLEAN THE MICROWAVE AGAIN.
GRRRGGHHH
GRRR
SNORT
12:00

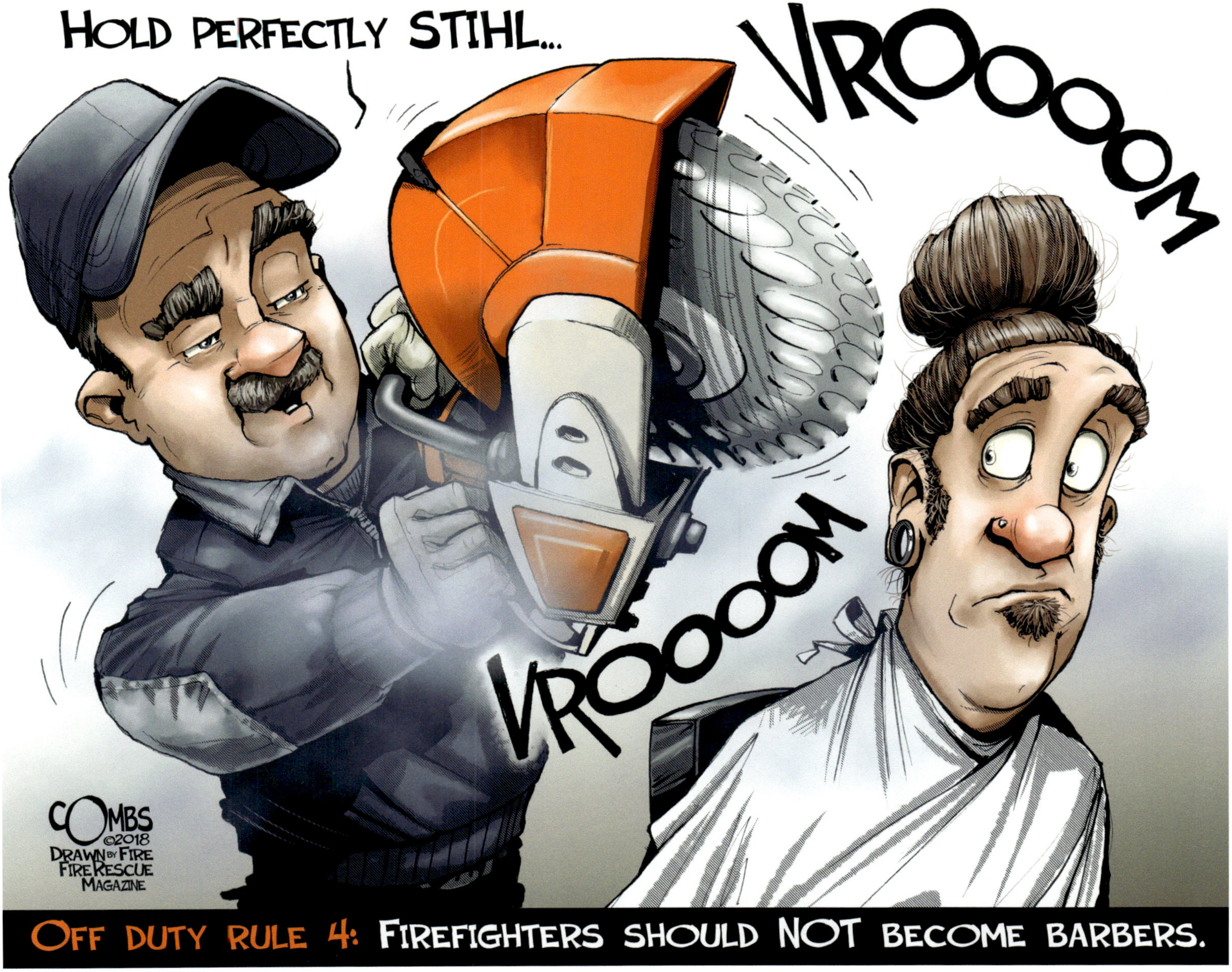

OFF DUTY RULE 4: FIREFIGHTERS SHOULD NOT BECOME BARBERS.

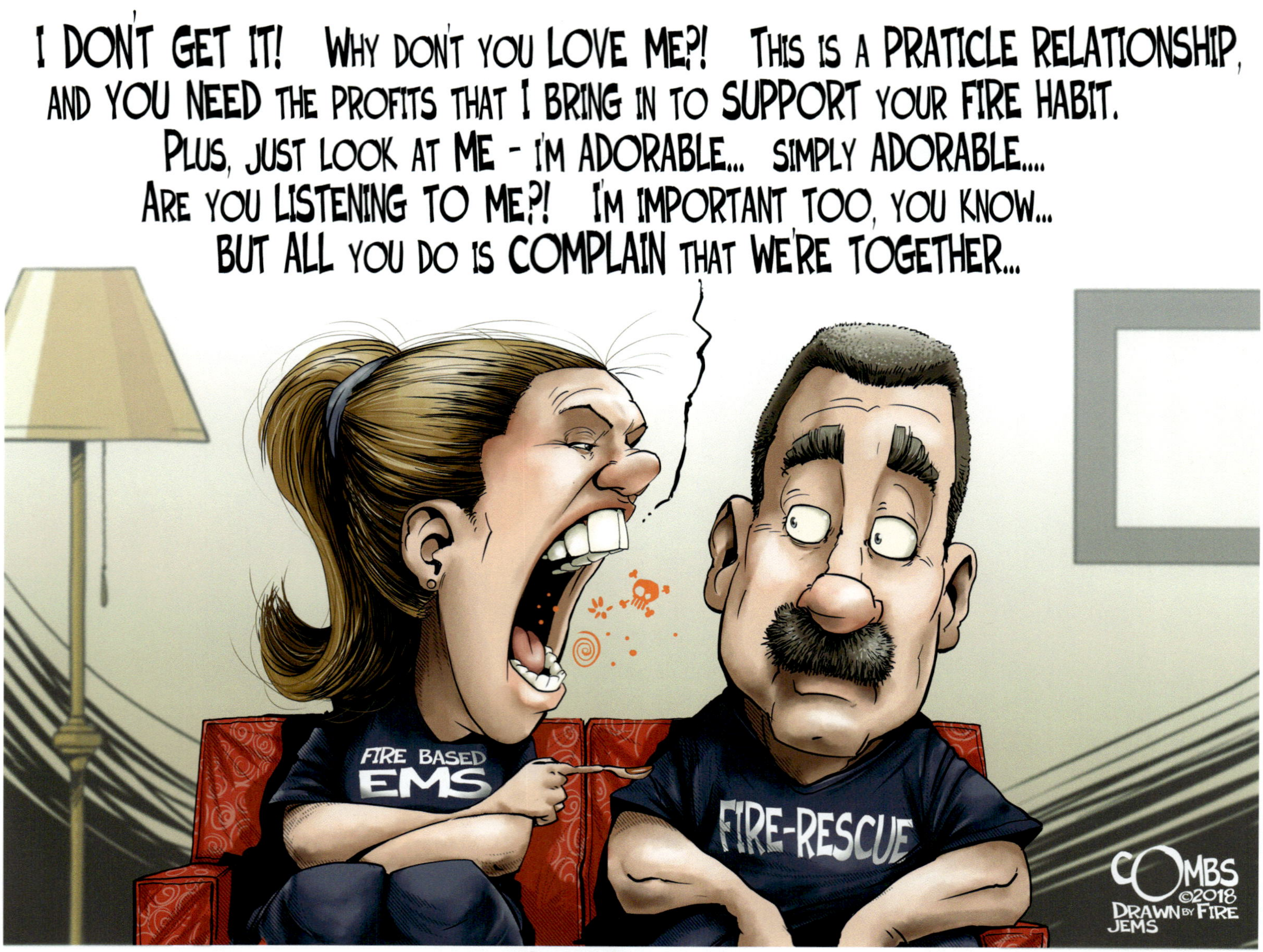

THE MARRIAGE

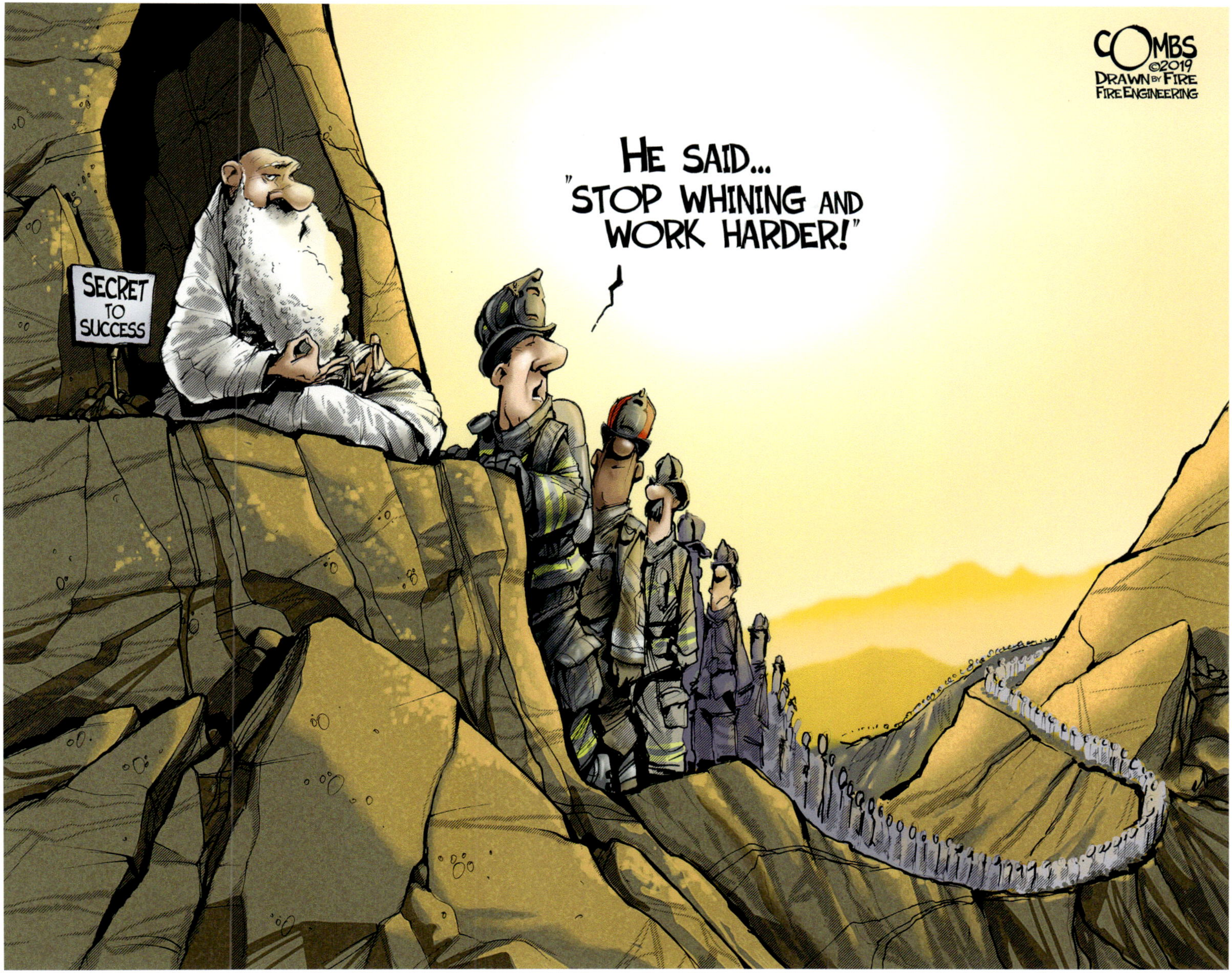
COMBS ©2019 DRAWN BY FIRE FIRE ENGINEERING
SECRET TO SUCCESS
HE SAID... "STOP WHINING AND WORK HARDER!"

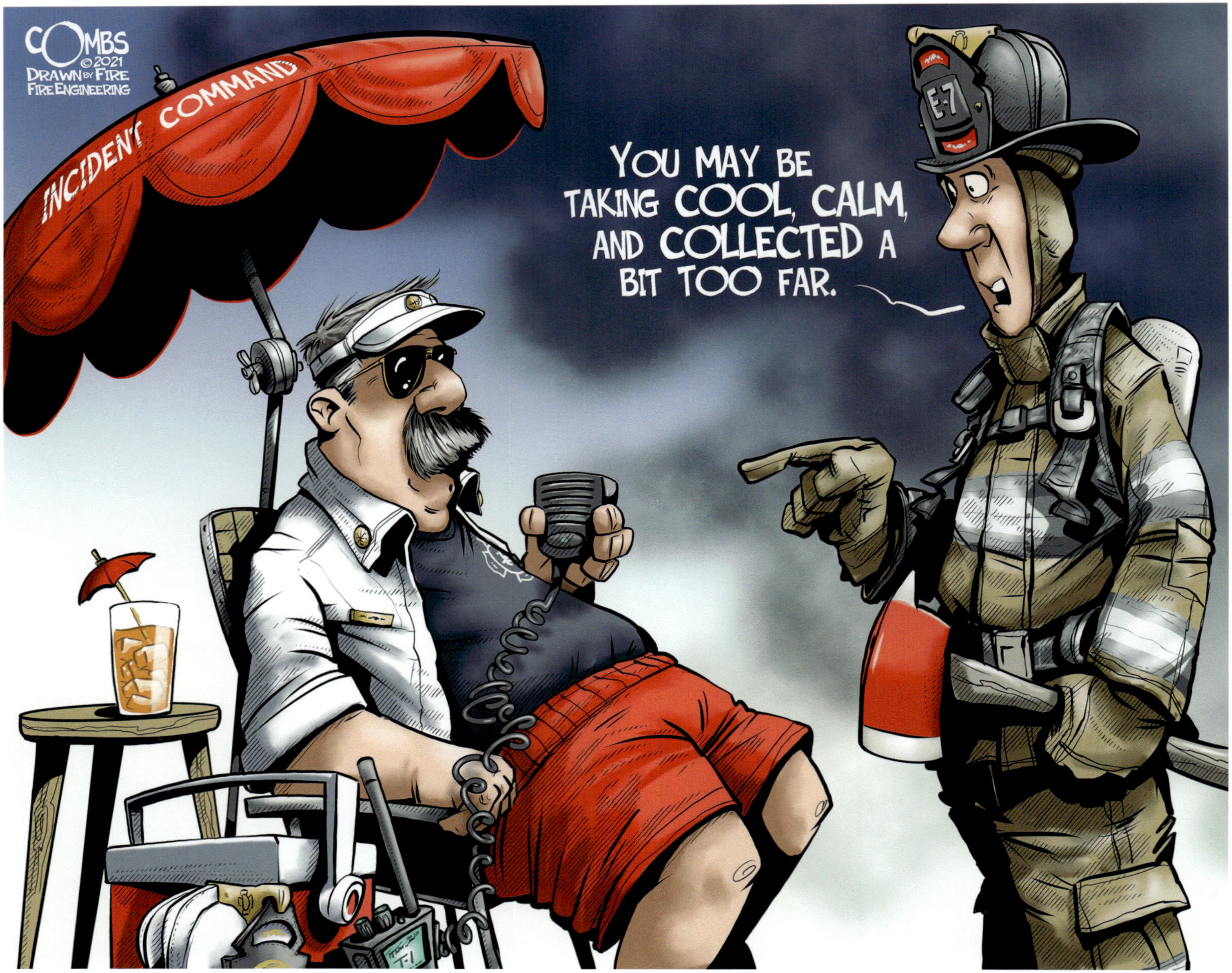
COMBS
©2021
DRAWN BY FIRE
FIRE ENGINEERING
INCIDENT COMMAND
YOU MAY BE TAKING COOL, CALM, AND COLLECTED A BIT TOO FAR.
E-7

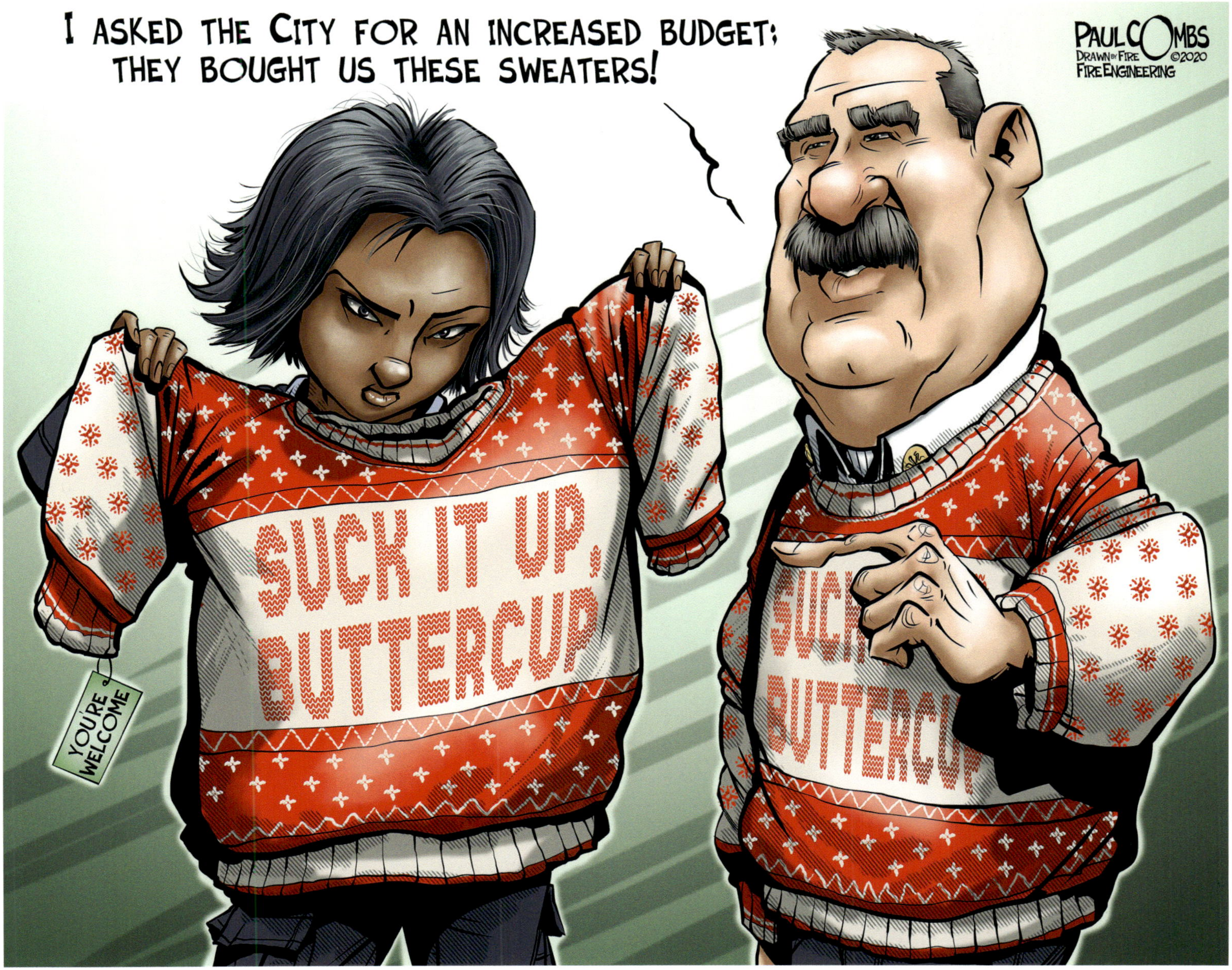
I ASKED THE CITY FOR AN INCREASED BUDGET; THEY BOUGHT US THESE SWEATERS!
PAUL COMBS
DRAWN BY FIRE ©2020
FIRE ENGINEERING
SUCK IT UP, BUTTERCUP
YOU'RE WELCOME

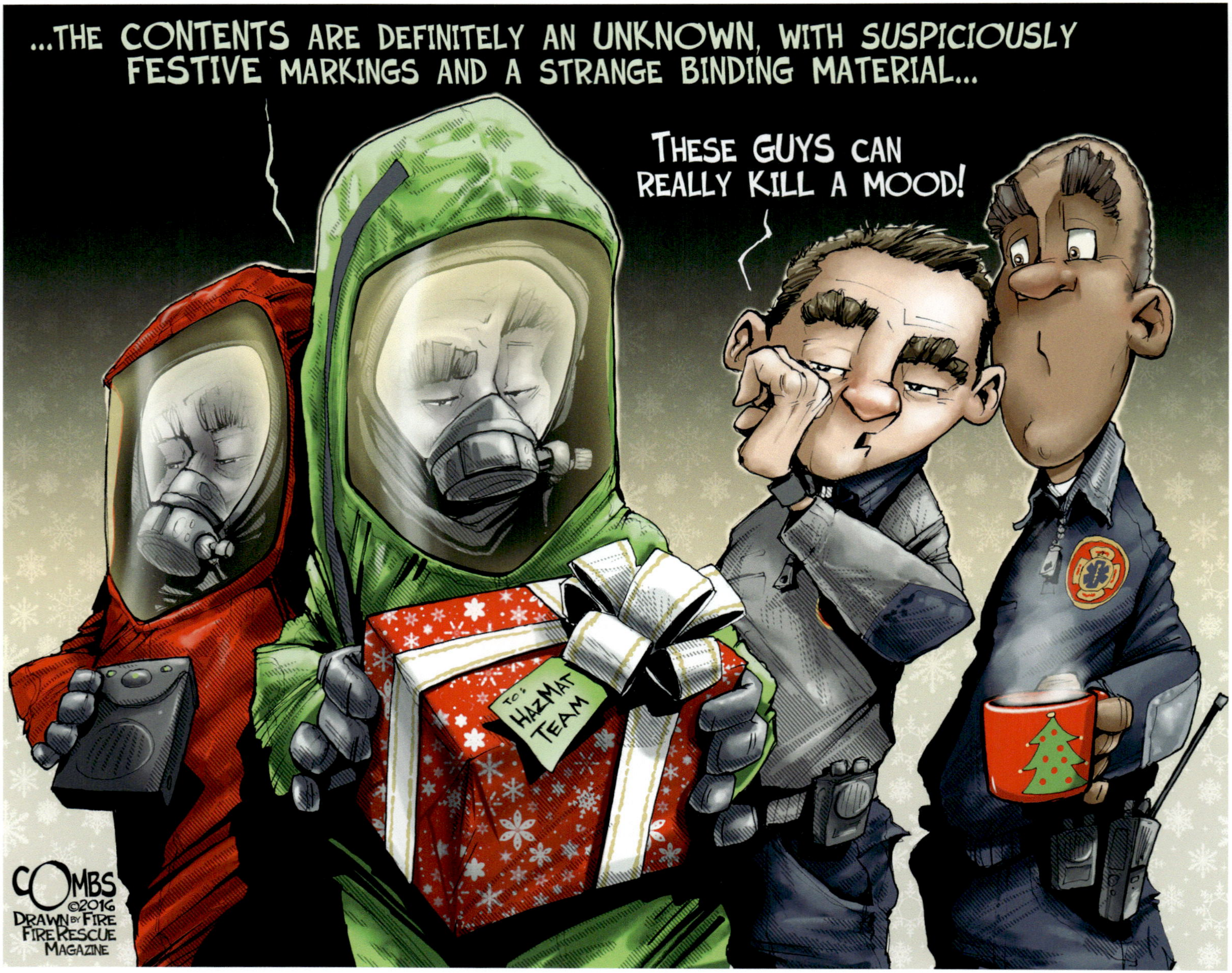
...THE CONTENTS ARE DEFINITELY AN UNKNOWN, WITH SUSPICIOUSLY FESTIVE MARKINGS AND A STRANGE BINDING MATERIAL...
THESE GUYS CAN REALLY KILL A MOOD!
TO: HAZMAT TEAM
COMBS ©2016 DRAWN BY FIRE FIRERESCUE MAGAZINE

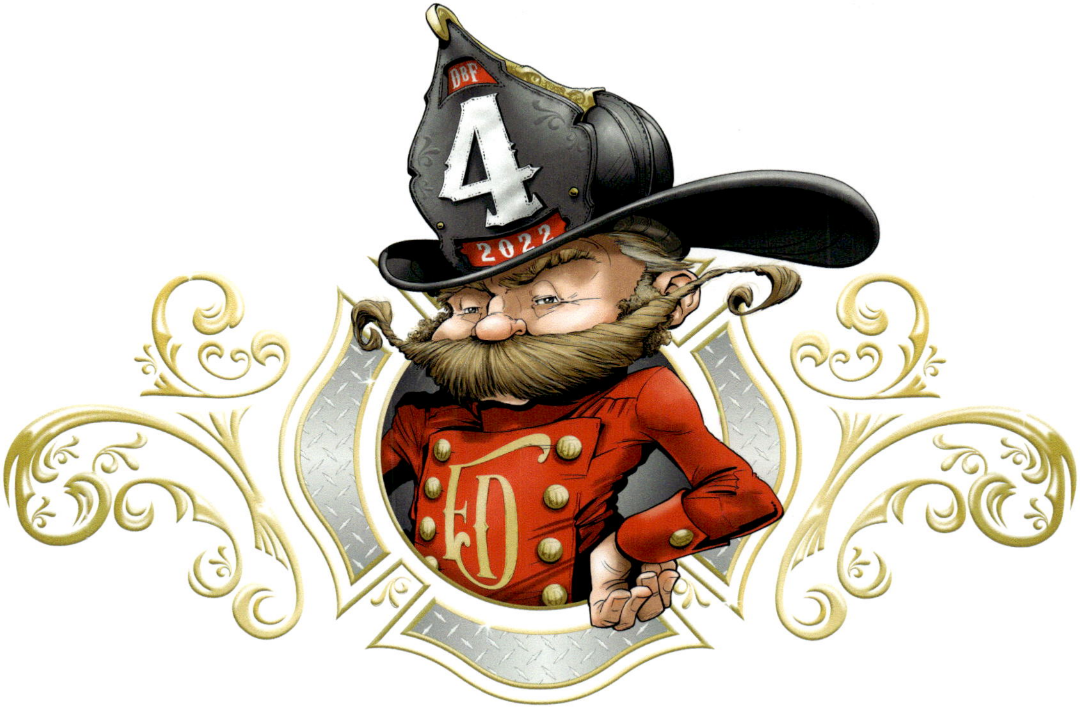

Mind, Body, Soul, and Fuzzy Slippers

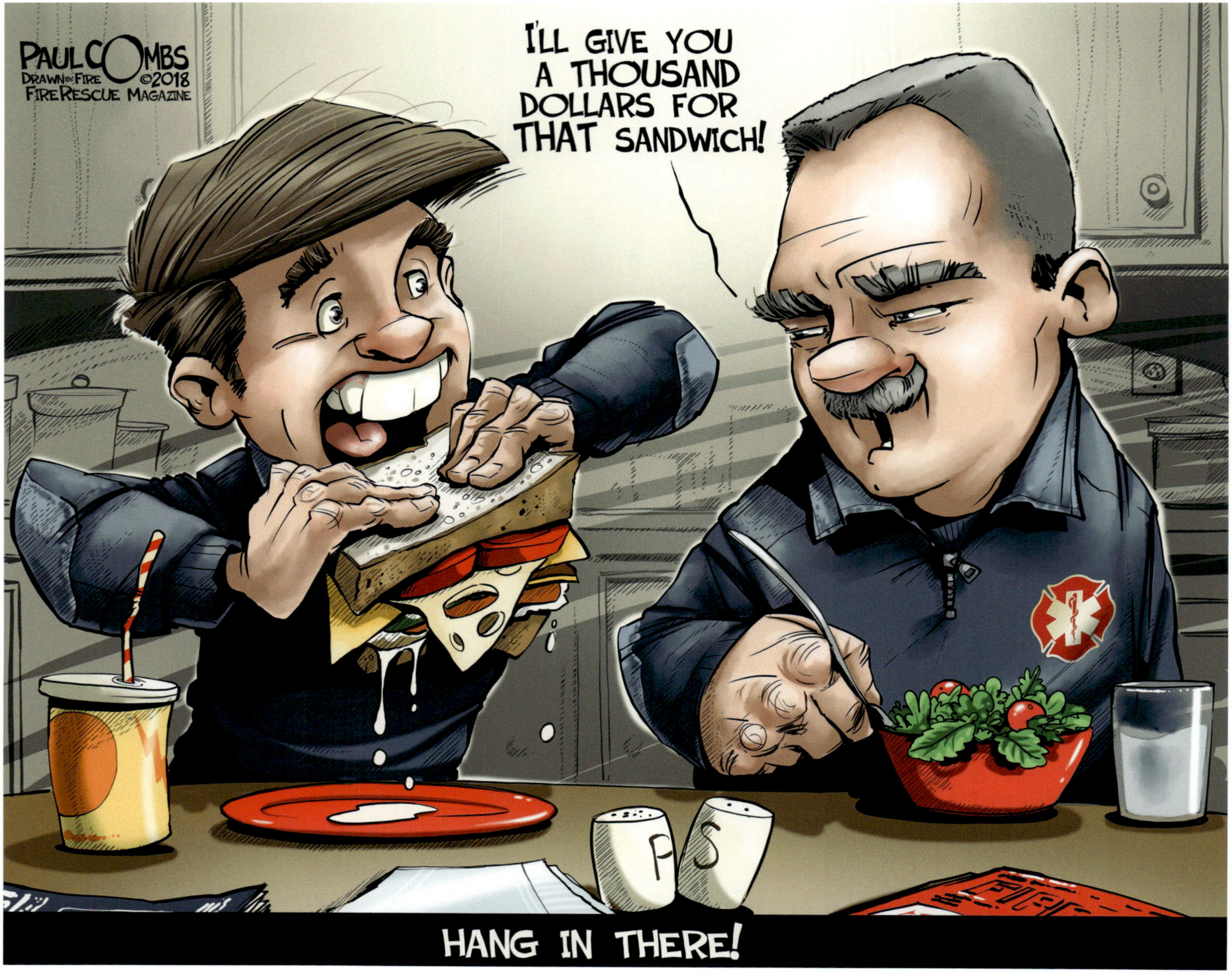
PAUL COMBS
DRAWN BY FIRE ©2018
FIRERESCUE MAGAZINE
I'LL GIVE YOU A THOUSAND DOLLARS FOR THAT SANDWICH!
HANG IN THERE!

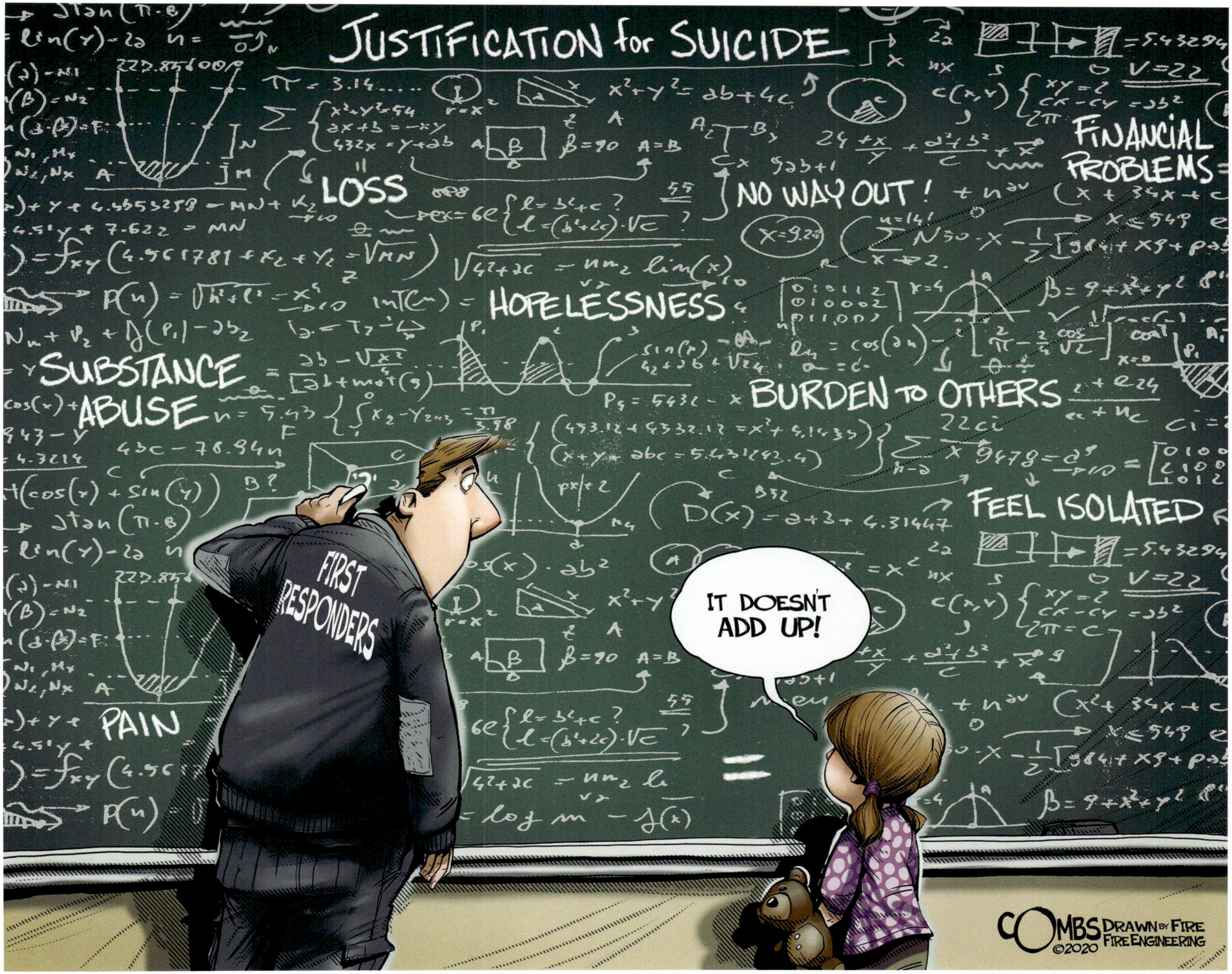
JUSTIFICATION for SUICIDE
FINANCIAL PROBLEMS
LOSS
NO WAY OUT!
HOPELESSNESS
SUBSTANCE ABUSE
BURDEN TO OTHERS
FEEL ISOLATED
PAIN
FIRST RESPONDERS
IT DOESN'T ADD UP!
COMBS ©2020 DRAWN BY FIRE FIRE ENGINEERING

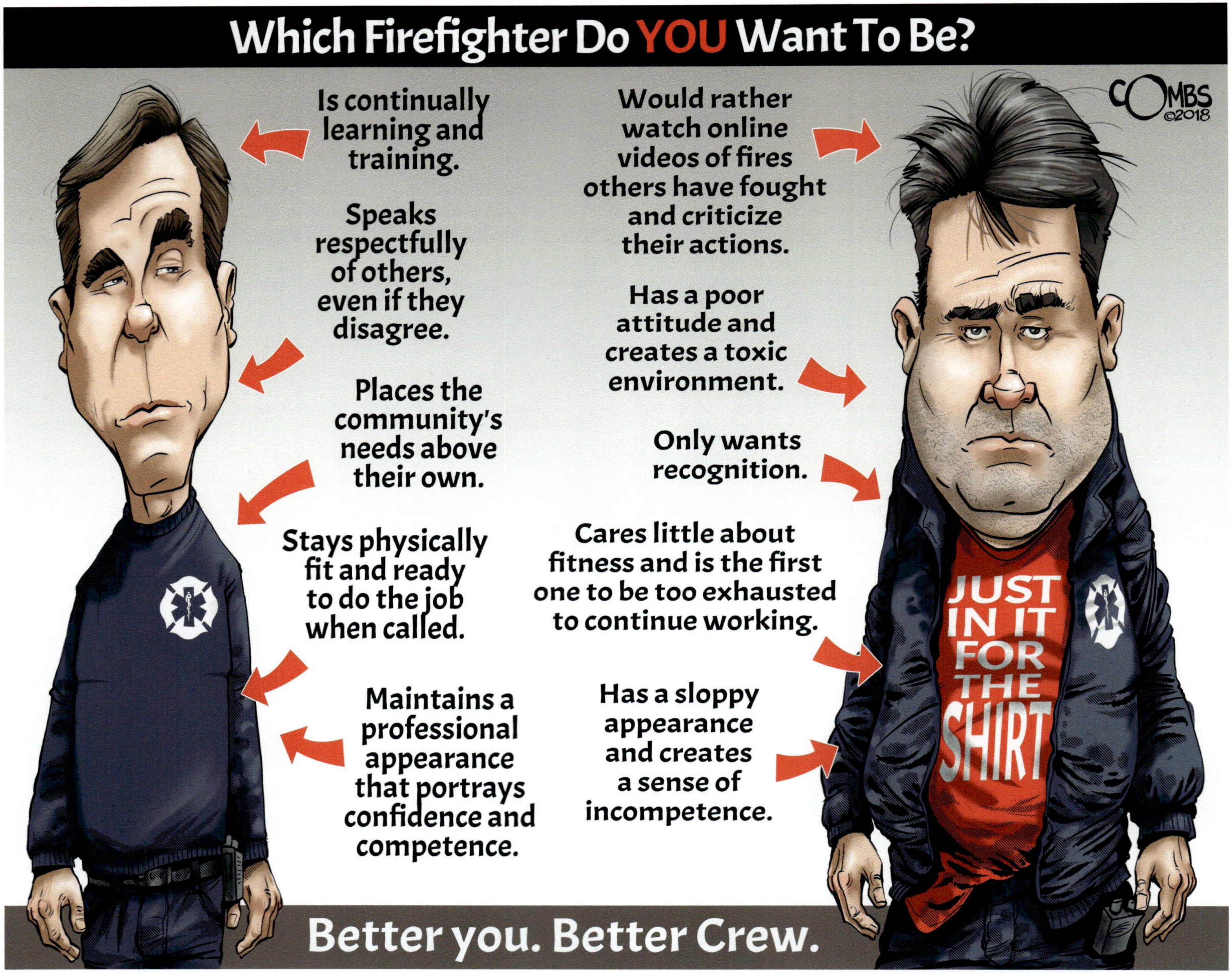
Which Firefighter Do YOU Want To Be?
COMBS ©2018
Is continually learning and training.
Speaks respectfully of others, even if they disagree.
Places the community's needs above their own.
Stays physically fit and ready to do the job when called.
Maintains a professional appearance that portrays confidence and competence.
Would rather watch online videos of fires others have fought and criticize their actions.
Has a poor attitude and creates a toxic environment.
Only wants recognition.
Cares little about fitness and is the first one to be too exhausted to continue working.
Has a sloppy appearance and creates a sense of incompetence.
JUST IN IT FOR THE SHIRT
Better you. Better Crew.

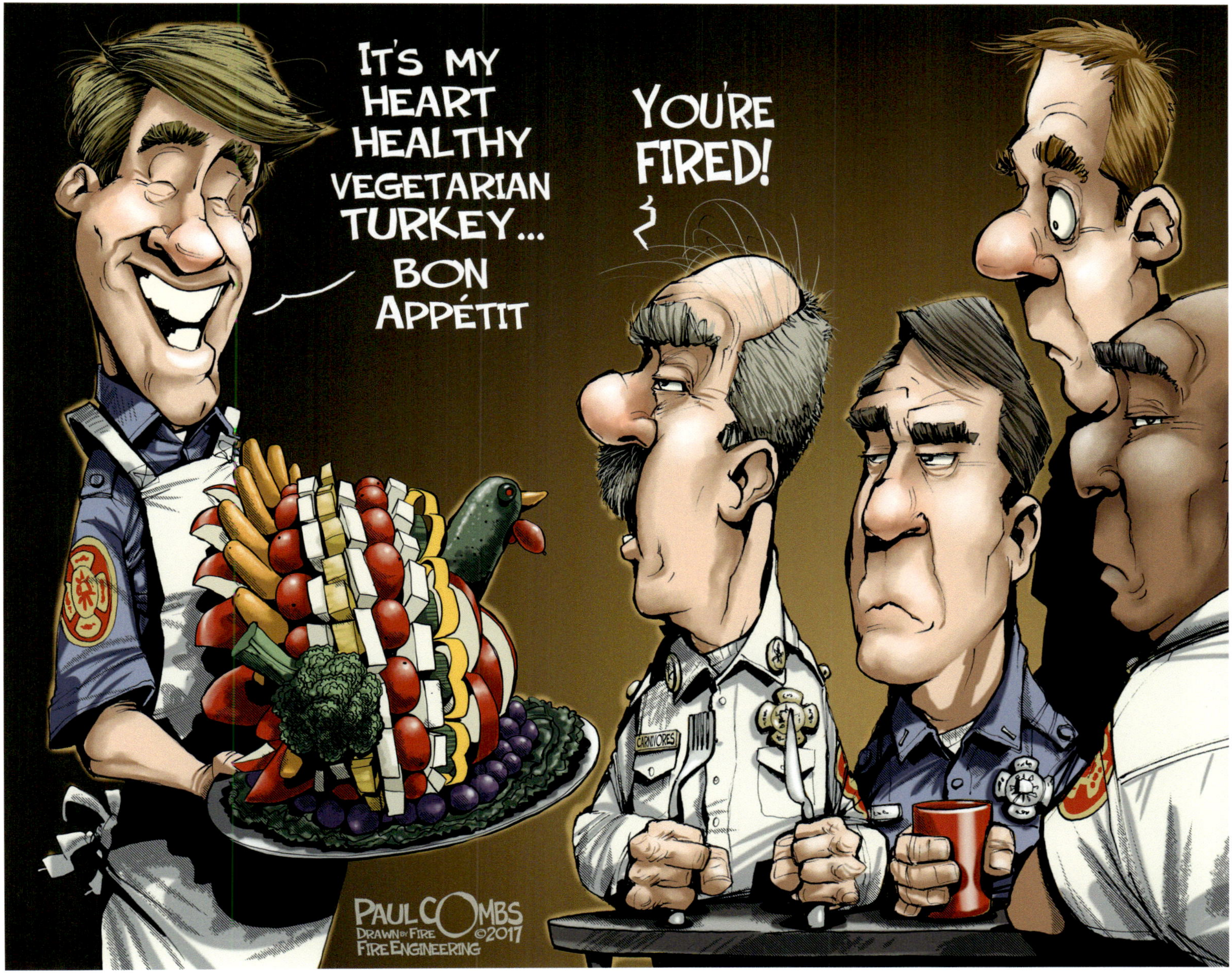
IT'S MY HEART HEALTHY VEGETARIAN TURKEY... BON APPÉTIT
YOU'RE FIRED!
CARNIVORES
PAUL COMBS
DRAWN BY FIRE ©2017
FIRE ENGINEERING

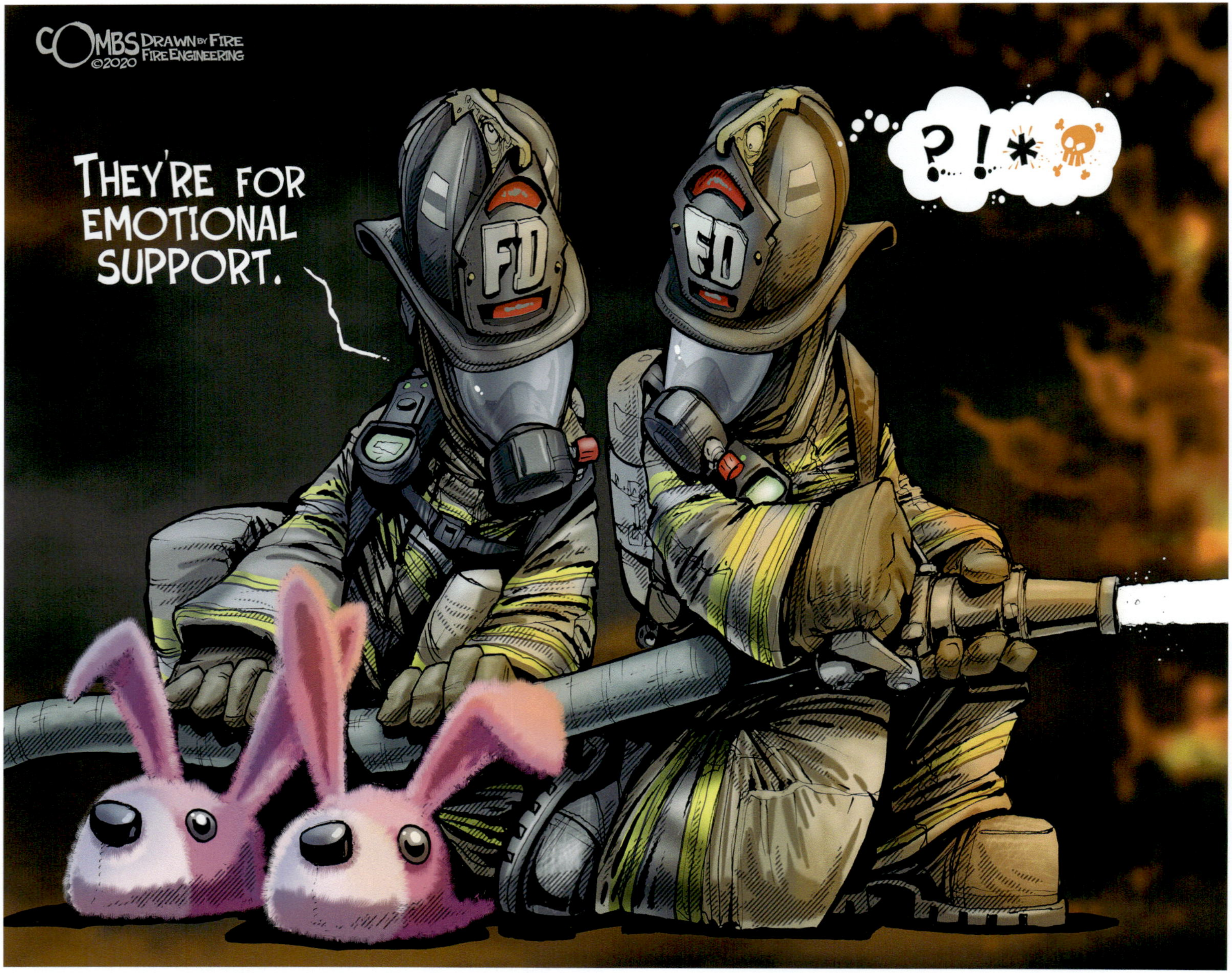
COMBS ©2020 DRAWN BY FIRE FIRE ENGINEERING
THEY'RE FOR EMOTIONAL SUPPORT.
?! *

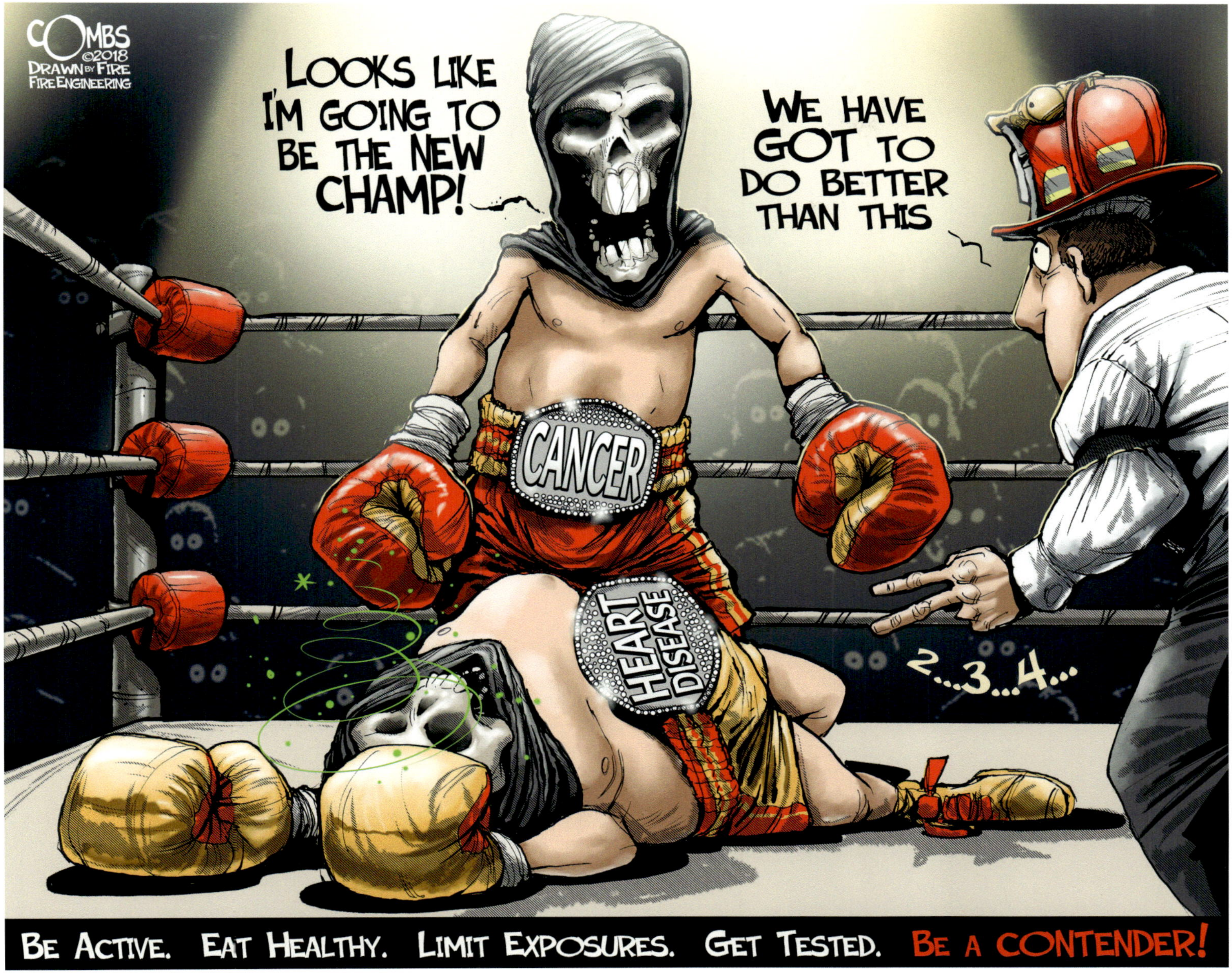
COMBS ©2018 DRAWN BY FIRE FIRE ENGINEERING
LOOKS LIKE I'M GOING TO BE THE NEW CHAMP!
WE HAVE GOT TO DO BETTER THAN THIS
CANCER
HEART DISEASE
2...3...4...
BE ACTIVE. EAT HEALTHY. LIMIT EXPOSURES. GET TESTED. BE A CONTENDER!

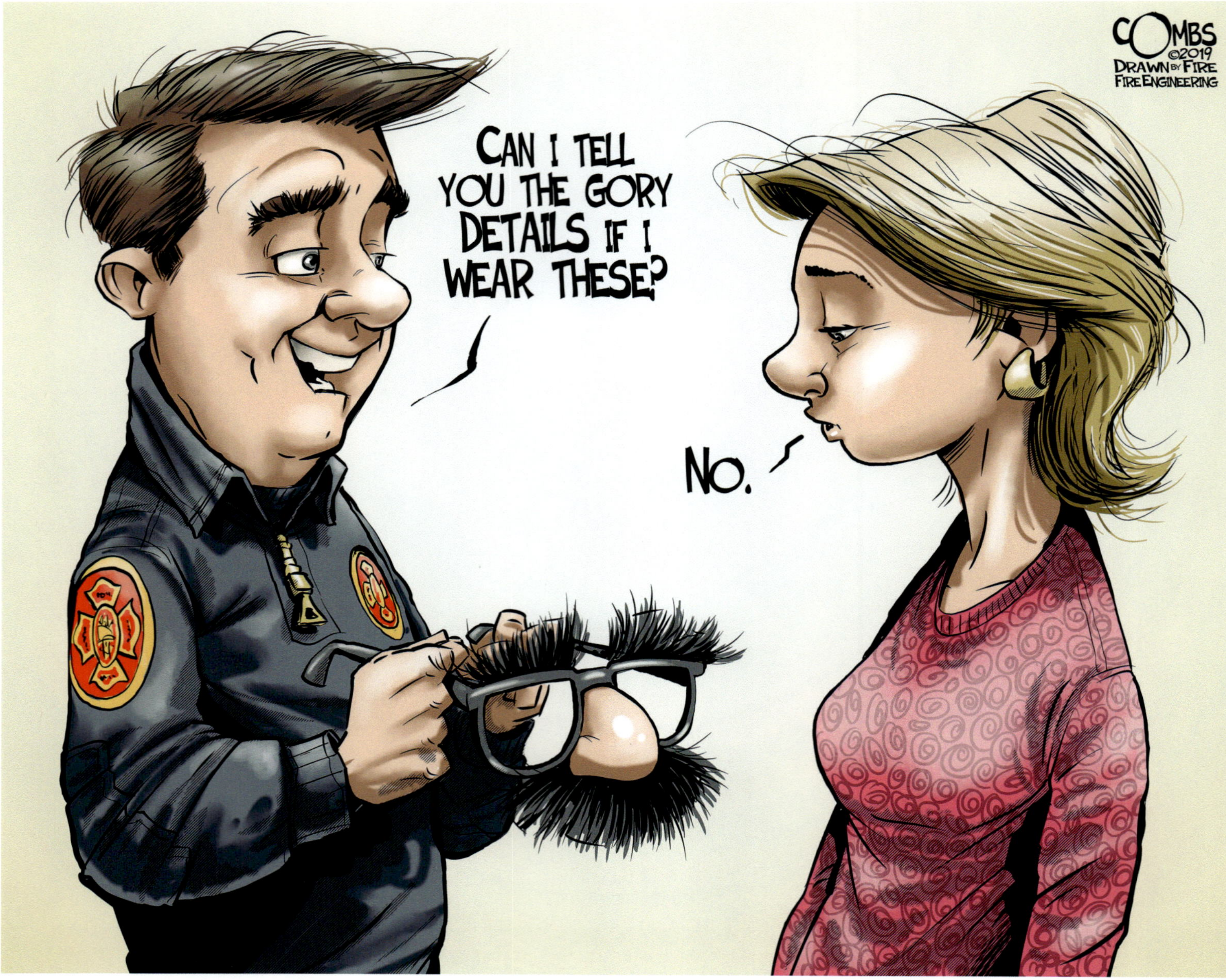
COMBS
©2019
DRAWN BY FIRE
FIRE ENGINEERING
CAN I TELL YOU THE GORY DETAILS IF I WEAR THESE?
NO.

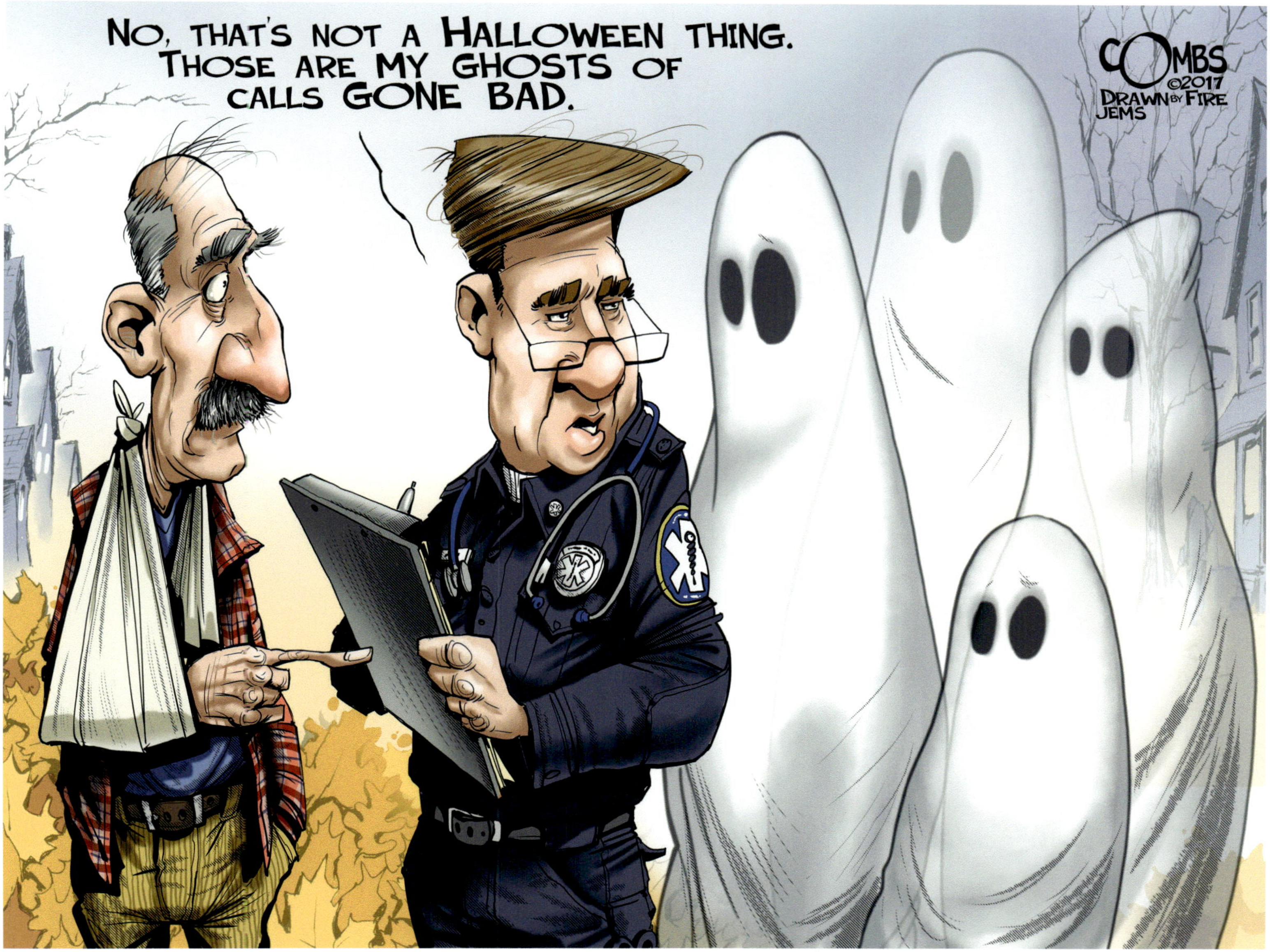

HOW DO YOU DEAL WITH YOUR GHOSTS?

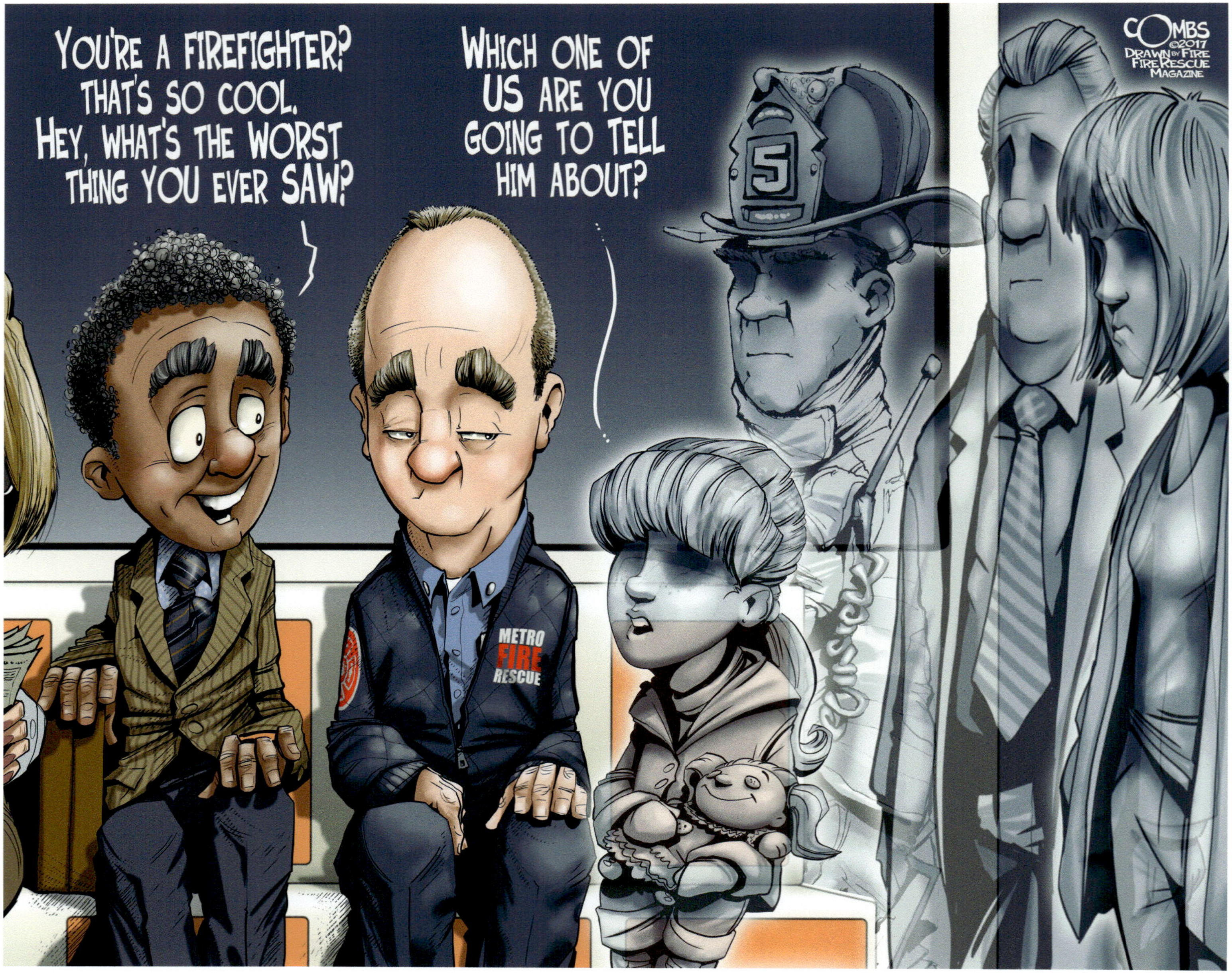
YOU'RE A FIREFIGHTER? THAT'S SO COOL. HEY, WHAT'S THE WORST THING YOU EVER SAW?
WHICH ONE OF US ARE YOU GOING TO TELL HIM ABOUT?
METRO FIRE RESCUE
COMBS ©2017 DRAWN BY FIRE FIRERESCUE MAGAZINE

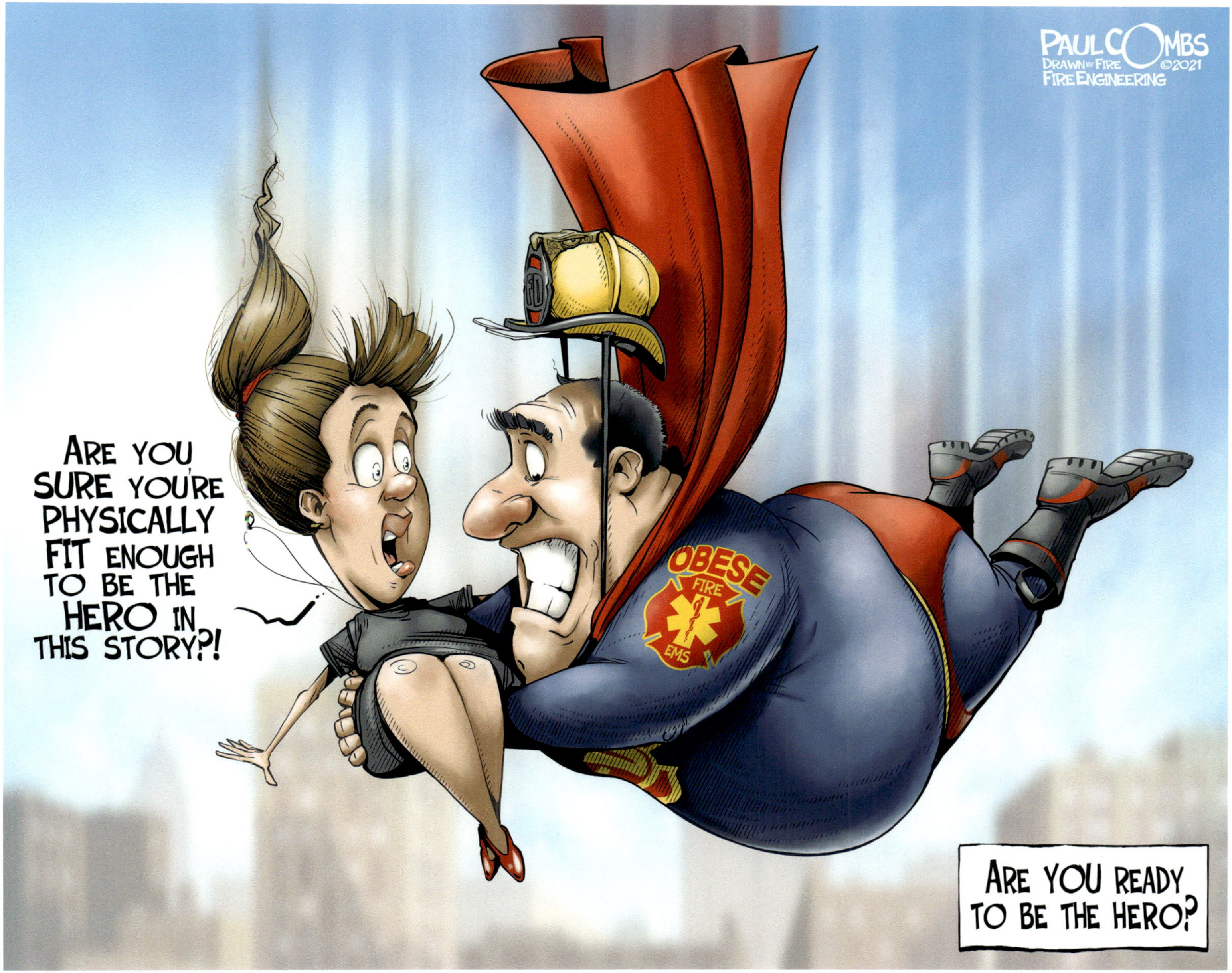
PAUL COMBS
DRAWN BY FIRE ©2021
FIRE ENGINEERING
ARE YOU SURE YOU'RE PHYSICALLY FIT ENOUGH TO BE THE HERO IN THIS STORY?!
OBESE FIRE EMS
ARE YOU READY TO BE THE HERO?

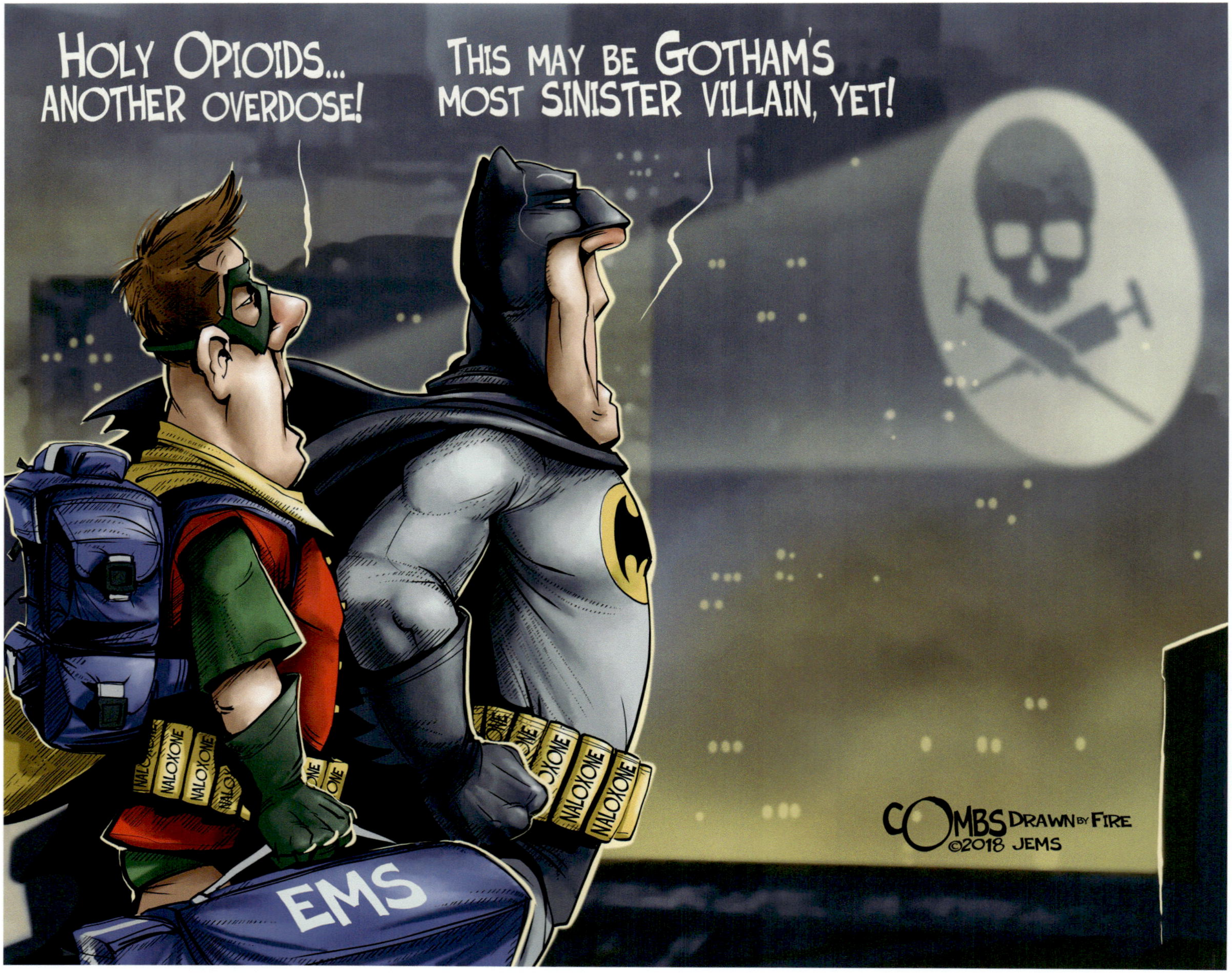
HOLY OPIOIDS... ANOTHER OVERDOSE!
THIS MAY BE GOTHAM'S MOST SINISTER VILLAIN, YET!
NALOXONE
EMS
COMBS DRAWN BY FIRE
©2018 JEMS

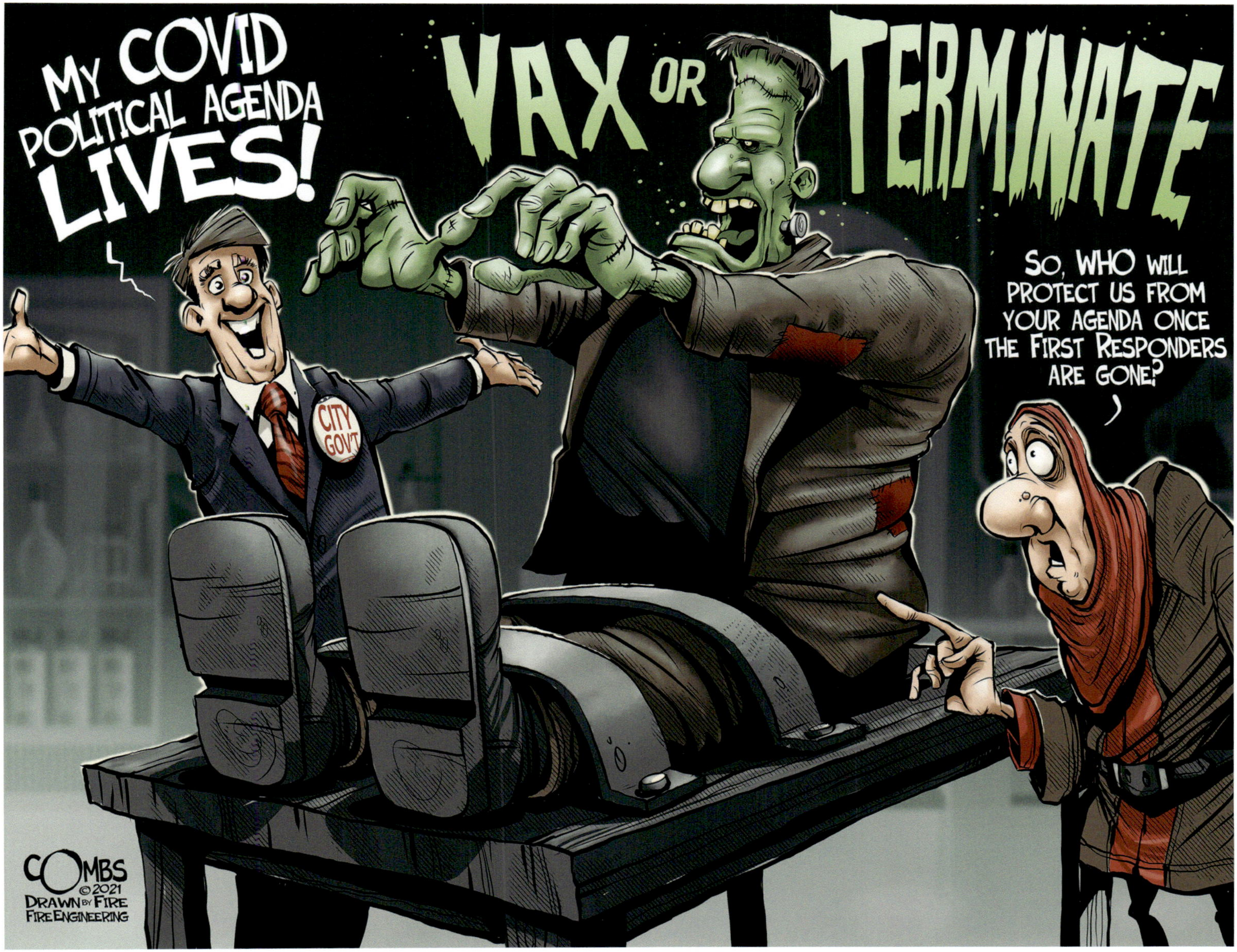
VAX OR TERMINATE
MY COVID POLITICAL AGENDA LIVES!
CITY GOV'T
SO, WHO WILL PROTECT US FROM YOUR AGENDA ONCE THE FIRST RESPONDERS ARE GONE?
COMBS ©2021 DRAWN BY FIRE FIREENGINEERING

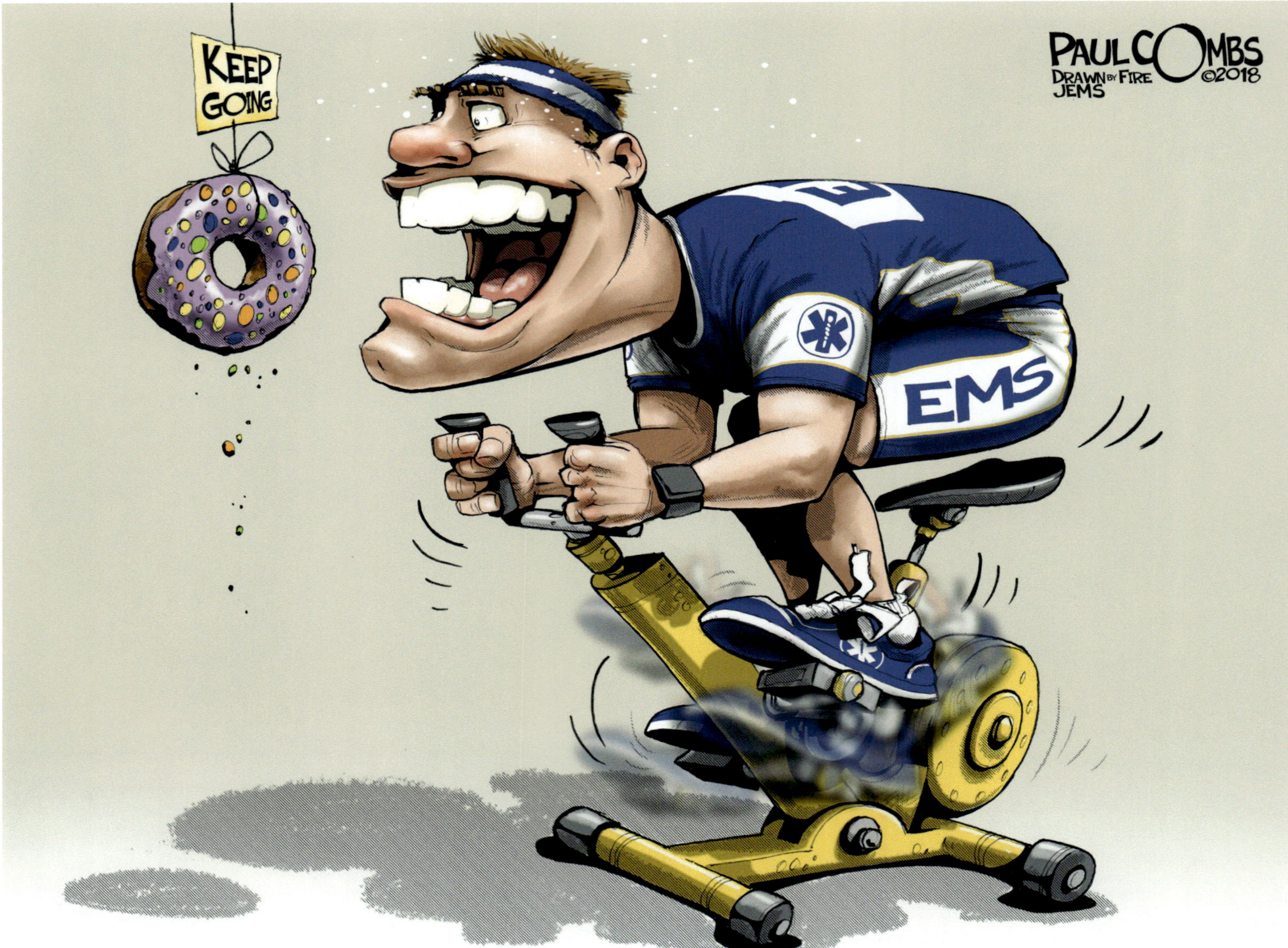

FIND MOTIVATION WHEREVER YOU CAN!

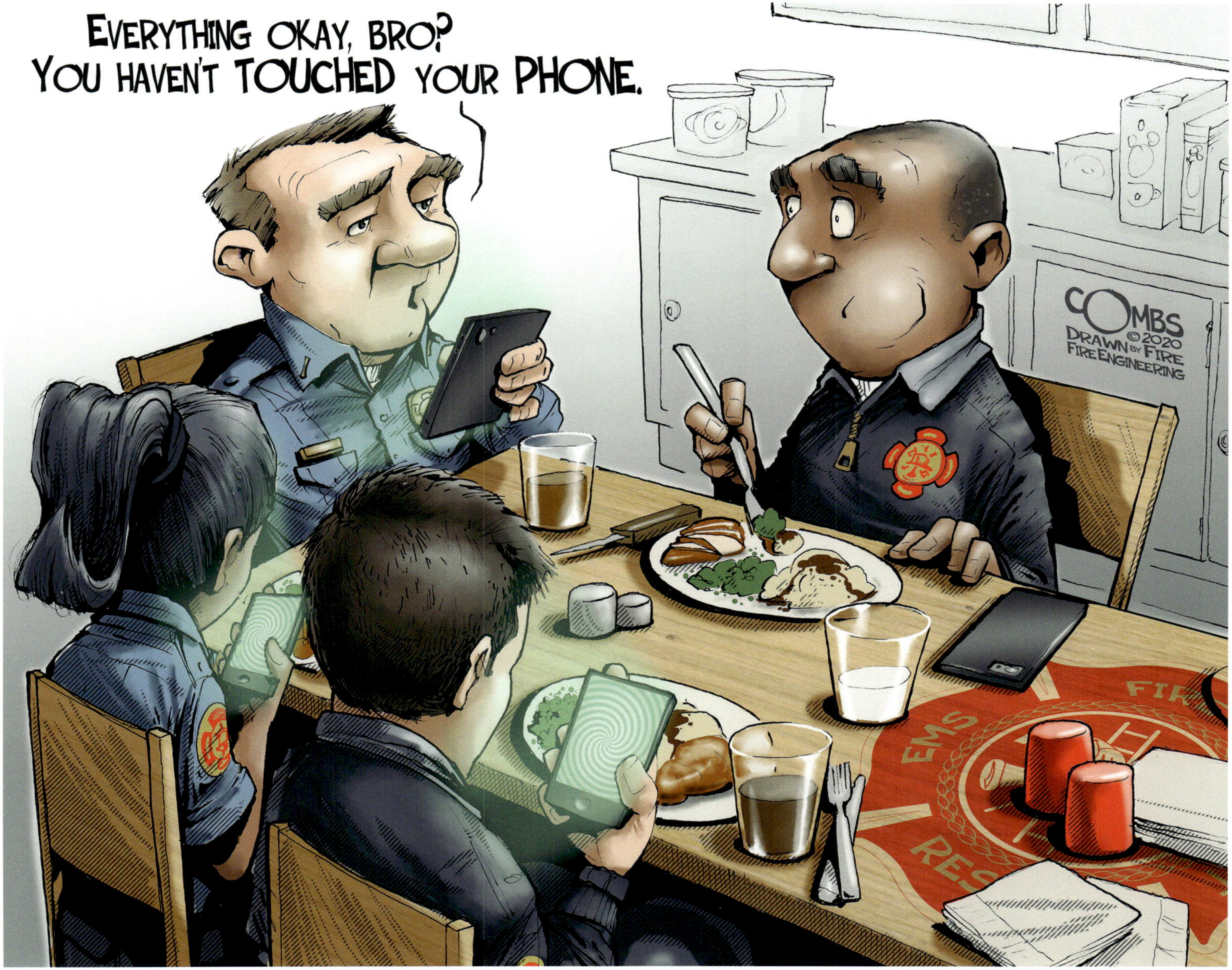
EVERYTHING OKAY, BRO? YOU HAVEN'T TOUCHED YOUR PHONE.
COMBS DRAWN BY FIRE ©2020 FIRE ENGINEERING
EMS
FIR
RES

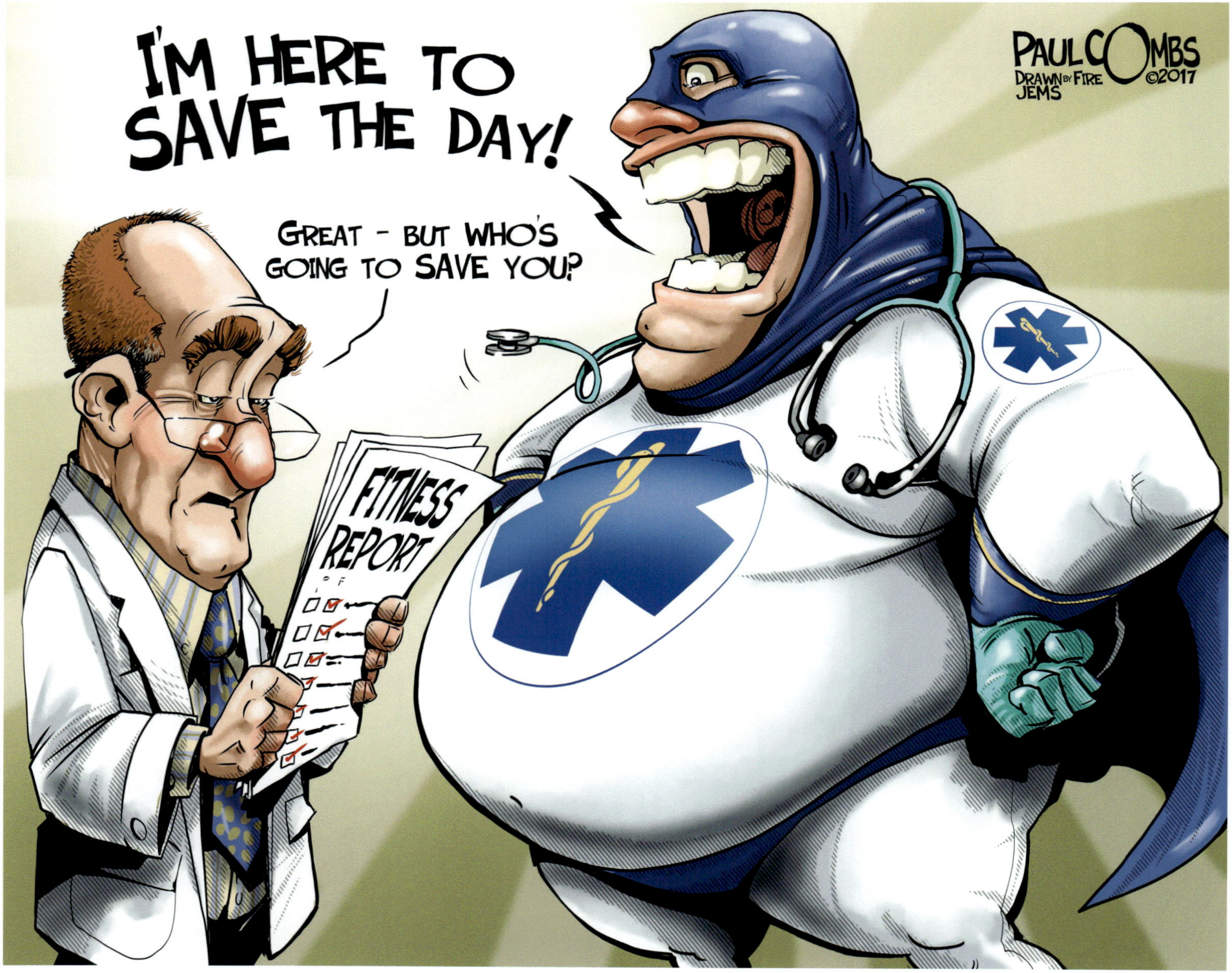
I'M HERE TO SAVE THE DAY!
GREAT - BUT WHO'S GOING TO SAVE YOU?
FITNESS REPORT
PAUL COMBS DRAWN BY FIRE JEMS ©2017

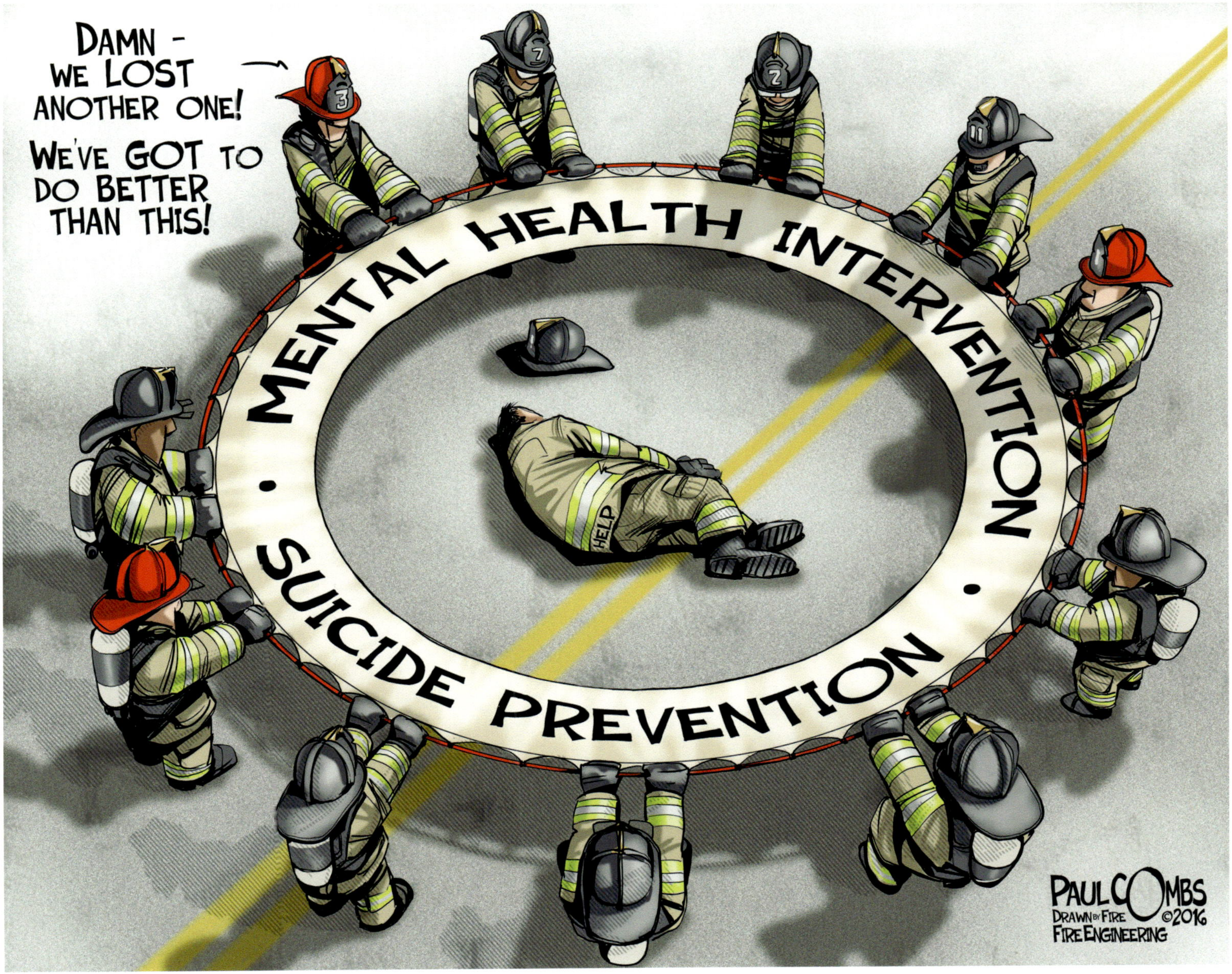
DAMN - WE LOST ANOTHER ONE!
WE'VE GOT TO DO BETTER THAN THIS!
MENTAL HEALTH INTERVENTION • SUICIDE PREVENTION •
HELP
PAUL COMBS
DRAWN BY FIRE ©2016
FIRE ENGINEERING

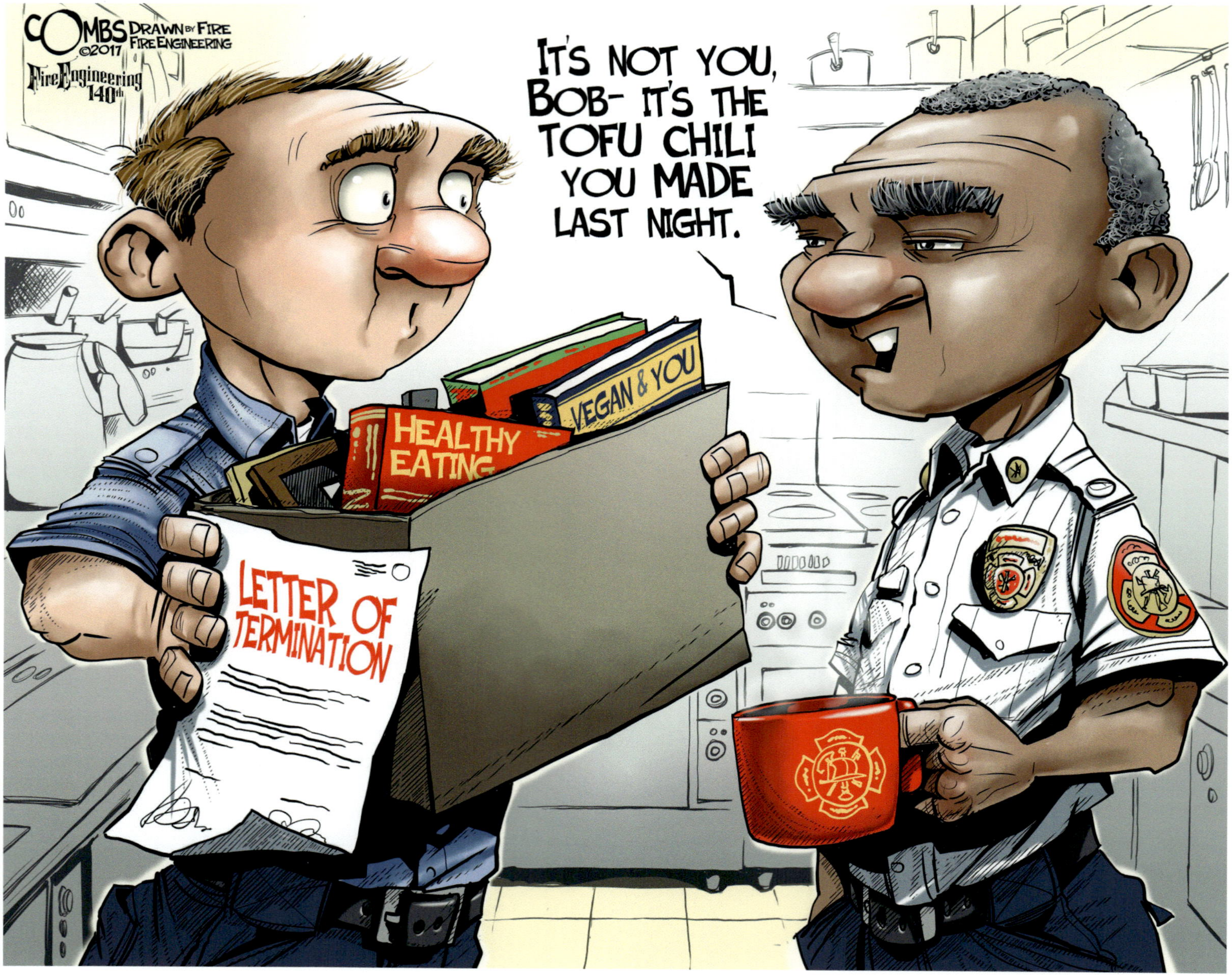
COMBS DRAWN BY FIRE FIRE ENGINEERING ©2017
Fire Engineering 140th
IT'S NOT YOU, BOB- IT'S THE TOFU CHILI YOU MADE LAST NIGHT.
VEGAN & YOU
HEALTHY EATING
LETTER OF TERMINATION

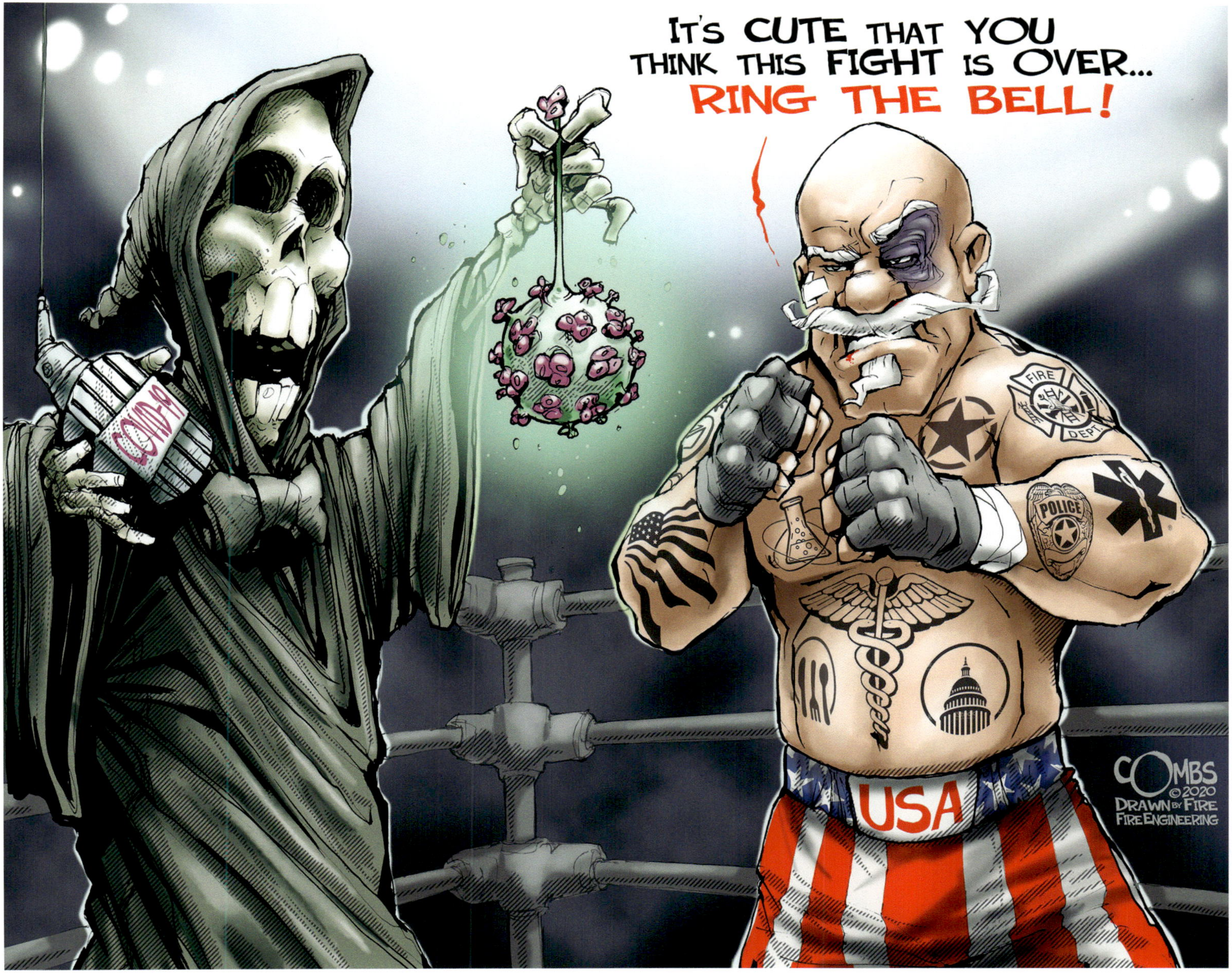
IT'S CUTE THAT YOU THINK THIS FIGHT IS OVER...
RING THE BELL!
COVID-19
FIRE DEPT.
POLICE
USA
COMBS ©2020 DRAWN BY FIRE FIREENGINEERING

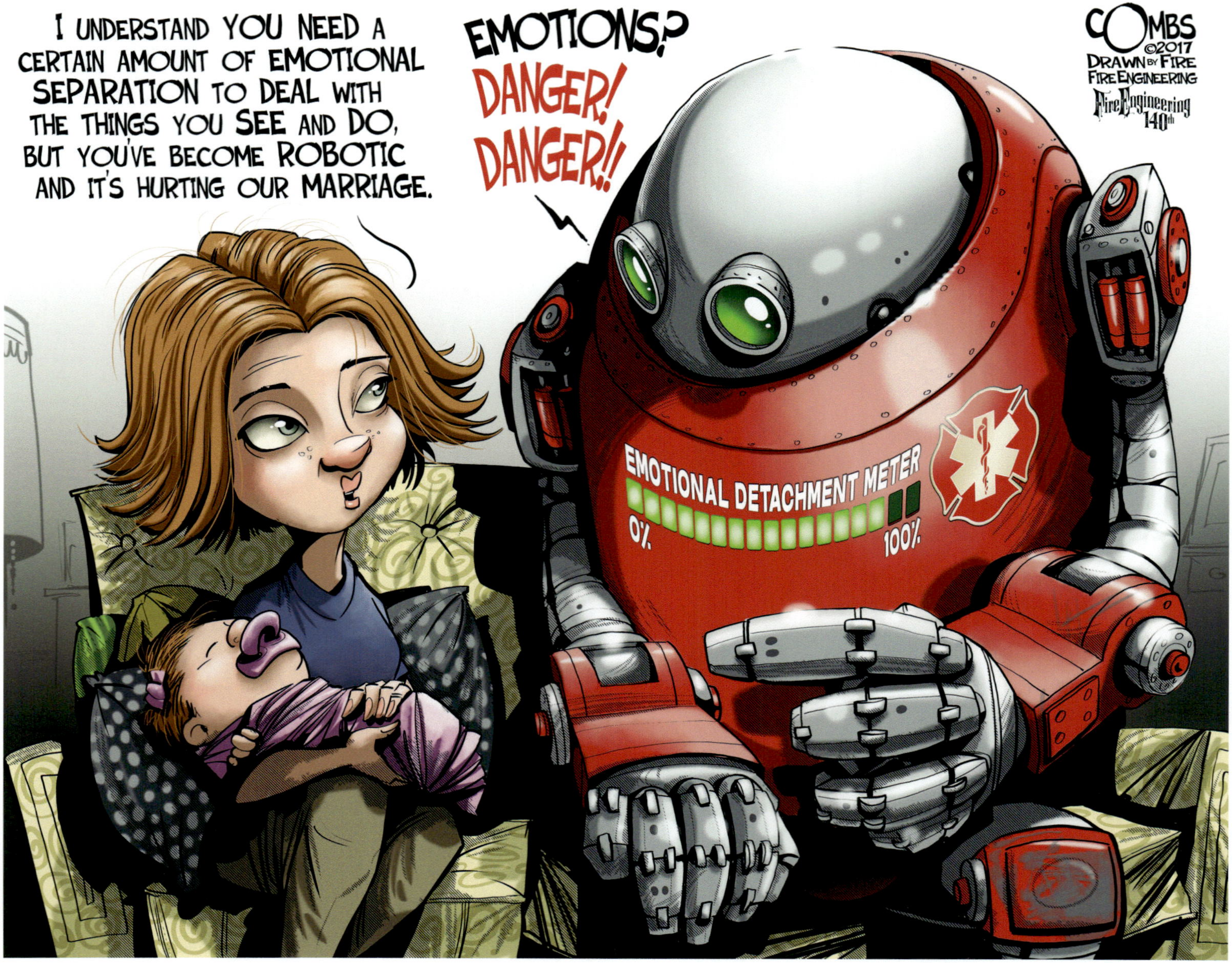
I UNDERSTAND YOU NEED A CERTAIN AMOUNT OF EMOTIONAL SEPARATION TO DEAL WITH THE THINGS YOU SEE AND DO, BUT YOU'VE BECOME ROBOTIC AND IT'S HURTING OUR MARRIAGE.
EMOTIONS?
DANGER! DANGER!!
EMOTIONAL DETACHMENT METER
0%
100%
COMBS ©2017 DRAWN BY FIRE FIRE ENGINEERING
Fire Engineering 140th

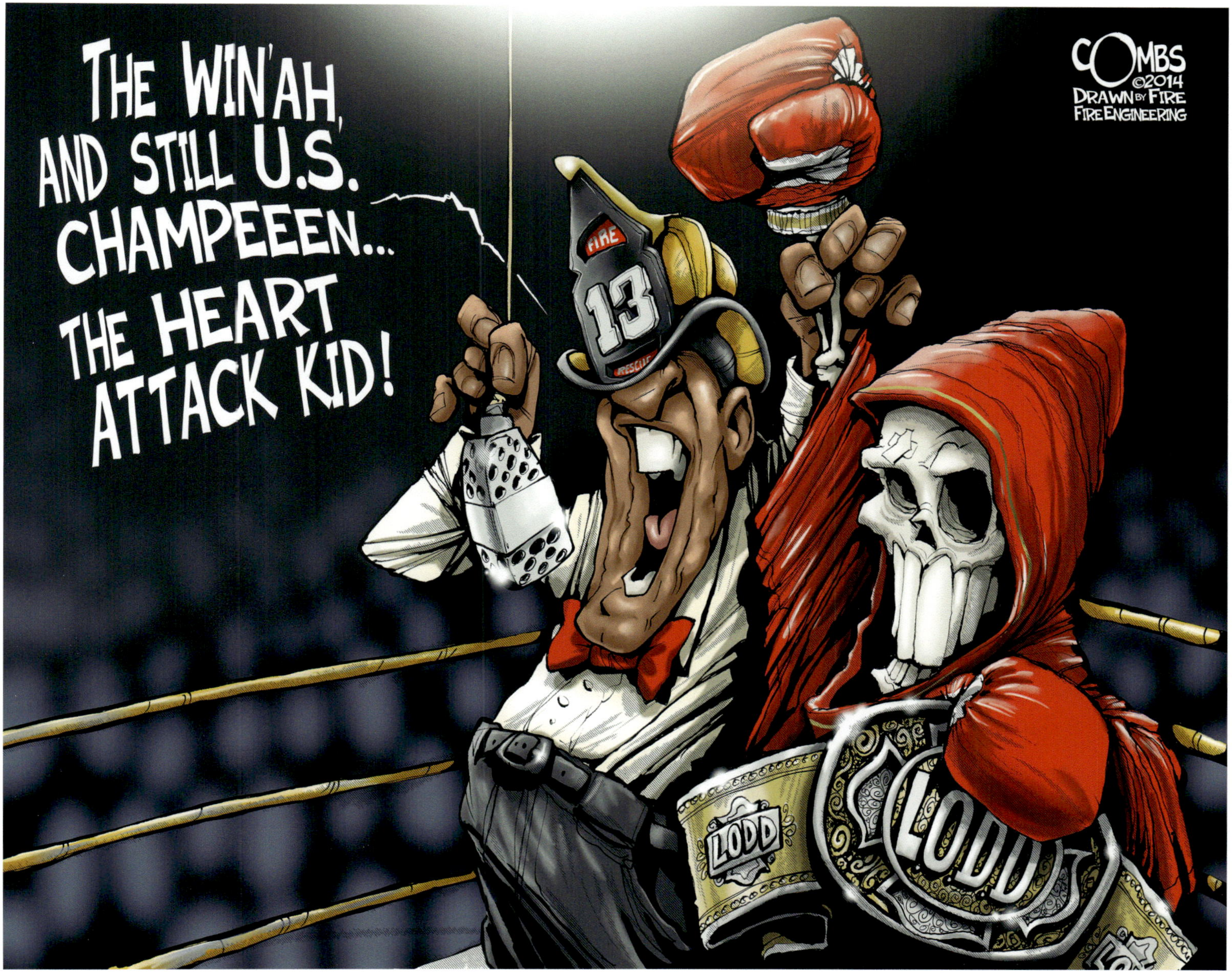
THE WIN'AH, AND STILL U.S. CHAMPEEEN... THE HEART ATTACK KID!
COMBS ©2014 DRAWN BY FIRE FIREENGINEERING
FIRE
13
RESCUE
LODD
LODD

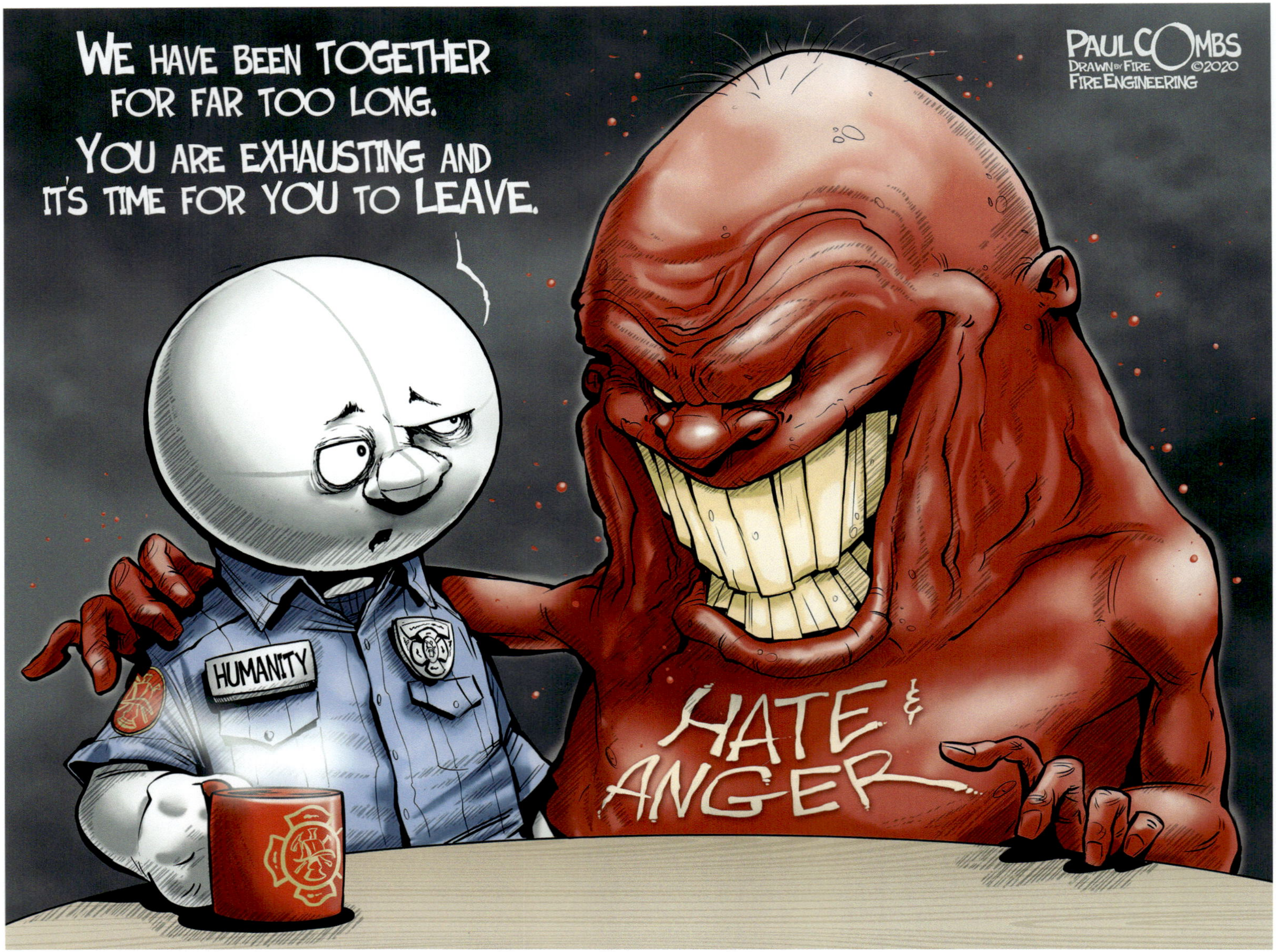
WE HAVE BEEN TOGETHER FOR FAR TOO LONG.
YOU ARE EXHAUSTING AND IT'S TIME FOR YOU TO LEAVE.
PAUL COMBS
DRAWN BY FIRE ©2020
FIRE ENGINEERING
HUMANITY
HATE & ANGER

THE BREAKUP

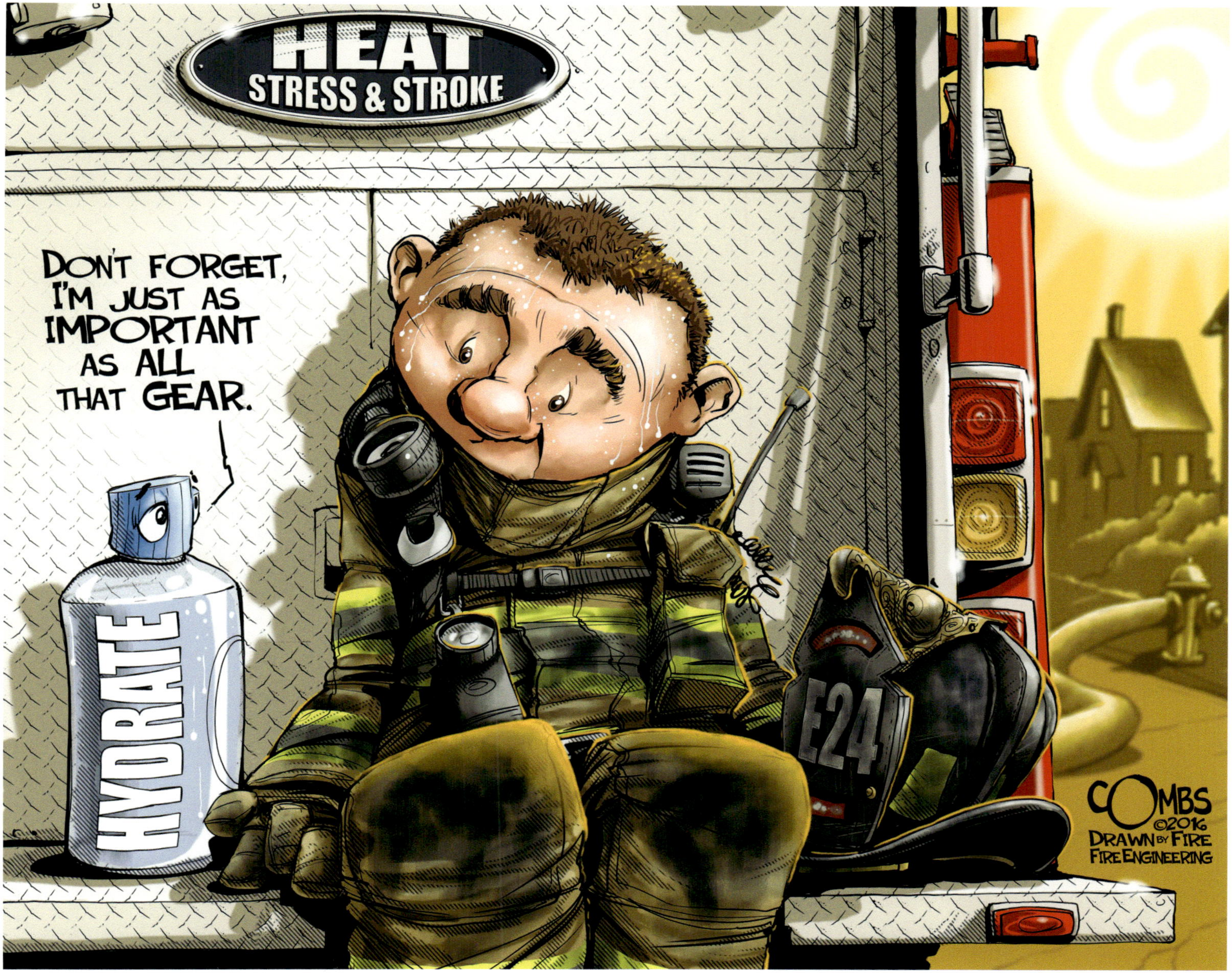
HEAT
STRESS & STROKE
DON'T FORGET, I'M JUST AS IMPORTANT AS ALL THAT GEAR.
HYDRATE
E24
COMBS
©2016
DRAWN BY FIRE
FIRE ENGINEERING

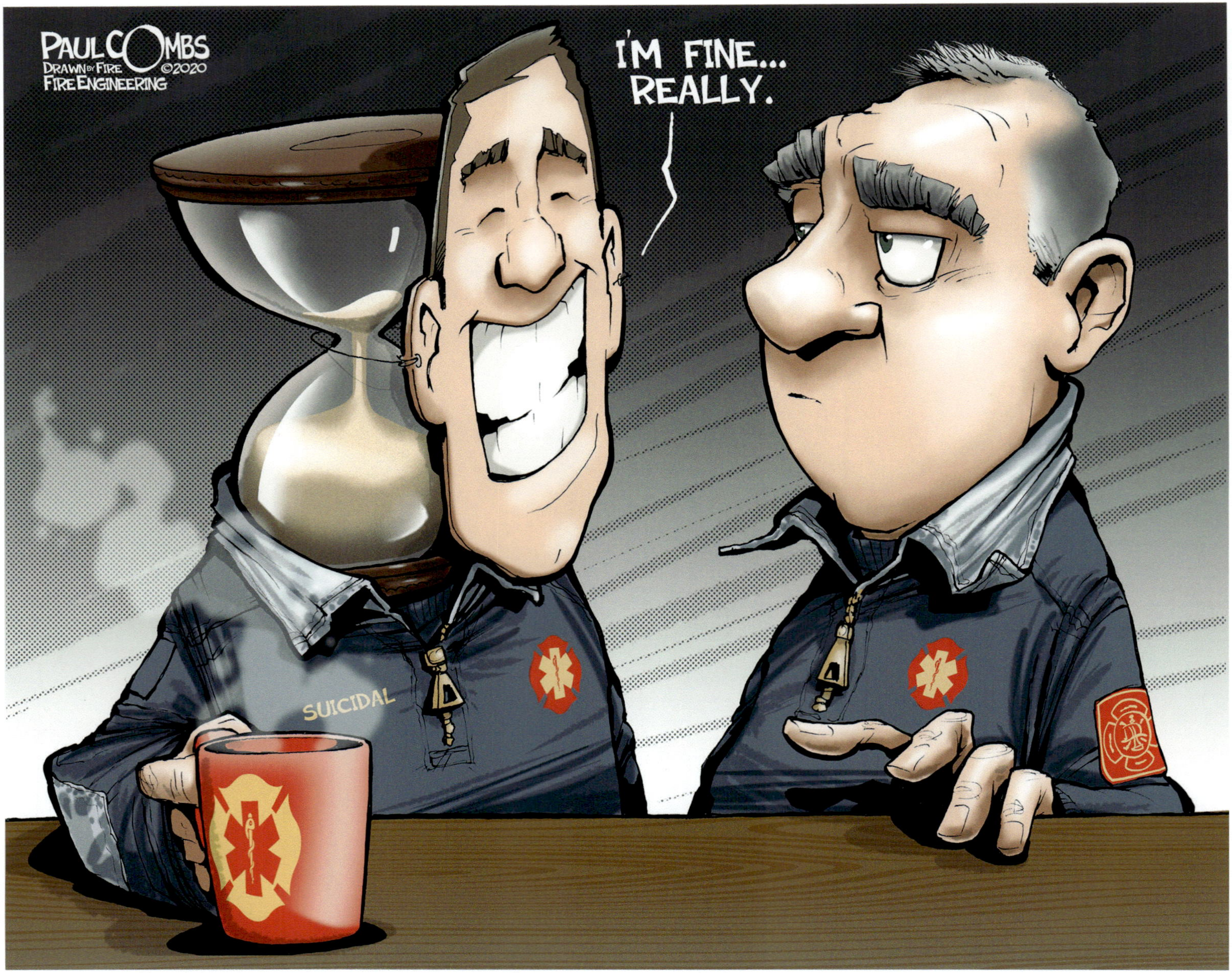
PAUL COMBS
DRAWN BY FIRE ©2020
FIRE ENGINEERING
I'M FINE... REALLY.
SUICIDAL

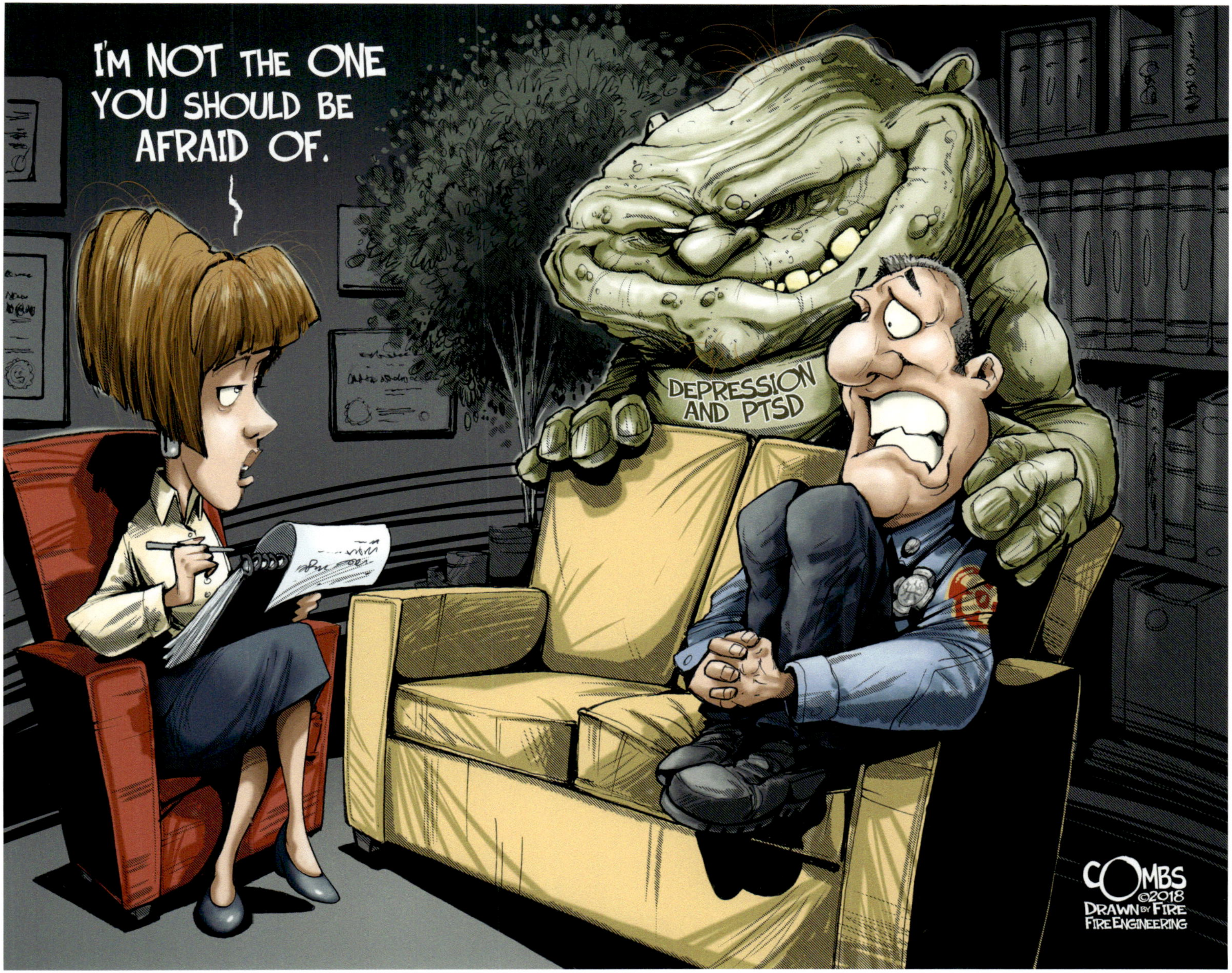
I'M NOT THE ONE YOU SHOULD BE AFRAID OF.
DEPRESSION AND PTSD
COMBS ©2018 DRAWN BY FIRE FIRE ENGINEERING

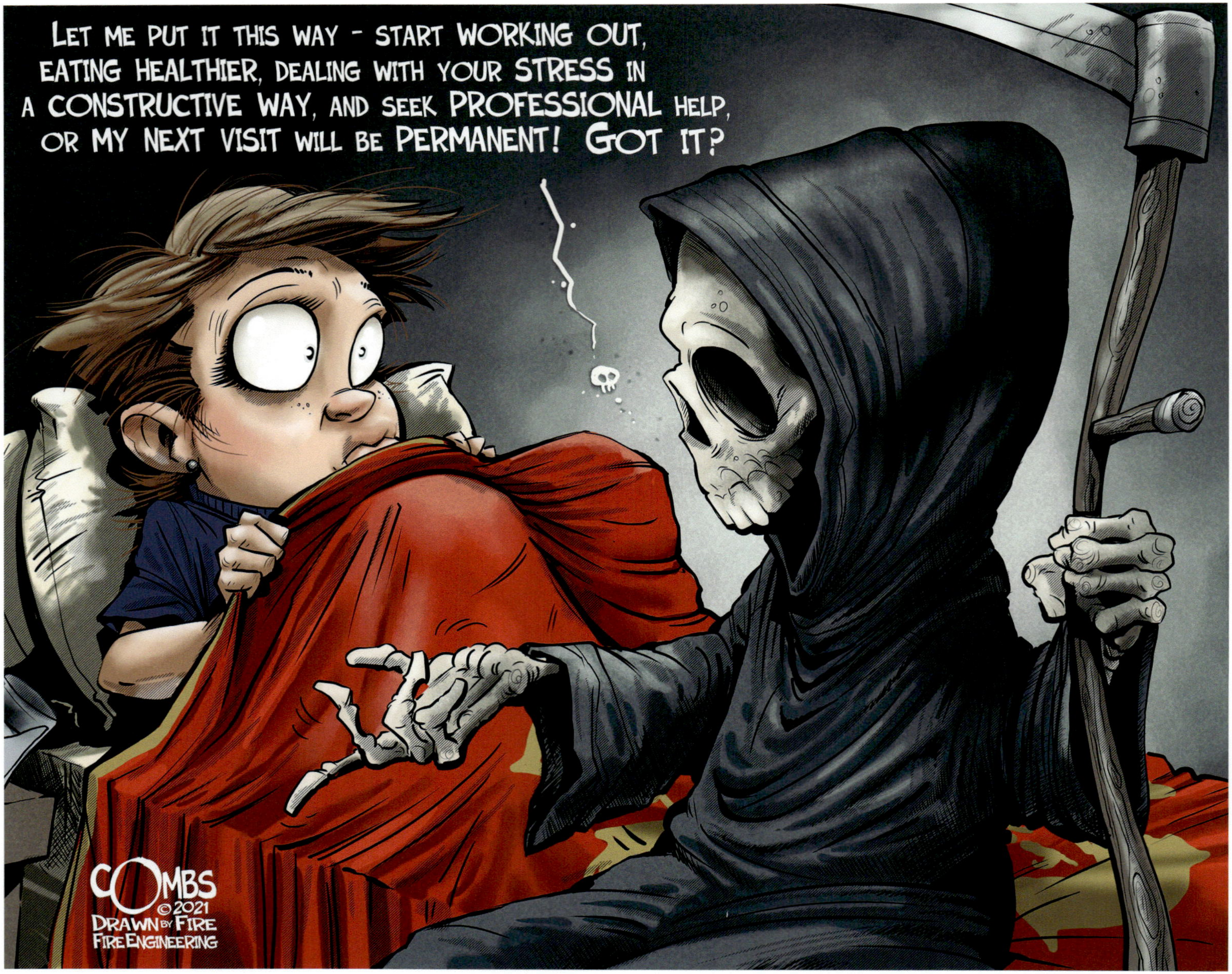
LET ME PUT IT THIS WAY - START WORKING OUT, EATING HEALTHIER, DEALING WITH YOUR STRESS IN A CONSTRUCTIVE WAY, AND SEEK PROFESSIONAL HELP, OR MY NEXT VISIT WILL BE PERMANENT! GOT IT?
COMBS ©2021 DRAWN BY FIRE FIRE ENGINEERING

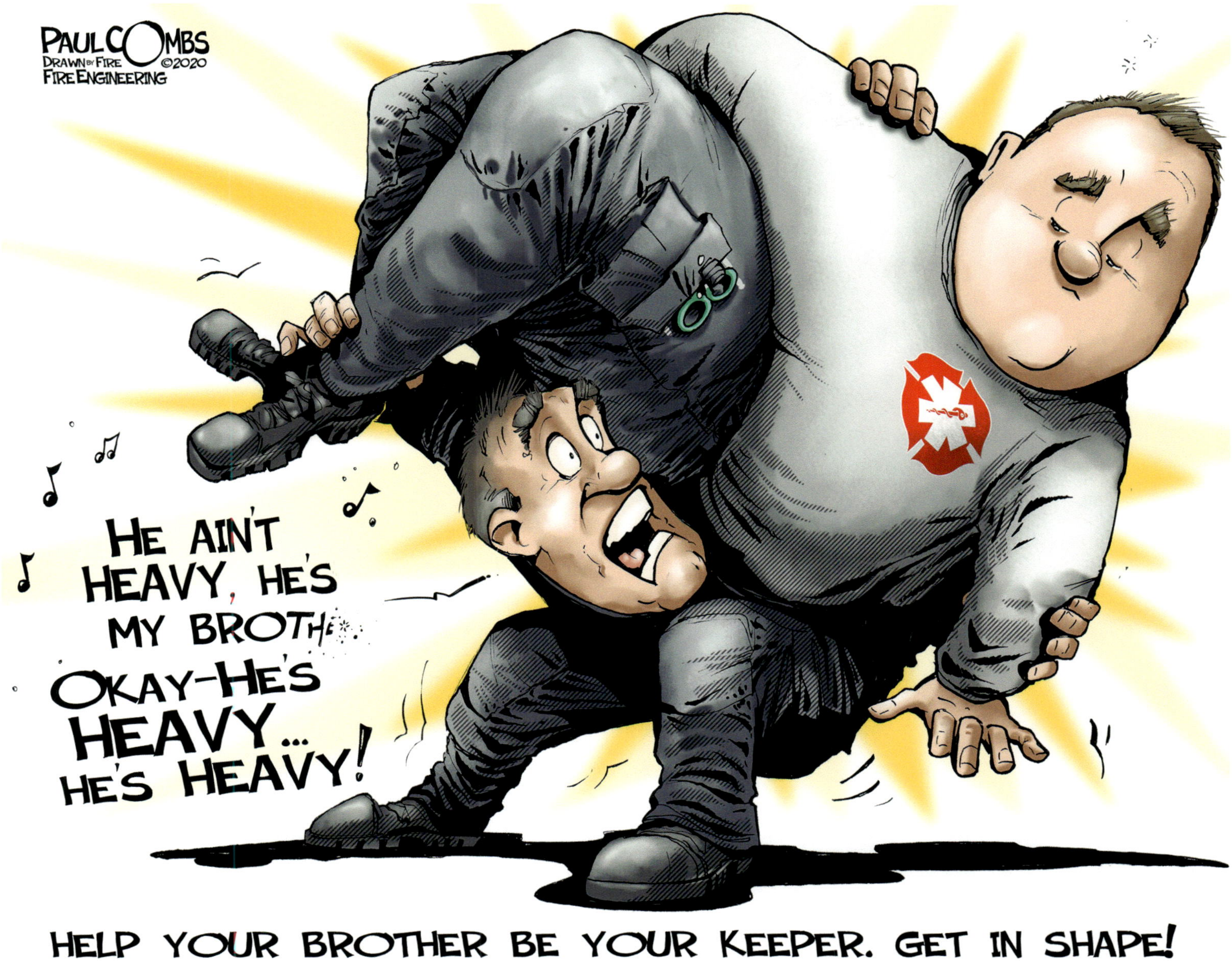
PAUL COMBS
DRAWN BY FIRE ©2020
FIRE ENGINEERING
HE AIN'T HEAVY, HE'S MY BROTH
OKAY-HE'S HEAVY... HE'S HEAVY!
HELP YOUR BROTHER BE YOUR KEEPER. GET IN SHAPE!

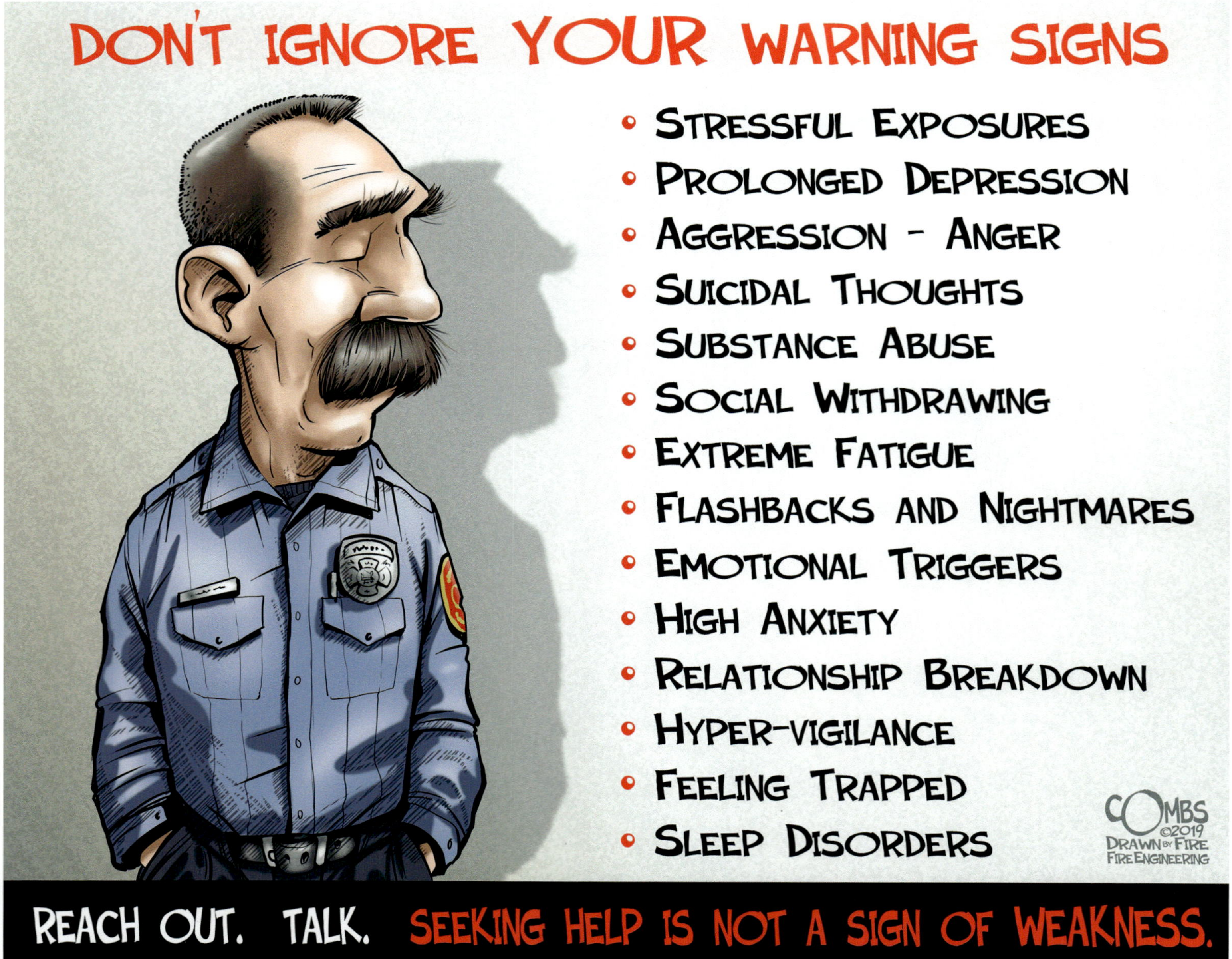
DON'T IGNORE YOUR WARNING SIGNS
• STRESSFUL EXPOSURES
• PROLONGED DEPRESSION
• AGGRESSION - ANGER
• SUICIDAL THOUGHTS
• SUBSTANCE ABUSE
• SOCIAL WITHDRAWING
• EXTREME FATIGUE
• FLASHBACKS AND NIGHTMARES
• EMOTIONAL TRIGGERS
• HIGH ANXIETY
• RELATIONSHIP BREAKDOWN
• HYPER-VIGILANCE
• FEELING TRAPPED
• SLEEP DISORDERS
COMBS ©2019 DRAWN BY FIRE FIRE ENGINEERING
REACH OUT. TALK. SEEKING HELP IS NOT A SIGN OF WEAKNESS.

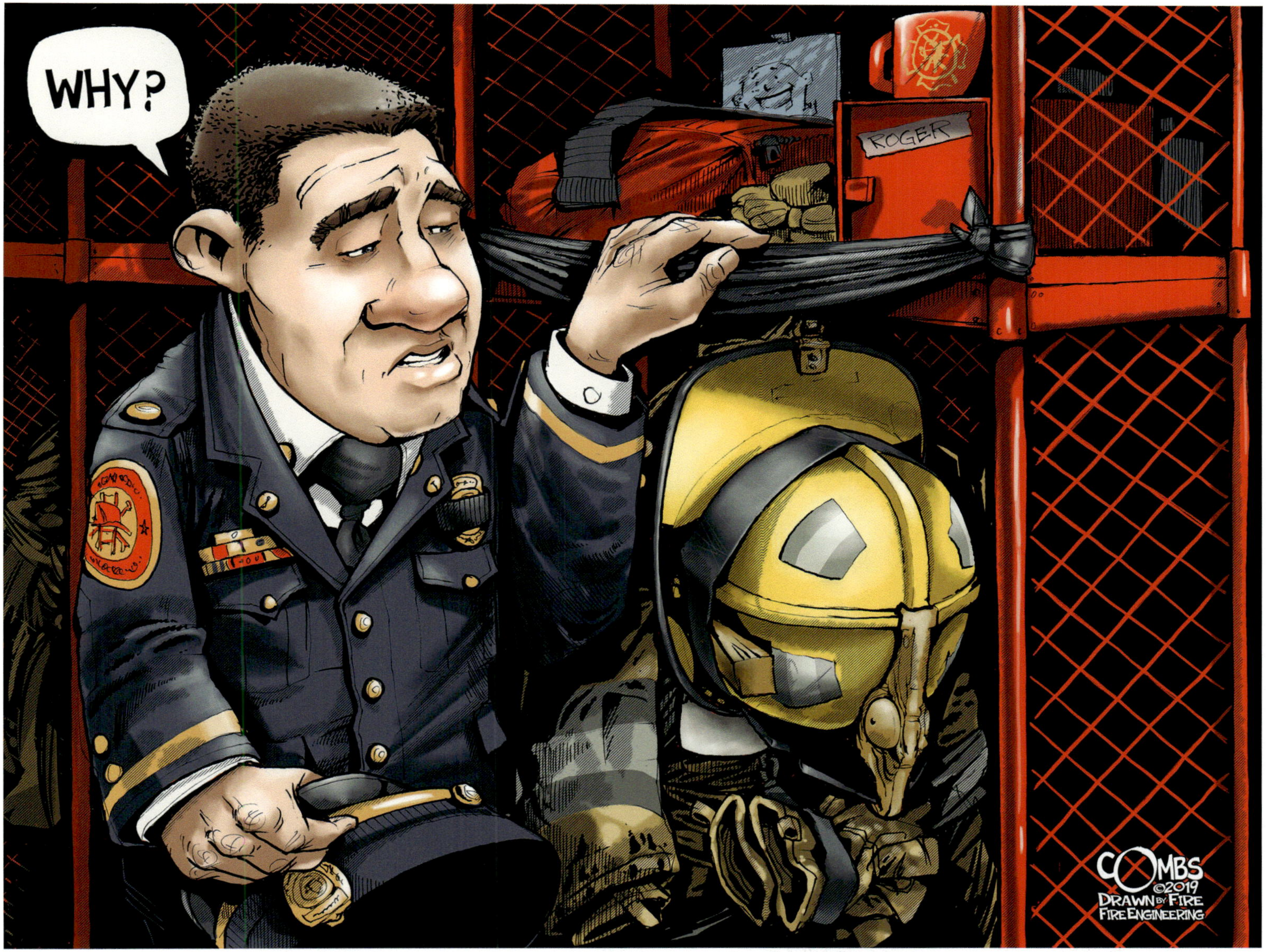

SEEKING HELP IS NOT WEAKNESS. YOUR LIFE MATTERS.

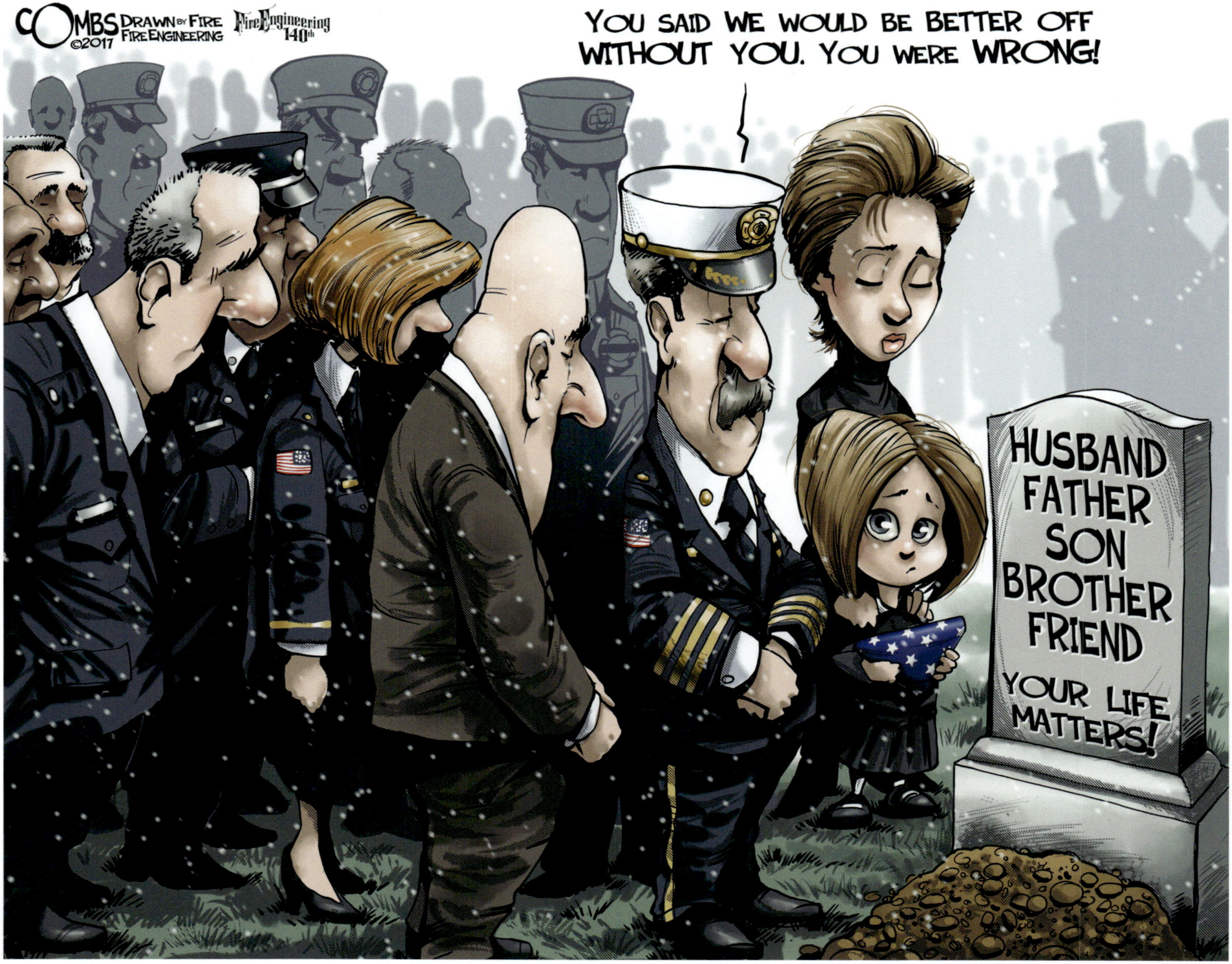
COMBS ©2017
DRAWN BY FIRE
FIRE ENGINEERING
Fire Engineering 140th
YOU SAID WE WOULD BE BETTER OFF WITHOUT YOU. YOU WERE WRONG!
HUSBAND
FATHER
SON
BROTHER
FRIEND
YOUR LIFE MATTERS!